Bioremediation of Energetics, Phenolics, and Polycyclic Aromatic Hydrocarbons

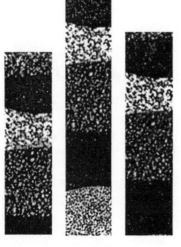

Editors

Victor S. Magar, Glenn Johnson, Say Kee Ong, and Andrea Leeson

The Sixth International In Situ and On-Site Bioremediation Symposium

San Diego, California, June 4–7, 2001

BATTELLE PRESS
Columbus • Richland

Library of Congress Cataloging-in-Publication Data

International In Situ and On-Site Bioremediation Symposium (6th : 2001 : San Diego, Calif.)
　　Bioremediation of energetics, phenolics, and polycyclic aromatic hydrocarbons : the Sixth International In Situ and On-Site Bioremediation Symposium : San Diego, California, June 4-7, 2001 / editors, V. Magar ... [et al.].
　　　p. cm. -- (The Sixth International In Situ and On-Site Bioremediation Symposium ; 3)
　　Includes bibliographical references and index.
　　ISBN 1-57477-113-2 (hc. : alk. paper)
　　1. Organic compounds--Biodegradation--Congresses. 2. Hazardous wastes--Biodegradation--Congresses. 3. Bioremediation—Congresses. I. Magar, V. (Victor), 1964-　. II. Title. III. Series: International In Situ and On-Site Bioremediation Symposium (6th : 2001 : San Diego, Calif.). Sixth International In Situ and On-Site Bioremediation Symposium ; 3.
TD192.5.I56 2001 vol. 3.
[TD1066.O73]
628.5 s--dc21
[628.5'2]

2001041198

Printed in the United States of America

Copyright © 2001 Battelle Memorial Institute. All rights reserved. This document, or parts thereof, may not be reproduced in any form without the written permission of Battelle Memorial Institute.

Battelle Press
505 King Avenue
Columbus, Ohio 43201, USA
614-424-6393 or 1-800-451-3543
Fax: 1-614-424-3819
Internet: press@battelle.org
Website: www.battelle.org/bookstore

For information on future environmental conferences, write to:
　Battelle
　Environmental Restoration Department, Room 10-123B
　505 King Avenue
　Columbus, Ohio 43201-2693
　Phone: 614-424-7604
　Fax: 614-424-3667
　Website: www.battelle.org/conferences

CONTENTS

Foreword   vii

Energetic Compounds

Biogeochemical Removal of RDX Using Iron Oxide and *Geobacter metallireducens* GS-15. *K.B. Gregory, M. von Arb, P.J.J. Alvarez, M.M. Scherer, and G.F. Parkin*   1

Field Demonstration of In Situ Biodegradation of HE in Soils. *K. Rainwater, C. Heintz, T. Mollhagen, C. Radtke, and L. Hansen*   9

Reductive Biotransformation of Nitrate and Explosives Compounds in Groundwater. *M.S. Heaston, P.W. Barnes, and K.R. Alvestad*   17

Treatment of Explosives-Contaminated Groundwater by In Situ Cometabolic Reduction. *P.W. Barnes, M.S. Heaston, and J.C. Compton*   25

Anaerobic Biological Treatment of RDX in Groundwater. *D.E. Jerger, T. Harris, A.H. Van Hout, and D.P. Leigh*   35

The Effects of Ubiquitous Electron Acceptors on the Initiation of RDX Biodegradation. *J. Davis, L.D. Hansen, and B. O'Neal*   43

Reductive Transformation of RDX in a Bench-Scale Simulated Aquifer. *L.D. Hansen, J.L. Davis, and L. Escalon*   51

Identification of Bottlenecks to the In Situ Bioremediation of Dinitrotoluene. *S.F. Nishino and J.C. Spain*   59

In Situ Bioremediation of TNT in Soil. *A. Gerth, A. Bohler, and H. Thomas*   67

Phenolic and Chlorophenolic Wastes

Application of Ligninolytic Fungi for Biotransformation of Chlorophenols. *A.A. Leontievsky, N.M. Myasoedeva, B.B. Baskunov, L.A. Golovleva, C. Bucke, and C.S. Evans*   75

Procedures for Composting Pentachlorophenol-Contaminated Soils. *M.K. Foget, J. Andrews, B. Vogt, and N. Sherman*   83

Biodegradation of Pentachlorophenol in Simulated Sand Sediments Co-Contaminated with Cadmium. *S.R. Kamashwaran and D.L. Crawford*   91

New Basidiomycetes on Bioremediation of Organochlorine-Contaminated Soil. *D.R. Matheus, V.L.R. Bononi, and K.M.G. Machado*   99

Evaluation of Aerobic and Anaerobic Degradation of Pentachlorophenol in Groundwater. S.J. MacEwen, F. Fadullon, and D. Hayes — 107

Biomass Influence on Bioavailability During Pentachlorophenol Transport Through Artificial Aquifers. Y. Dudal, R. Samson, and L. Deschenes — 115

Anaerobic In Situ Bioremediation of Pentachlorophenol. A.J. Frisbie and L. Nies — 125

Arthrobacter chlorophenolicus A6 -- The Complete Story of a 4-Chlorophenol-Degrading Bacterium. K. Westerberg, C. Jernberg, A.M. Elvang, and J.K. Jansson — 133

Polynuclear Aromatic Hydrocarbons (PAHs)

Bacterial Cell Surface Hydrophobicity Affecting Degradation of PAHs in NAPLS. P.A. Willumsen and U. Karlson — 141

Sequential Reactivity Studies of B[a]P and B[e]P by Chemical and Biological Processes. V. Librando, F. Castelli, M.G. Sarpietro, and M. Aresta — 149

Assesssment of Mass-Transfer Limitations During Slurry Bioremediation of PAHs and Alkanes in Aged Soils. M.H. Huesemann, T.S. Hausman, and T.J. Fortman — 157

Biodegradation of Anthracene and Pyrene in a Laboratory-Scale Biobarrier. C. Sartoros, L. Yerushalmi, P. Beron, and S.R. Guiot — 165

The Study of Polycyclic Aromatic Hydrocarbon Bioavailability in Soils. D.M. Lorton, S. Smith, A.D.G. Jones, and J.R. Mason — 173

Biological Treatment of PAH-Contaminated Soil Extracts. B. Guieysse, B. Mattiasson, S. Lundstedt, and B. van Bavel — 181

Availability and Bioslurry Treatment of PAHs in Contaminated Dredged Materials. J.W. Talley, U. Ghosh, and R.G. Luthy — 189

The Effect of Nutrient Amendments on PAH Degradation in a Creosote-Contaminated Soil. K.A. Pennie, G.W. Stratton, and G.B. Murray — 197

Anaerobic Polycyclic Aromatic Hydrocarbon (PAH)-Degrading Enrichment Cultures Under Methanogenic Conditions. W. Chang, T.N. Jones, and T.R. Pulliam Holoman — 205

Acute Phytotoxicity Bioassay (APB) Application on PAH-Contaminated Soil. C. Mollea, F. Bosco, and B. Ruggeri — 211

Biological Removal of Iron and PAHs from MGP-Impacted Groundwater. D.E. Richard and D.F. Dwyer — 219

Ecotoxicological Characterization of Metabolites Produced During PAH Biodegradation in Contaminated Soils. *F. Haeseler, D. Blanchet, P. Werner, and J.-P. Vandecasteele* — 227

Effects of Aging, Bacterial Source and Desorption on PAH Biodegradation. *S. Hwang and T.J. Cutright* — 235

Effect of a Rhamnolipid Biosurfactant on the Phenanthrene Desorption from a Clay-Loam Soil. *C. Amezcua-Vega, R. Rodriguez-Vazquez, H. Poggi-Varaldo, and E. Rios-Leal* — 243

Bacterial Strategies to Optimize the Bioavailability of Solid PAH. *L. Y. Wick and H. Harms* — 251

Field-Scale Studies on Removal of Creosote From Contaminated Soil. *H.L. Allen, L. Francendese, G. Harper, and T.F. Miller* — 259

Isolation of a Soil Bacterial Strain Capable of Dibenzothiophene Degradation. *S. Di Gregorio, D. Lizzari, O. Massella, C. Zocca, and G. Vallini* — 267

Author Index — 275

Keyword Index — 303

FOREWORD

The papers in this volume correspond to presentations made at the Sixth International In Situ and On-Site Bioremediation Symposium (San Diego, California, June 4-7 2001). The program included approximately 600 presentations in 50 sessions on a variety of bioremediation and supporting technologies used for a wide range of contaminants.

This volume focuses on *Bioremediation of Energetics, Phenolics, and Polycyclic Aromatic Hydrocarbons.* Remediation of these recalcitrant contaminants continues to pose challenges to environmental engineers and scientists. Energetics, phenolics, and PAHs are common and challenging groups of contaminants that affect soils, sediments, and groundwater. Methods for treating these compounds undergo continuing change and refinement, to overcome recalcitrance, toxicity, microbial bioavailability, and degradation endpoints. This volume covers in situ and ex situ, aerobic and anaerobic processes, from sorption and bioavailability issues to physical/chemical pre-treatment processes to enhance contaminant degradation.

The author of each presentation accepted for the symposium program was invited to prepare an eight-page paper. According to its topic, each paper received was tentatively assigned to one of ten volumes and subsequently was reviewed by the editors of that volume and by the Symposium chairs. We appreciate the significant commitment of time by the volume editors, each of whom reviewed as many as 40 papers. The result of the review was that 352 papers were accepted for publication and assembled into the following ten volumes:

Bioremediation of MTBE, Alcohols, and Ethers — 6(1). Eds: Victor S. Magar, James T. Gibbs, Kirk T. O'Reilly, Michael R. Hyman, and Andrea Leeson.

Natural Attenuation of Environmental Contaminants — 6(2). Eds: Andrea Leeson, Mark E. Kelley, Hanadi S. Rifai, and Victor S. Magar.

Bioremediation of Energetics, Phenolics, and Polycyclic Aromatic Hydrocarbons — 6(3). Eds: Victor S. Magar, Glenn Johnson, Say Kee Ong, and Andrea Leeson.

Innovative Methods in Support of Bioremediation — 6(4). Eds: Victor S. Magar, Timothy M. Vogel, C. Marjorie Aelion, and Andrea Leeson.

Phytoremediation, Wetlands, and Sediments — 6(5). Eds: Andrea Leeson, Eric A. Foote, M. Katherine Banks, and Victor S. Magar.

Ex Situ Biological Treatment Technologies — 6(6). Eds: Victor S. Magar, F. Michael von Fahnestock, and Andrea Leeson.

Anaerobic Degradation of Chlorinated Solvents — 6(7). Eds: Victor S. Magar, Donna E. Fennell, Jeffrey J. Morse, Bruce C. Alleman, and Andrea Leeson.

Bioaugmentation, Biobarriers, and Biogeochemistry — 6(8). Eds: Andrea Leeson, Bruce C. Alleman, Pedro J. Alvarez, and Victor S. Magar.
Bioremediation of Inorganic Compounds — 6(9). Eds: Andrea Leeson, Brent M. Peyton, Jeffrey L. Means, and Victor S. Magar.
In Situ Aeration and Aerobic Remediation — 6(10). Eds: Andrea Leeson, Paul C. Johnson, Robert E. Hinchee, Lewis Semprini, and Victor S. Magar.

In addition to the volume editors, we would like to thank the Battelle staff who assembled the ten volumes and prepared them for printing: Lori Helsel, Carol Young, Loretta Bahn, Regina Lynch, and Gina Melaragno. Joseph Sheldrick, manager of Battelle Press, provided valuable production-planning advice and coordinated with the printer; he and Gar Dingess designed the covers.

The Bioremediation Symposium is sponsored and organized by Battelle Memorial Institute, with the assistance of a number of environmental remediation organizations. In 2001, the following co-sponsors made financial contributions toward the Symposium:

Geomatrix Consultants, Inc.
The IT Group, Inc.
Parsons
Regenesis
U.S. Air Force Center for Environmental Excellence (AFCEE)
U.S. Naval Facilities Engineering Command (NAVFAC)

Additional participating organizations assisted with distribution of information about the Symposium:

Ajou University, College of Engineering
American Petroleum Institute
Asian Institute of Technology
National Center for Integrated Bioremediation Research & Development (University of Michigan)
U.S. Air Force Research Laboratory, Air Expeditionary Forces Tec hnologies Division
U.S. Environmental Protection Agency
Western Region Hazardous Substance Research Center (Stanford University and Oregon State University)

Although the technical review provided guidance to the authors to help clarify their presentations, the materials in these volumes ultimately represent the authors' results and interpretations. The support provided to the Symposium by Battelle, the co-sponsors, and the participating organizations should not be construed as their endorsement of the content of these volumes.

Andrea Leeson & Victor Magar, Battelle
2001 Bioremediation Symposium Co-Chairs

BIOGEOCHEMICAL REMOVAL OF RDX USING IRON OXIDE AND *GEOBACTER METALLIREDUCENS* GS-15

Kelvin B. Gregory, Michelle von Arb, Pedro J. J. Alvarez, Michelle M. Scherer, and Gene F. Parkin (The University of Iowa, Iowa City, Iowa)

ABSTRACT: Biogeochemical interactions between iron-reducing bacteria (e.g., *Geobacter metallireducens* GS-15) and iron oxides could promote the removal of redox-sensitive pollutants, such as RDX (hexahydro-1,3,5-trinitro-1,3,5-triazine) from contaminated soils and groundwater. RDX, a common contaminant at military installations, is a suspected carcinogen that is relatively recalcitrant to microbial degradation. RDX, however, is prone to undergo chemical reduction. Our results suggest that iron-reducing bacteria may enhance RDX removal not only through biotransformation but also by the production of reactive, biogenic iron oxides such as magnetite that can reduce RDX abiotically. Interestingly, biogenic magnetite was more reactive than chemically synthesized magnetite. Consequently, iron-reducing bacteria could improve the performance of iron barriers and increase the reactivity of some mineral surfaces for enhanced natural attenuation of priority pollutants that are prone to be abiotically reduced.

INTRODUCTION

RDX (hexahydro-1,3,5-trinitro-1,3,5-triazine) is the British acronym for Research Department Explosive (Testud et al., 1996). RDX is toxic to humans and a variety of organisms, and is classified as a Class C (possible human) carcinogen by the U.S. Environmental Protection Agency (McLellan et al., 1988). Due to its recalcitrance to microbial degradation, low tendency to volatilize (dimensionless Henry's constant, $H' = 2 \times 10^{-11}$), and high mobility in aquifers ($\log K_{ow} = 0.8$) (Gorontzy et al., 1994), clean-up of RDX contaminated sites is a challenging problem. RDX is also thermodynamically susceptible to reduction reactions, making it a good model to investigate the fate of redox-sensitive, priority pollutants.

Encouraging results in laboratory and field experiments have recently stimulated a rapid increase in the use of zero-valent iron (Fe(0)) as a reactive material to remove redox-sensitive contaminants from groundwater. Up to now, research on Fe(0) systems has focused primarily on abiotic processes. Nevertheless, we recently found that Fe(0) (via cathodic hydrogen production during corrosion) may serve as the sole exogenous electron donor for the biotransformation of reducible contaminants, including chlorinated methanes (Gregory et al., 2000; Novak et al., 1998; Weathers et al., 1997), chlorinated ethanes (Gregory, et al., 2000), nitrate (Till et al., 1998), and RDX (Wildman and Alvarez, 2001). These experiments showed that hydrogen-utilizing organisms could significantly enhance the treatment efficiency of permeable reactive barriers (PRBs). However, there are numerous niches in Fe(0) systems, besides those created by cathodic hydrogen, that could be exploited by microorganisms for enhanced pollutant removal.

For example, dissimilatory iron-reducing bacteria (DIRB) could enhance the reactivity of iron in PRBs by reductive dissolution of passivating layers of Fe(III)-oxides (Gerlach, et al., 2000). Alternatively, DIRB may establish more reactive surfaces in PRBs and in aquifer minerals through the formation of Fe(II)-oxides or surface associated Fe(II) (Fredrickson and Gorby, 1996; McCormick et al., 2000). Since numerous Fe(II) and Fe(III) oxides may be formed as Fe(0) oxidizes in anoxic groundwater, the geochemistry of PRBs is largely unknown (Scherer et al., 2000). Many iron oxides found in and around PRBs may serve as electron acceptors for DIRB. Goethite, magnetite, hematite, and hydrous ferric oxide(s) (HFO), among others, have been identified both as oxides in PRBs (Phillips et al., 2000) and as electron acceptors for DIRB (Nealson and Little, 1997). Therefore, these oxides could provide a niche for contaminant transformation by DIRB. In addition, many oxidized iron species may be found downgradient from PRBs that could be reduced by DIRB for enhanced attenuation at mineral surfaces.

This study was undertaken to determine if DIRB could affect the reactivity of iron oxides that passivate Fe(0), and whether we could exploit such biogeochemical interactions to remove RDX as a model redox-sensitive pollutant. Emphasis was placed on identifying the reactive oxide species produced during DIRB respiration to make the findings of this study also relevant for natural attenuation processes at mineral surfaces.

MATERIALS AND METHODS

Biotic and abiotic RDX transformations were studied separately and interactively to evaluate potential synergistic effects. Batch degradation assays were conducted in amber, 100-mL serum vials with 25 mL of HEPES buffer (10 mM, 4-(2-hydroxyethyl)-1-piperazineethanesulfonic acid, pH=7) under a 20% CO_2/80% N_2 headspace. RDX was added from an acetone stock solution and dried with N_2 gas before addition of liquid medium, cells, or iron oxide. The reactors were sealed with Teflon-lined, grey-butyl septa and aluminum crimp caps. All incubations took place on a rotary shaker at 30°C.

Analytical. Reactors were periodically analyzed for RDX, and some transformation products (i.e., 1,3-dinitro-5-nitroso-1,3,5-triazacyclohexane [MNX], 1,2-dinitroso-5-nitro-1,3,5-triazacyclohexane [DNX] and 1,3,5-trinitroso-1,3,5-triazacyclohexane [TNX]). Anaerobic, 1 mL liquid samples were removed from reactors for analysis by High Performance Liquid Chromatography (HPLC). Fe(II) determinations were performed using the colorimetric complexing agent, Ferrozine, and measuring absorbance at 562 nm.

Cell Cultures. *Geobacter metallireducens* GS-15 was routinely cultured under strict anaerobic conditions as described elsewhere (Lovley and Phillips, 1988). 1-L batches of cells were cultured in liquid medium (Lovely et al., 1993) using acetate (10 mM) as carbon source and electron donor, and Fe(III) citrate (50 mM) as the electron acceptor. For biological experiments, cells were grown to late log phase, washed and resuspended, at 3× concentration, in deoxygenated HEPES buffer. Microbial activity was monitored using Fe(II) production as an indicator. Cell reactivity towards RDX was observed in the presence and absence of HFO. In addition, all cell treatments were performed under fed (with 10 mM acetate) and unfed conditions.

DIRB-Modified HFO. Acetate-fed GS-15 was allowed to respire HFO for two days. Reactors spiked with RDX were then amended with the DIRB-reduced HFO. A biocide (Kathon CG/ICP or 300 mg/L $HgCl_2$) was added also to remove the potential for biological transformation to confound the analysis of abiotic RDX removal by DIRB-modified HFO.

Biogenic Magnetite. Biogenic magnetite was produced by growth of GS-15 in 10 mL batches of liquid medium with HFO as an electron acceptor (Lovley et. al., 1993). Typically, a black, magnetic phase was observed after three weeks of incubation at 30°C. This magnetic phase was separated using a neodymium-iron-boron magnet (McMaster-Carr, Elmhurst, IL). It was washed and resuspended 6 times in 10 mM, pH=7 HEPES buffer. Confirmation of magnetite particles was obtained through X-ray diffraction (XRD) and Mossbauer Spectroscopy. The specific surface area was determined by the method of Brunauer, Emmett and Teller (BET).

Synthetic Magnetite. Chemically synthesized magnetite with a 5-μm average particle diameter was obtained from Cerac, Inc. (Milwaukee, WI). X-ray diffraction and spectrographic analysis (by Cerac, Inc.) confirmed a 99% pure magnetite. Varying mass concentrations of magnetite were tested for reactivity towards RDX. The specific surface area was later determined by BET.

RESULTS AND DISCUSSION

Biotransformation. Figure 1 summarizes results obtained with washed cell suspensions of GS-15 under acetate-fed and unfed conditions and in the presence and absence of HFO. GS-15 was capable of removing RDX. Removal was not affected by the presence of HFO in the absence of acetate (i.e., under unfed conditions. However, under acetate-fed conditions, removal was slightly faster when HFO was present. Removal by fed cells with HFO indicated a biogeochemical transformation of RDX. In addition these results suggest that biotransformation was faster than the abiotic reaction with the oxides.

Under all conditions, less than 10% of the initial molar concentration of RDX was observed as the heterocyclic reduction products, MNX, DNX and TNX. The low percentage of RDX that was recovered as heterocyclic nitroso intermediates may be explained by the formation of an unknown intermediate that was observed to build up in all reactors containing GS-15. Figure 2 shows the buildup of this unknown metabolite in a reactor containing washed GS-15 and HFO. Other researchers have also reported the accumulation of a variety of products that are formed during the anaerobic biodegradation of RDX (Hawari et.

al., 2000; Oh and Alvarez, 2001). Identification of this metabolite is currently under way using ^{14}C-RDX and Liquid Chromatography-Mass Spectrometry.

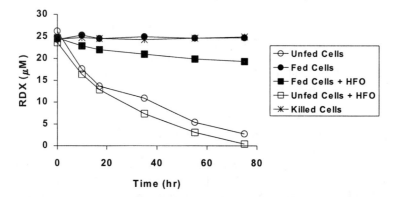

FIGURE 1. Biotransformation of RDX by washed GS-15 cells in the presence and absence of HFO and under acetate-fed and unfed conditions.

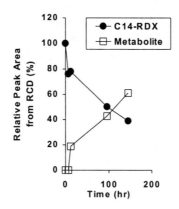

FIGURE 2. Removal of ^{14}C-RDX and appearance of unknown ^{14}C-metabolite in a treatment with fed cells of GS-15 and HFO. This figure is also representative of treatments that contained unfed GS-15 without HFO.

DIRB-Modified HFO. Results from modified HFO treatments are shown in Figure 3. HFO that had been exposed to GS-15 prior to experimentation removed RDX from solution. Unexposed HFO was not reactive towards RDX, as well as HFO that had been exposed to killed cells of GS-15. Since it has been demonstrated that GS-15 may produce magnetite from its growth on HFO as an electron acceptor (Lovley et al., 1987), our results indicate that this biogenic magnetite may transform RDX. Indeed, similar results have been obtained with carbon tetrachloride (McCormick et al., 2000) and nitroaromatics (Heijman et al., 1995).

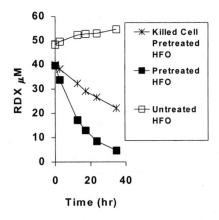

FIGURE 3. Interactive treatments with GS-15 and HFO.

Biogenic Magnetite. The removal of RDX in the presence of pretreated HFO suggests that GS-15 was producing a reactive surface associated Fe(II)-oxide from HFO. Results obtained from incubations with biogenic magnetite are shown in Figure 4. RDX (50 µM) was rapidly removed in less than 80 hours. MNX was not removed rapidly, and represented approximately 20% of the initial molar concentration of RDX. DNX was transiently observed with concomitant production of TNX, which built up and only slowly and began to decay after 150 hr of incubation (data not shown).

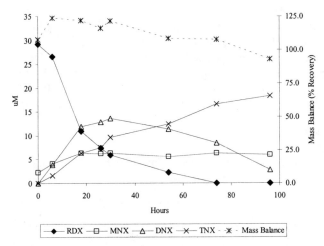

FIGURE 4. Transformation of RDX with Biogenic Magnetite.

The MNX depicted in Figure 4 is suspicious since it was detected since the beginning of the experiment, at t=0. The MNX peak on chromatograms may have been slightly interfered by a co-eluting peak. We are currently attempting LC-MS techniques to confirm the MNX data.

The initial surface area of magnetite was determined to be 224 m^2/g. This area is approximately twice that observed by McCormick et al. (2000). The surface area concentration of magnetite was therefore 215 m^2/L in these treatments. These data confirm that biologically produced magnetite from GS-15 respiring HFO was responsible for removal of RDX.

Synthetic Magnetite. The initial surface area of the synthetic magnetite was determined to be 44 m^2/g. Synthetic magnetite was not reactive towards RDX over 100 hours of incubation. Incubations with several surface area concentrations (88, 176, 352, and 880 m^2/L) of synthetic magnetite were unsuccessful in removing RDX within 80 hours of exposure.

CONCLUSIONS

Geobacter metallireducens GS-15, a ubiquitous DIRB, is capable of transforming RDX by at least two mechanisms: 1) biotransformation either as a fortuitous reaction, or potentially as a nutrient coupled to growth; and 2) the production of reactive magnetite through its respiration of HFO. This biogenic magnetite appears to be more reactive towards redox sensitive contaminants than a chemically synthesized form.

Since iron(III) is one of the most abundant anaerobic electron acceptors in the natural environment and because human intervention creates environments where iron is plentiful, the significance of DIRB in contaminant removal may not be fully appreciated. Biogeochemical removal of RDX may play a significant role in the detoxification of RDX contaminated groundwater, soil and sediment.

ACKNOWLEDGEMENTS

The authors thank Aaron G. B. Williams for assistance with oxide characterizations. Kelvin Gregory was supported through fellowships from the National Science Foundation and The University of Iowa Center for Biocatalysis and Bioprocessing. Partial funding of this work was provided by the Strategic Environmental Research and Development Program (SERDP) (Project No. DACA72-00-P-0057) and Iowa Biotechnology Byproducts Consortium.

REFERENCES

Fredrickson, K.J. and Y.A. Gorby. 1996. "Environmental Processes Mediated by Iron-reducing Bacteria". *Cur. Opin. Biotech.* 7:287-294.

Gerlach, R., A.B. Cunningham, and F. Caccavo. 2000. "Disimilatory Iron-reducing Bacteria Can Influence The Reduction of Carbon Tetrachloride by Iron Metal". *Env. Sci. Technol.* 34:2461-2464.

Gorontzy T., O., M.W. Drzyzga, D. Kahl, J. Bruns-Nagel, E. Breitung, K.H. von Leow, and Blotevogel. 1994. "Microbial Degradation of Explosives and Related Compounds" *Crit. Rev. in Microbiol. 20*:265-284.

Gregory, K.B., M.G. Mason, H.D. Picken, L.J. Weathers, and G.F. Parkin. 2000. "Bioaugmentation of Fe(0) for the Remediation of Chlorinated Aliphatic Hydrocarbons". *Env. Eng. Sci. 17*:169-181.

Hawari J., A. Halasz, T. Sheremata, S. Beaudet, C. Groom, L. Paquet, C. Rhofir, G. Ampleman, and S. Thiboutot. 2000. "Characterization of Metabolites During Biodegradation of Hexahydro-1,3,5-trinitro-1,3,5-triazine (RDX) with Municipal Anaerobic Sludge". *Appl. Env. Microbiol. 66*:2652-2657.

Heijman, C.G., E. Grieder, C. Holliger, R.P. Schwarzenbach. 1995. "Reduction of Nitroaromatic Compounds Coupled to Microbial Iron Reduction in Laboratory Aquifer Columns". *Env. Sci. Technol. 29*:775-783.

Lovley, D.R., J.F. Stolz, G.L. Nord, and E.J.P. Phillips. 1987. "Anaerobic Production of Magnetite by a Dissimilatory Iron-reducing Microorganism". *Nature 330*:252-254.

Lovley, D.R. and E.J.P. Phillips. 1988. "Novel Mode of Microbial Energy Metabolism: Organic Carbon Oxidtion Coupled to Dissimilatory Reduction of Iron or Manganese". *Appl. Env. Microbiol. 54*:1472-1480.

Lovely, D.R., S.J. Giovannoni, D.C. White, J.E. Campine, E.J.P. Phillips, Y.A. Gorby and S. Goodwin. 1993. "*Geobacter metallireducens* gen. Nov. sp. Nov., a Microorganism Capable of Coupling the Complete Oxidation of Organic Compounds to the Reduction of Iron and Other Metals". *Arch. Microbiol. 159*:336-344.

McCormick, M.L., H.S. Kim, and P. Adriaens. 2000. "Transformation of Tetrachloromethane in a Defined Iron Reducing Culture: Relative Contributions of Cell and Mineral Mediated Reactions" In *Preprints of Extended Abstracts Presented at the 219th ACS National Meeting*, San Francisco, CA. Division of Environmental Chemistry, Inc. American Chemical Society.

McLellan, W., W.R. Hartley, and M. Brower. 1988. *Health Advisory For Hexahydro-1,3,5-Trinitro-1,3,5-Triazine*. Technical Report PB90-273533. Office of Drinking Water, U.S. EPA, Washington, DC.

Nealson, K.H. and B. Little, (1997). Breathing Manganese and Iron: Solid-State Respiration. In Neidleman, S.L. and A.I. Laskin (Ed.), *Adv. Appl. Microbiol.* Volume 45. Academic Press: San Diego, CA.

Novak, P., L. Daniels, and G. Parkin. 1998. "Enhanced Dechlorination of Carbon Tetrachloride and Chloroform in the Presence of Elemental Iron and *Methanosarcina barkeri*, *Methanosarcina thermophila*, or *Methanosaeta concillii*". *Env. Sci. Technol. 32*:1438-1443.

Oh, B-T. and P.J.J. Alvarez. 2001. "RDX degradation with bioaugmented Fe^0 filings: implications for enhanced PRB performance". In *Proceedings from In Situ and On-Site Bioremediation*, The Sixth International Symposium. June 4-7, 2001. San Diego, CA. This volume.

Phillips D.H., B. Gu, D. Watson, Y. Roh, L. Lianf, and S.Y. Lee. 2000. "Performance Evaluation of a Zero-valent Iron Reactive Barrier: Mineralogical Characteristics". *Env. Sci. Technol. 34*:4169-4176.

Scherer M.M., S. Richter R.L. Valentine, and P.J.J. Alvarez. 2000. "Chemistry and microbiology of permeable reactive barriers for in situ groundwater cleanup". *Crit. Rev. Env. Sci. Technol. 30*:363-411.

Testud, F., J. Glanclaude, and J. Descotes. 1996. *Clinical Toxicology 34*:109-111.

Till, B., L. Weathers, and P. Alvarez. 1998. "Fe(0)-supported autotrophic denitrification". *Env. Sci. Technol. 32*:634-639.

Weathers, L.J., G.F. Parkin, and P.J. Alvarez. 1997. "Utilization of Cathodic Hydrogen as Electron Donor for Chloroform Cometabolism by a Mixed Methanogenic Culture". *Env. Sci. Technol. 31*:880-885.

Wildman, M. and P. Alvarez. 2001. "RDX Degradation Using an Integrated Fe(0)-Microbial Treatment Approach". *Water. Sci. Technol. 43*: in press

FIELD DEMONSTRATION OF *IN SITU* BIODEGRADATION OF HE IN SOILS

Ken Rainwater, Caryl Heintz, and Tony Mollhagen (Texas Tech University Water Resources Center, Lubbock, Texas)
Corey Radtke (INEEL, Idaho Falls, Idaho)
Lance Hansen (U.S. Army COE-ERDC, Vicksburg, Mississippi)

ABSTRACT: The first field pilot-scale demonstration of a technology for *in situ* remediation of vadose zone soils contaminated with high explosives (HE) has been performed at the Department of Energy's Pantex Plant. The HE of concern at the demonstration site were RDX and 1,3,5-TNB, with concentrations ranging from 5 to 35 ppm, above the risk reduction clean-up criteria of 2.6 and 0.51 ppm, respectively. The shallow (<10 m depth) soils at the site could not be excavated due to the presence of buried utilities. Based on previous laboratory studies, it was found that the impacted soils had indigenous microbial populations that could be encouraged to reduce the RDX and TNB. A five-spot well pattern, with injection at the central well and extraction at the four outer wells (each 4.6 m from the injection well), was used to flood the target vadose zone soils with nitrogen gas with the intent of stimulating the HE degraders. The system was monitored periodically for gas composition as well as HE concentrations and microbial activity in retrievable soil samples. After 295 days of treatment, the target HE concentrations dropped by 40 percent from the initial site. Operation of the site continues.

INTRODUCTION

In the last two decades, contamination of soil and groundwater by high explosives (HE) has been found at many government and private facilities. These facilities typically were involved with the missions of the Department of Defense or the Department of Energy. The HE contaminants are remnants of past or current manufacture, testing, or training with conventional ordnance or nuclear weapons (Ramsey et al., 1995). Under ambient environmental conditions, HE are highly persistent in soil and groundwater and exhibit a resistance to naturally-occurring volatilization or biodegradation (Craig et al., 1995). The HE exhibit relatively low water solubilities that contribute to both significant residual concentrations in soil and significant concentrations in groundwater. Efficient and cost-effective techniques for remediating the HE contamination problems are now being developed and implemented at the affected sites. Unfortunately, due to the different site conditions and facility missions, no single remedial approach has yet been found appropriate at all locations. Soil contamination is typically dealt with by excavation followed by treatment and/or disposal, making this approach useful only for relatively shallow soils. To date no *in situ* treatment method has been demonstrated to allow reduction of HE concentrations in place.

The Pantex facility, located 27 km northeast of Amarillo, Texas, has utilized HE in the production of weapons since September 17, 1942. The facility began production of conventional munitions shortly after World War II started and still remains as the Department of Energy's final assembly and disassembly plant for all nuclear weapons in the United States. Today, the 6,500-hectare (16,000-acre) facility, composed mostly of farmland, is operated by BWXT Pantex. During World War II, several buildings were used to process and mold HE in the production of munitions. From 1952 to the present, Pantex has performed casting of machining of HE use in nuclear weapons. Any spills or excess HE were washed into concrete troughs that emptied into unlined ditches and flowed north, west, and south into playa lakes located on the facility. As a result, the HE-contaminated wastewaters have infiltrated into and contaminated the vadose zone, as well as a perched aquifer located 82 m (270 ft) below the Pantex facility. Only since the late 1980s have the HE waste streams been reworked to reduce contaminant discharges.

The primary HE contaminating the soil and groundwater are octahydro-1,3,5,7-tetranitro-1,3,5,7-tetrazocine (HMX), hexahydro-1,3,5-trinitro-1,3,5-triazine (RDX), 2,4,6-trinitrotoluene (TNT), and 1,3,5-trinitrobenzene (TNB) (Ramsey et al., 1995). In 1996, the Texas Natural Resource Conservation Commission (TNRCC) negotiated subsurface cleanup criteria for these compounds and set the Risk Reduction Standard 2 (RRS2) values for HMX, RDX, TNT, and TNB in soil at 511, 2.6, 5.1, and 0.511 mg/ kg of soil, respectively. The RRS2 requires the removal and/or decontamination of HE to levels such that any substantial present or future threat to human health or the environment is eliminated. Concentrations of RDX and TNB above the RRS2 have been observed at several locations in Zone 12 and at depths up to 82 m.

To achieve the required cleanup criteria set by the TNRCC, two techniques have been explored, *ex situ* and *in situ* remediation of the soil. *Ex situ* remediation can be used to treat shallow soils that can be easily excavated from the facility. However, some areas at the Pantex facility have a large number of buried utility lines thus preventing the excavation of soil below these lines. In the area surrounding Buildings 12-43 and 12-24, extensive HE contamination has been reported in surface soils as well as subsurface soils. Extensive soil contamination occurred here because HE-contaminated wastewater was discharged onto the soil and ditches near building 12-43. The presence of many buried utilities precludes excavation. *In situ* bioremediation can offer a feasible approach to treating the HE-contaminated soil while avoiding any buried utility lines.

The purpose of this project was to develop an *in situ* method to biodegrade high explosives in the vadose zone. The project involved the construction of an experimental field site to force an anaerobic treatment zone and thus stimulate indigenous microorganisms to biodegrade the HE. The desired level of treatment is to reduce the HE concentrations to below the RRS2 values. The specific objectives in developing the *in situ* treatment method included [1] location of a site with high levels (greater than 20 mg HE/kg soil) of HE to remediate, [2] determination of distributions of HE contamination and microbial activity present within the soil, [3] design, construct, and operate the field site and control buildings, and [4] evaluate the effectiveness of the process.

BACKGROUND

Previous research at the Texas Tech University Water Resources Center (TTUWRC), the University of Texas, and INEEL demonstrated the potential for in situ degradation of the target HE in Pantex soils. Medlock (1998) examined the relationships among HE concentration, metabolic activity within the soil, and microbial population. The soil samples were collected from the first 10 m below ground surface in the Building 12-43 vicinity. Medlock used EPA Method 8330 to determine the concentrations of HE in the soil. Microbial activity was determined using the Rapid Automated Biological Impedance Technique (RABIT), and microbial populations were quantified with a spiral plate method. Aerobic and anaerobic microbial activity were present in all samples taken from the field site and that metabolic activity levels were similar at all soil depths below 1.5 m. HE concentrations did not affect the amount of anaerobic metabolic activity present within the soil, thus showing the anaerobic organisms remain viable in the presence of HE. Shaheed (1998) conducted multiple tests on soil samples containing HMX and RDX to evaluate the ability of indigenous anaerobic microorganisms to respond to various carbon, nitrogen, and phosphorus amendments. The RABIT method was used to evaluate the response of microorganisms to the nutrients present in the RABIT test cells. Solutions with over one percent concentrations of nutrients did produce positive metabolic responses when used individually or in combinations. McKinney and Speitel (1998) investigated the feasibility of *in situ* bioremediation by determining the environmental conditions needed for RDX to be degraded by the indigenous microorganisms present in soil samples from the cores taken near Building 12-43. The experiments involved the addition of varying amounts of oxygen, nitrogen, and organic carbon source to the headspace of closed vials containing 2-g portions of soil inoculated with ^{14}C-radiolabeled RDX. The samples were incubated at 20°C, and the radiolabeled RDX and its derivatives were monitored at regular intervals by a liquid scintillation counter. The results showed that anoxic (little or no oxygen) environmental conditions should be present for microorganisms to be able to degrade RDX. In addition, biodegradable organic carbon increased biodegradation rates.

Radtke and Roberto (1998) of INEEL investigated the addition of organic vapors in a nitrogen atmosphere to laboratory soil columns to stimulate the indigenous anaerobic microorganisms and encourage biodegradation of RDX and TNB. The soil samples were provided by the TTUWRC from samples collected near Building 12-43. The addition of nitrogen gas by itself created the anaerobic atmosphere that was needed by the indigenous microorganisms to biodegrade HE. In addition to the nitrogen gas, ethanol, acetone, acetic acid, and isobutyl acetate vapors were selected as possible carbon sources. Three soil column replicates for each organic solvent addition and three nitrogen only soil columns were used in the experimental apparatus. In the soil column setup, solvent-laden nitrogen and humidified nitrogen were combined and injected through the soil column. Each of the 15 soil columns were injected with nitrogen gas and the four organic vapors for 98 days. At the end of the 98-day test period, the soil columns were sampled. The samples were then analyzed at INEEL and the U.S. Army Corps of Engineers Cold Regions Research and Engineering Laboratory (CRREL) to determine the final concentrations of RDX and TNB. Average RDX and TNB concentrations decreased under all five test conditions. The use of nitrogen gas alone also resulted in the degradation of both RDX (20 percent) and TNB (60 percent). It is likely that organic carbon available in the clayey sand soil was degraded in the nitrogen only atmosphere while the RDX and TNB were cometabolized. The columns with the added vapors also showed reductions in RDX (all near 20 percent) and TNB (85 to 95 percent).

MATERIALS AND METHODS

The field site for the demonstration was set up north of Building 12-43 following characterization of the HE distribution in the vicinity. The area was investigated by geoprobe, with soil samples collected continuously to a depth of 9.2 m. The soils were analyzed at 0.6-m intervals using Method 8330. The intents was to identify a site with RDX and TNB concentrations above 20 mg/kg and large enough to allow construction of a modified five-spot injection and extraction well system. Figure 1 describes the layout of the system.

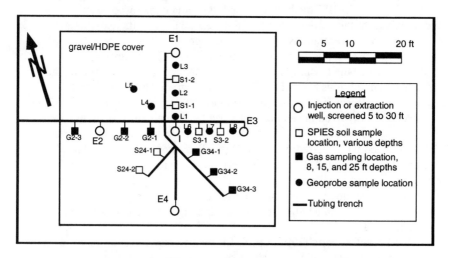

FIGURE 1. Field Site Well and Sampling Positions

The injection (I in Figure 1) and extraction (E1 to E4) wells were bored using a geoprobe rig provided by Sandia National Laboratory (SNL). Each well extended to a depth of 9.2 m (30 ft) and had a diameter of 5.1 cm (2 in). The wells were cased with 2.5-cm (1-in) schedule 80 PVC pipe and were screened from 1.5 to 9.2 m (5 to 30 ft); the slot size of the screen was 0.51 mm (0.020 in). The screened section of each well was sand-packed to allow for adequate gas flow into and out of the wells. Finally, bentonite chips were used in the annulus of the

0.9 m (3 ft) above the sand pack in each well to prevent water from infiltrating down into the well. The surface completion of the well included concrete in the annulus of the last 0.6 m (2 ft) to the ground surface and a steel 25-cm (10-in) diameter manhole to allow access to the connections to the 0.64-cm (0.25-in) copper injection-extraction tubing.

Six gas sampling wells (shown as "Gx-x" in Figure 1) were placed in the field site so that the gas composition over the entire field site could be monitored. Placement of the gas sampling ports at three different depths was intended to show the gas composition in the shallow and deep regions of the treatment zone. These wells were bored to their desired depths using a 7.6-cm (3-in) diameter auger. In each well, the gas sampling ports were set at depths of 2.4, 4.6, and 7.6 m (8, 15, and 25 ft). The gas sampling ports were constructed from a 20-cm (8-in) wire mesh screen and had a geoprobe drive tip placed on the end. A 0.64-cm (0.25-in) O.D. plastic tube carried the sampled gas to the top of the well where it was then joined to 0.64-cm (0.25-in) O.D. copper tube using a compression fitting. The three gas sampling ports were sand packed to allow for adequate gas flow into the sampling port. In addition, bentonite chips were placed in between each gas sampling port to prevent gas from moving vertically within each well. The surface completions were the same as the injection and extraction wells.

A set of specially designed strategically placed in-situ environmental samplers (SPIES) were used to monitor the amount of HE degradation that was occurring in the treatment zone. The SPIES soil samples were removed from the actual field site using a geoprobe rig, and the initial HE concentrations were determined for each end of the soil sample. Each soil sample was initially 0.6 m (2 ft) long with a diameter of 2.5 cm (1 in). The SPIES soil samples were each housed in a plastic tube with small holes drilled around the circumference to allow adequate movement of gas into the soil sample. The SPIES wells (shown as "S" in Figure 1) were constructed to have soil sample depths of 2.4, 4.6, and 7.6 m, (8, 15, or 25 ft, depth noted on Figure 1). The six SPIES wells were cased using 6.4-cm (2.5-in) PVC pipe with screened sections in the bottom 0.76 m (2.5 ft) of the well. The screened section of each well was sand packed and bentonite chips were used just above the screened section. A 0.64-cm (25-in) copper tube was installed from the middle of the screened section to the manhole to monitor the composition of gasses in the well. The surface completions were the same as the injection and extraction wells. The SPIES soils were retrieved at selected times and analyzed for HE and microbial activity.

In order to keep water from infiltrating into the soil and limit communication with the atmosphere, a 12 m by 12 m (40 ft x 40 ft), 60-mil high-density polyethylene (HDPE) geomembrane was placed over the entire field site. A soil and gravel cover was placed on top of the HDPE for aesthetics and to hold the geomembrane in place. Plumbing work in the finished field site was completed using 0.64 cm (0.25-in) O.D. copper tubing, and compression fittings were used to join all copper tubes. The tubing lines were run from the finished field site to the small control buildings located approximately 46 m (150 ft) southwest of the finished site. The location of the two control buildings was chosen because of the close proximity of the field site to a sensitive building. For safety purposes, the control buildings were placed approximately 150 ft southwest of the field well site. One control building housed most of the plumbing work, valves, flow meters, and extraction pumps. The other control building housed the liquid nitrogen cylinder, water column, and gas monitoring devices.

The purpose of the nitrogen injection system was to create anaerobic conditions within the soil in the treatment zone. The injection flow rate was 4.8 L/min (nominal 0.17 cfm). The residual pressure in the liquid nitrogen cylinder was used to inject nitrogen gas into the injection well. The nitrogen was bubbled through a water column to maintain a relative humidity of approximately 30 percent in the nitrogen gas and keep from drying the soil. Each of the four 1/3 hp extraction vacuum pumps were set with rotameter flow controllers to target flow rates of 4.8 L/min (nominal 0.17 cfm).

A single pump was used to extract gas from the 18 gas sampling ports and the 6 soil sampling gas ports. Each of the 24 gas sampling lines had a two-way valve placed on its end so that the lines could be sampled individually. When a gas port was sampled, the extracted gas left the extraction pump and was injected into a one liter, tedlar gas sampling bag so that it could be analyzed with a landfill gas monitor.

The gas composition from the extraction wells and gas sampling ports was analyzed with a LANDTEC™ GA-90 landfill gas analyzer. The gas analyzer was used to determine the percent oxygen, methane, and carbon dioxide in the extracted gas. Gas composition monitoring was performed weekly on all extraction wells, gas sampling ports, and gas ports in the soil sampling wells.

RESULTS AND DISCUSSION

The system was put in operation on May 24, 1999. Operation continued relatively continuously for 295 days, with occasional maintenance shutdowns. Experience showed that some air and runoff water leaks occurred near some of the manholes, and the leaks were repaired. Most of the gas sampling ports did not allow gas flow for sampling, so it was concluded that the plastic tubes from the ports were most likely blocked by squeezing by the swelled bentonite annular packing. The gas sampling ports in the SPIES were then used as the primary indicators of the target zone oxygen levels. It was found that only very low capacity (1/8-hp) vacuum pumps provided gas samples that showed low oxygen percentages at these six locations. Larger suction encouraged leakage of atmospheric air into the SPIES, and raised the measured concentrations. Application of the low capacity pump showed the target oxygen levels of 5 percent or less were achieved in four of the six SPIES holes during the treatment period. Methane was never detected above 0.1 percent at any of the extraction or SPIES holes. Carbon dioxide levels in the extraction and SPIES holes ranged from 0.0 to over 2.0 percent during the test duration.

At the end of the treatment period, soil cores were collected with a geoprobe unit provided by the Pantex Environmental Restoration group. The cores were 5-cm (2-in) diameter, 1.2-m (4-ft) long, collected continuously to depths of 0 to 9.2 m (30 ft) at the locations L1-L8 shown on Figure 1. These locations were selected to be in the target treatment zone between the injection and extraction wells, to avoid the buried plumbing, and to allow for future sample locations elsewhere. The samples were collected as "continuous" core, and soil was analyzed every 0.6 m (2 ft) in the cores. The average (with 95 percent confidence interval) RDX and 1,3,5-TNB concentrations in these 117 samples were 10.8±1.9 and 10.3±2.1 mg/kg, respectively, which were 40 percent lower than the initial average values of 18.2±2.8 and 17.1±3.3 mg/kg, respectively, taken from the original five well locations at the site as shown in Figure 2.

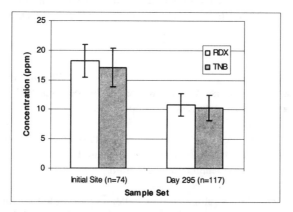

FIGURE 2. Comparison of Initial and Final HE Concentrations with 95 percent Confidence Intervals

Operation of the system continued until 333 d after commencement. The final average SPIES soils RDX and TNB concentrations at 333 d were 5.6±5.8 mg/kg and 1.4±1.0 mg/kg, respectively, which were lower than the initial average values of 12.9±4.3 mg/kg and 11.6±4.8 mg/kg. Simple first order decay equations were fitted to the RDX and TNB vs. time data for the eight SPIES sampling dates. The rate coefficients were 0.0025 d^{-1} and 0.0071 d^{-1} for the RDX and TNB, respectively. Based on the RABIT analyses, metabolic activity remained high, while variable, within the SPIES samples during the entire treatment period. Metabolic activity in the samples from the eight boreholes taken at day 295 was distributed more deeply in the target treatment zone than that seen in the initial conditions.

CONCLUSIONS

Based on these encouraging results, the treatment process was successful in reducing the RDX and TNB concentrations at the site. Pursuit of a process patent is underway. The demonstration is continuing at this field site. During the summer of 2000, the system was modified in two ways. First, the liquid nitrogen tank source was replaced with a membrane nitrogen generator for continuous, dependable nitrogen supply with much higher flow rates than previously available. Second, the flow regime was reversed, with injection at the four outer wells (known in this report at E1, E2, E3, and E4) and extraction at the central well (I). This scheme has lead to more uniform reductions in oxygen content within the treatment system. This approach is being applied and monitored for several months, then followed with a geoprobe sampling event in spring 2001.

ACKNOWLEDGMENTS

The authors wish to acknowledge the support of the BWXT Pantex Environmental Restoration soil remediation group lead by Jay Childress. The project was also supported by the Amarillo National Research Center, the DOE's Innovative Technology Remediation Demonstration Program, and the U.S. Army Corps of Engineers Environmental Research and Development Center.

REFERENCES

Craig, H.D., Sisk, W.E., Nelson, M.D., Dana, W.H. 1995. "Bioremediation of Explosive-Contaminated Soils: A Status Review." Proceedings of the 10th Annual Conference on Hazardous Waste Research. pp. 164-179.

McKinney, D.C., and Speitel, G.E. 1998. "Feasibility of In-Situ Remediation of Residual High Explosives in the Vadose Zone Beneath the Pantex Plant." Amarillo National Resource Center for Plutonium.

Medlock, W.N. 1998. "Laboratory Studies Indicating the Potential for Bioremediation of High Explosives in Soil at the Pantex Plant." M.S. Thesis, Texas Tech University, Lubbock, TX.

"Nitroaromatics and Nitramines by High Performance Liquid Chromatography (HPLC)." 1994. EPA Method SW-846-8330. Revision 0.
Radtke, C.W. and Roberto, F.F. 1998. "Solvent Screening for the Enhancement of Vapor-Phase Anaerobic Degradation of RDX at the PANTEX Facility." Idaho National Engineering and Environmental Laboratory Biotechnologies Department.

Ramsey, R.H., Rainwater, K.A., and Mollhagen, T.R. 1995. "Investigation of Historic Discharges to the Ditches and Playas at the Pantex Plant." Water Resources Center, Texas Tech University.

Shaheed, M.S. 1998. "Use of Impedance Microbiology in Assessing Metabolic Responses of Microorganisms in High Explosive Contaminated Soil." M.S. Thesis, Texas Tech University, Lubbock, TX.

REDUCTIVE BIOTRANSFORMATION OF NITRATE AND EXPLOSIVES COMPOUNDS IN GROUNDWATER

Mark S. Heaston (Earth Tech, Pueblo, Colorado)
Paul W. Barnes (Earth Tech, Pueblo, Colorado)
Kimberly R. Alvestad (Earth Tech, Pueblo, Colorado)

ABSTRACT: Earth Tech has successfully completed two bench-scale studies demonstrating the treatment of explosives contaminated groundwater by amendment with Hydrogen Release Compound® (HRC®). The first of these two bench scale studies was intended to demonstrate the removal of explosive contaminants, with the second designed to assess transformation product formation and accumulation. Groundwater used in these bench scale studies was obtained from groundwater monitoring wells located at the Pueblo Chemical Depot, in Pueblo, Colorado. Groundwater from this portion of the site contains various explosive compounds including, hexahydro-1,3,5-trinitro-1,3,5-triazine (RDX), 1,3,5-trinitrobenzene (TNB), 2,4,6-trinitrotoluene (TNT), 4-amino-2,6-dinitrotoluene (4-Am-2,6-DNT), 2-amino-4,6-dinitrotoluene (2-Am-4,6-DNT), and 2,4-dinitrotoluene (2,4-DNT), as well as nitrate. Biotransformation of these compounds is achieved by cometabolic reduction resulting from the addition of HRC to the groundwater.

Results of the first bench scale study showed that treatment was achieved as quickly as 14 days in test reactors, with strongly reducing conditions (<-400mV) being achieved within 7 days. The efficacy of removal was greater than 95% for most compounds, as determined by Earth Tech's on-site laboratory performing U.S.E.P.A. Method SW8330 analysis.

Since early attempts to describe the reduction pathway of RDX included byproducts of environmental concern such as the mono-, di-, and trinitroso deriviatives (MNX, DNX, and TNX) of RDX, n-nitrosodimethylamine (NDMA), and hydrazine, a second study was performed to evaluate the potential for accumulation of these compounds. Samples were collected at a point believed to represent the end of the reductive treatment process and analyzed for these potential byproducts. Post treatment analyses for these byproducts confirm that these compounds are either not produced or are further reduced under continued reducing conditions. These results indicate that reductive biotransformation may be suitable for *in situ* treatment of explosives contaminated groundwater and that HRC maybe a suitable electron donor for this process. Successful *in situ* treatment of explosives contaminated groundwater by HRC could provide a cost-effective alternative to other *in situ* treatment technologies such as chemical oxidation, or *ex situ* 'pump and treat' strategies.

INTRODUCTION

The Pueblo Chemical Depot was home to an ammunition washout facility, which operated intermittently from the early 1940s until 1974. Wastewater from this operation was discharged from the washout facility to an unlined ditch leading to an unlined evaporative lagoon. Over time, residual explosives contained within this wastewater have migrated through the soil and into the shallow aquifer beneath. Investigative samples have shown explosives contamination in groundwater ranging from levels high as 10,300 parts per billion (ppb) total explosives in the source area to low ppb levels near the installation's boundary.

As a part of an *in situ* treatment technology evaluation, Earth Tech conducted two bench-scale treatment studies using Hydrogen Release Compound® (HRC®). HRC is a proprietary polylactate ester compound bonded to a glycol carrier. HRC is intended to serve as an electron donor for the reduction of nitroaromatic and explosive compounds.

The goals of these studies were; to determine if HRC could effectively reduce nitrate and nitroaromatic explosives compounds, to determine if the kinetics of reduction were favorable at low starting concentrations, and to determine if postulated, potentially harmful, transformation products of hexahydro-1,3,5-trinitro-1,3,5-triazine (RDX) are formed and/or accumulated during the reductive transformation process. Earlier work conducted in this area suggested the potential for the formation of several byproducts of environmental concern. (McCormick *et* al., 1981) These potential byproducts of concern included the mono-, di-, and tri- nitorso derivatives (MNX, DNX, and TNX) of RDX, as well as n-nitrosodimethylamine (NDMA), and hydrazine.

METHODS AND MATERIALS

Sample Collection. For the first bench-scale study, samples were collected from two separate areas of the site. Ground water was collected from two wells located in direct proximity to the source area, and from one well located near the facility boundary. Historical sample results from wells near the source area indicated the presence of high-levels (>2000 µg/L) of 1,3,5-trinitrobenzene (TNB) and 2,4,6-trinitrotoluene (TNT). Historical data from wells near the facility boundary indicate the presence of TNB, 2,4-dinitrotoluene (2,4-DNT), RDX, and nitrate at levels 2-3 orders of magnitude lower than the source area.

Prior to sample collection, 3-5 well volumes were removed from each well until ground water pH, dissolved oxygen, turbidity, temperature, and conductivity were stable. Samples were collected from each well using a peristaltic pump. Approximately 100 gallons of groundwater were collected at both the source area and boundary locations.

In order to more accurately represent *in situ* conditions, clean soil, similar in composition to saturated aquifer material was collected from the dry bed of nearby Chico Creek. Soil collected from Chico Creek was used to create twelve laboratory reactors.

Study Design. For the first bench-scale study twelve reactors were constructed from 45-gallon polyethylene drums. A perforated, filter-cloth wrapped pipe was installed horizontally through the sidewall of the drum approximately 6 inches from the bottom of the reactor for sample collection. A similar tube was installed vertically through the top of the reactor to allow for parameter monitoring by direct-read electrodes. These reactors were filled to approximately 90% capacity with the soil collected from Chico Creek. Two reactors were then filled with groundwater collected from the source area (high contaminant levels), and two reactors were filled with groundwater collected near the facility boundary (low contaminant levels. These reactors were filled to minimal headspace and sealed to serve as control reactors for the demonstration. Three test reactors each for both high and low contaminant levels were then filled. A dose of approximately 5,000 mg/L HRC was added to groundwater from each of the sample areas. The HRC product and groundwater were continuously cycled through a pump until all of the HRC product was either in solution or suspended within the groundwater. Three reactors were filled to a minimal headspace with high and low-level groundwater amended with HRC. These reactors were sealed to serve as test reactors.

Upon completion of loading the reactors, temperature, pH, and ORP measurements were collected at intermittent intervals to track the effects of the HRC addition. Samples collected from each of the reactors and were also analyzed for nitrate and explosives concentration periodically. Samples for nitrate were analyzed using a Hach field kit, and samples for explosives were analyzed by method SW8330. Earth Tech's on-site laboratory conducted all analyses.

Reactors for the second bench-scale study were assembled in the same way, with triplicate control and test reactors for low-level concentrations only. Analyses for the second study were specific to RDX and its transformation products. Explosives analyses were conducted using method SW8330 and conducted at Earth Tech's on-site laboratory. Samples for MNX, DNX, and TNX were analyzed using a modified method SW8330 at the United States Army Corps of Engineer's Waterways Experiment Station, and samples for hydrazine and NDMA were analyzed by methods ASTM D1385 (colorimetric)/SW8315 (chromatographic) and SW1625, respectively at off-site contract laboratory facilities.

RESULTS AND DISCUSSION
Discussion of the first bench-scale study results will focus mainly on the low-level test and control reactors. In the high-level study, abiotic removal in both test and control reactors made differences related to treatment indistinguishable. Similar results were noted in the low-level control reactors for several compounds, but not to the extent observed in the high-level control reactors. This phenomenon, although poorly understood, is well documented in previously published soil column studies. (Ainsworth *et al.*, 1993; Pennington *et al.*, 1995) Discussion of the second bench-scale study results will focus only on the reduction of RDX and the production and/or accumulation of potential transformation products.

ORP and pH. ORP measurements were collected intermittently throughout both studies to verify microbial activity. The average initial (Day 0) ORP levels were measured to be 52 mV and 119 mV for the control and test reactors respectively.

By day 7 of the study the average ORP readings of the test reactors had dropped to less than –400 mV while ORP levels in control reactors remained at or above baseline levels. ORP levels in the test reactors remained below –400 mV on average through Day 12, before gradually returning to oxidative levels. Minimal change was observed in the ORP levels of the control reactors throughout the duration of the study. This indicates that groundwater from PCD amended with the HRC product is capable of achieving strongly reducing conditions, increasing the potential for nitroaromatic explosive and nitrate reduction.

Similar changes in ORP readings were observed in the second study, as depicted in Figure 1, with the initial (Day 0) average ORP readings for the control and test reactors measuring 180 mV and 269 mV respectively. ORP levels in the test reactors were measured to be as low as –489 mV, with little to no change observed in the control reactors. ORP levels in the treated test reactors remained below –300 mV for 5 days before beginning to climb towards oxidative conditions.

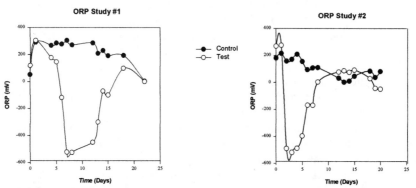

FIGURE 1. ORP (mV) measurements over time in test and control reactors

Measurements of pH were also collected at the same interval as ORP to assess the effect that HRC and cometabolic processes on the pH of the system. Control reactors for both studies displayed little to no change, maintaining a pH of approximately 7.5. Addition of the HRC product reduced the initial pH to 5.5; microbial activity caused further reduction in pH to approximately 5.0, before re-stabilizing at approximately 5.5. Microbial activity was apparently unaffected by the lower pH, as reducing conditions were achieved very quickly.

Nitrate. Nitrate is an analyte of significant concern when considering *in situ* treatment of explosive compounds. Numerous studies have been conducted using oxidative technologies for the treatment of explosives. However, oxidative technologies are not effective in the treatment of nitrate, which is often a co-contaminant of nitroaromatic explosives.

Nitrate samples were collected and analyzed at various points throughout this study to assess the potential of HRC to reduce nitrate concentrations. Initial (Day 0) concentrations of nitrate were measured to be 9.61 mg/L and 9.88 mg/L on average in the low-level test and control reactors respectively. After 21 Days of exposure to the HRC product, average nitrate levels in the test reactors measured 1.22 mg/L as compared to 8.03 mg/L in the control reactors. Total reduction in the test reactor was 87% as compared to only 19% reduction in control reactor concentrations. These results indicate that HRC served as an effective electron donor for the reduction of nitrate. Figure 2 illustrates the reduction of nitrate concentration in the test reactors as compared to the control reactors

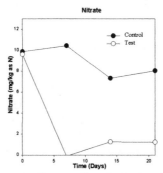

FIGURE 2. Nitrate concentration vs. time in test and control reactors.

1,3,5-Trinitrobenzene. Day 0 average concentrations of TNB were determined to be 3.9029 µg/L and 3.7809 µg/L in the test and control reactors respectively. After 14 Days of study, results from the test reactors were non-detect at the analytical reporting limit of 0.25 µg/L and results from the control reactors were also non-detect. TNB reduction in the test reactors could not be attributed to amendment with HRC due to the removal of TNB from the control reactors apparently by abiotic processes such as adsorption or abiotic reduction.

2,4-Dinitrotoluene. Average Day 0 concentrations of 2,4-DNT were 1.6049 µg/L in test reactors, and 1.3087 µg/L in the control reactors. By Day 14 of the study, concentrations of 2,4-DNT had been reduced to less than the analytical reporting limit of 0.1 µg/L in the test reactors. Control reactors also displayed significant reduction in DNT concentrations, with the average Day 14 concentration being measured at 0.2238 µg/L. Similar to TNB, 2,4-DNT concentrations in the control reactors were significantly reduced. This indicates that while HRC may have increased the rate of 2,4-DNT reduction, removal cannot be attributed solely to amendment with HRC.

RDX. Samples collected at Day 0 returned average concentrations of 10.1634 µg/L and 9.2830 µg/L of RDX in the test and control reactors, respectively. RDX did not appear to be as susceptible to abiotic removal as TNB and 2,4-DNT. Samples collected after 14 days of study revealed that the test reactors were all non-detect for RDX at the analytical reporting limit of 0.25 µg/L, as compared to an average concentration of 8.0413 µg/L in the control reactors. The treated reactors displayed a reduction >97% as compared to only 13% in the control group, thus indicating that amendment with HRC produced conditions that let to the reduction of RDX concentrations.

During the second study, Day 0 samples were collected to establish a baseline concentration for monitoring the reduction of RDX expected during the formation of the transformation products of concern. The average Day 0 concentrations of RDX observed in the control and test reactors were 8.783 µg/L and 9.590 µg/L respectively. Samples collected from the three test reactors were non-detect for RDX by Day 8 and were again confirmed to be non-detect at day 20. Evidence of treatment in the treated test reactors is offered by the results of untreated control reactors collected simultaneously. By day 8, untreated control reactors displayed an average RDX concentration of greater than 5.0 µg/L, significantly greater than the treated test reactors. Figure 3 shows the reduction in RDX concentrations over time in test reactors as compared to control reactors. Reduction in RDX concentrations in untreated control reactors was believed to be attributed to an abiotic mechanism. One of the three control reactors did however display evidence of microbial activity as seen in the ORP readings for this reactor. This was believed to be the result of residual HRC retained in the reactor from a previous HRC study. If results from this reactor are discarded, the average untreated control reactor concentrations were 6.0 µg/L and 4.7 µg/L at Day 8 and Day 20 respectively, displaying a gross reduction of >97% in the treated test reactors as compared to 32% (Day 8) and 47% (Day 20) observed in the control reactors.

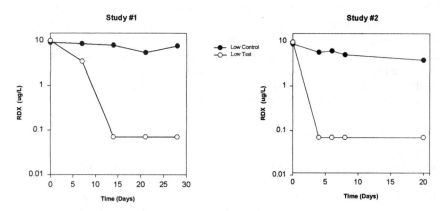

FIGURE 3. RDX concentrations over time in test and control reactors

MNX, DNX, and TNX. MNX, DNX, and TNX have been shown to be the first step transformation products in the reduction of RDX. Day 0 samples analyzed for MNX, DNX, and TNX were a found to be non-detect for all three compounds. Samples collected at Day 8 and Day 20 also returned values of non-detect. This would tend to support the theory that the intermediate nitroso- derivatives of RDX are relatively unstable, and susceptible to further reduction in strongly reducing environments such as those observed in this study.

Hydrazine. Hydrazine, although once postulated as a potential component of the RDX reduction pathway, has never been detected in a significant quantity during laboratory testing. Day 0 samples collected for hydrazine were analyzed by ASTM D1385. Later investigation into this method revealed that it was a colorimetric method subject to interferences such as high turbidity and amine compounds, which could be produced under strong reducing conditions such as those observed in this study. Day 0 samples analyzed by this colorimetric method provided a result of 7.0 µg/L of hydrazine in the control reactors, which could not be reproduced in the Day 0 test reactor result, which provided a non-detect value. Similarly, Day 8 samples provided conflicting data. The test reactors produced a second non-detect value, while the control reactors produced a result of 30.0 µg/L. Day 20 samples from test and control reactors produced detections at 12.0 µg/L and 7.0 µg/L respectively. Although neither of these values exceeds the highest observed concentration in the control reactors, concern over the validity of these sporadic and fluctuating detections led to the collection of a sample at Day 27. This sample was analyzed by an alternative chromatographic method, SW8315, recommended by the laboratory. Results from this analysis were non-detect for both test and control reactors. Discussions with the laboratory performing the chromatographic analysis revealed that there were numerous matrix peaks in the Day 27 samples and that hydrazine was not present. Detections reported by the colorimetric method were most likely the result of matrix interference, high turbidity, or the presence of amine compounds produced under reducing conditions. Results from the Day 27 sample indicate that hydrazine, if produced, does not persist under the conditions of this study.

NDMA. Like hydrazine, NDMA has also been postulated as a potential product of the cometabolic reduction of RDX. Day 0 samples from test and control reactors produced results of 0.0037 µg/L and 0.0043 µg/L respectively. Day 8 samples collected from both control and test reactors returned non-detect values for NDMA, suggesting that NDMA did not accumulate during the reductive biotransformation of RDX under the conditions observed in this study. These results were further validated by Day 20 samples, in which the test reactors were again confirmed to be non-detect and the control reactors returned an estimated result (less than the reporting limit but greater than the method detection limit) of 0.0011J µg/L.

CONCLUSION

Although significant abiotic removal of 2,4-DNT and TNB made results for these compounds difficult to interpret, RDX and nitrate results provided strong evidence that anaerobic treatment using HRC as an electron donor facilitates the reductive transformation of these contaminants in groundwater. Removal rates for RDX and nitrate suggest that the kinetics associated with this type reductive transformation are favorable even at low starting concentrations. Non-detect results displayed by test reactors for explosives suggest that HRC is capable of facilitating contaminant reduction to levels below risk based standards for explosives as well as nitrate. It is also evident that the reductive transformation products of RDX considered for in this study are either not formed, or do not persist under the conditions of this study. This combination of reduction rate and efficacy of treatment make this technology a favorable candidate for *in situ* field pilot demonstration at the Pueblo Chemical Depot. Areas still in need of development for this technology include transformation product accumulation analysis for other explosive compounds, and those not well demonstrated in this study, as well as *in situ* demonstration.

REFERENCES

Ainsworth, C. C., Harvey, S. D., Szewcsody, J.E., Simmons, M. A., Cullinan, V. I., Resch, C. T., and Mong, G. H. 1993. "Relationship between the leachability characteristics of unique energetic compounds and soil properties," Final Report, Project Order No. 91PP180, U.S. Army Biomedical Research and Development Laboratory, Fort Detrick, Frederick, MD.

Hawari, J., A. Halasz, T. Sheremata, S. Beaudet, C. Groom, L. Paquet, C. Rhofir, G. Ampleman, and S. Thiboutot. 2000. Characterization of Metabolites during Biodegradation of Hexahydro-1,3,5-Trinitro-1,3,5-Triazine (RDX) with Municipal Anaerobic Sludge. *Appl. Environ. Microbiol.*. 66:2652-2657.

Hawari, J. 2000. Biodegradation of RDX and HMX: From Basic Research to Field Application. p. 277-310. In J. C. Spain, J. B. Hughes and H-J. Knackmuss (ed.), *Biodegradation of Nitroaromatic Compounds and Explosives*, CRC Press, Boca Raton, Florida.

McCormick, N. G., J. H. Cornell, and A.M. Kaplan. 1981. Biodegradation of hexahydro-1,3,5-trinitro-1,3,5-triazine. *Appl. Environ. Microbiol.*. 42:817-823.

Pennington, J. C., Myers, T. E., Davis, W. M., Olin, T. J., McDonald, T. A., Hayes, C. A., and Townsend, D. M. 1995. "Impacts of sorption on in situ bioremediation of explosives-contaminated soils," *Technical Report IRRP-95-1*, U.S. Army Engineer Waterways Experiment Station, Vicksburg, MS.

TREATMENT OF EXPLOSIVES-CONTAMINATED GROUNDWATER BY IN SITU COMETABOLIC REDUCTION

Paul W. Barnes (Earth Tech, Pueblo, Colorado)
Mark S. Heaston (Earth Tech, Pueblo, Colorado)
Joanne C. Compton (REACT Environmental Engineers, Saint Louis, Missouri)

ABSTRACT: Earth Tech Inc. has completed the first field-scale demonstration of *in situ* explosives-contaminated groundwater treatment by cometabolic reduction using Hydrogen Release Compound® (HRC®) at Pueblo Chemical Depot (PCD) in Pueblo, Colorado. The study was conducted to address contaminants originating from a former munitions washout facility including; 2,4-dinitrotoluene (2,4-DNT), hexahydro-1,3,5-trinitro-1,3,5-triazine (RDX), 1,3,5-trinitrobenzene (1,3,5-TNB), and nitrate. High rates of removal were observed for all contaminants of concern, and risk-based treatment standards were achieved in multiple monitoring wells within 105 days of injection of the HRC product. RDX and 2,4-DNT, the primary contaminants of concern, were reduced from average concentrations 9.3 µg/L and 3.7 µg/L, respectively to less than their analytical method detection limits of 0.0616 µg/L and 0.0509 µg/L, respectively in the most significantly affected wells. 1,3,5-TNB was reduced from an average concentration of 231 µg/L to less than the method detection limit of 0.0811 µg/L, and nitrate was reduced from an average concentration of 6.1 mg/L to less than the method detection limit of 0.2 mg/L in the same wells. The study also evaluated indirect methods of tracking treatment progress, and demonstrated both the *in situ* movement of HRC product and the diffusion of low redox conditions.

INTRODUCTION

Pueblo Chemical Depot, located approximately 15 miles East of Pueblo Colorado, was once the site of significant munitions storage and reprocessing operations. The U.S. Army operated a TNT washout facility on-site, intermittently discharging explosives-contaminated wash water to an unlined trench and leach bed over a period of approximately thirty years. These activities resulted in the contamination of groundwater in a shallow surficial aquifer with a number of explosives and explosives-related compounds including 2,4,6-trinitrotoluene (2,4,6-TNT), 2,4-dinitrotoluene (2,4-DNT), hexahydro-1,3,5-trinitro-1,3,5-triazine (RDX), 1,3,5-trinitrobenzene (1,3,5-TNB), and nitrate. The site hydrogeology, extent of contamination, and terrain make *ex situ* remediation challenging and expensive at PCD. Further, the combination of nitroaromatic contaminants and nitrate present at some locations would require two separate *ex situ* treatment technologies. Hydrogen Release Compound® (HRC®) is an injectable polylactate glycol ester distributed by Regenesis Inc. that offers the potential for simultaneously treating both nitroaromatic explosives and nitrate *in situ*. When injected into the subsurface, HRC serves as a slow-releasing carbon substrate for indigenous microorganisms, which in turn consume oxygen and lower the Oxidation-Reduction Potential (ORP) to reducing or strongly reducing levels. Contaminants are transformed to reduced forms by reduction of their susceptible functional groups.

Objective. This study was conducted to evaluate the potential for treating groundwater contaminated by nitrate and the nitroaromatic explosives 2,4-dinitrotoluene (2,4-DNT), hexahydro-1,3,5-trinitro-1,3,5-triazine (RDX), and 1,3,5-trinitrobenzene (1,3,5-TNB) *in situ* by cometabolic reduction using HRC. Secondary goals included verifying bench-scale results and identifying surrogate measurements that could accurately indicate treatment progress more quickly and at lower cost than actual explosives analyses. The study was also intended to confirm that low redox conditions could be achieved and maintained *in situ* despite minimal saturated thickness and a high rate of groundwater flow.

METHODS
Location. The location for the study was selected to be an area where the hydrogeology was well understood and representative of the conditions at the site boundary, where full-scale remediation would ultimately be required. The area had been well characterized by previous investigative drilling and geophysical efforts and was known to include a narrow, meandering subsurface channel with a relatively high rate of groundwater flow. The channel was estimated to be 140' wide and found to contain one to three feet of sand to silty/clayey sands capped by several feet of confining clay or silty clay. The hydraulic conductivity in the channel aquifer was estimated to be 51 feet per day and the hydraulic gradient was 0.02 foot per foot. Groundwater in the channel has historically displayed elevated levels of the explosives and related compounds typically found near the site boundary including nitrate ($\approx$ 6.8 mg/L), 2,4-DNT ($\approx$ 3.9 µg/L), RDX ($\approx$ 9.9 µg/L), and 1,3,5-TNB ($\approx$260 µg/L). Average concentrations of 2,4-DNT and RDX were well above the risk-based levels calculated for the site (0.0885 µg/L and 0.55 µg/L for 2,4-DNT and RDX, respectively).

Study Area Layout. The location selected for the study was the narrowest point of the subsurface channel through which most of the groundwater in that portion of the surficial aquifer flows. Injections of HRC were made in three parallel rows perpendicular to the estimated direction of groundwater flow, crossing the full width of the channel. Injections in each row were spaced at 10-foot intervals. Rows were 5 feet apart with injections symmetrically offset from adjacent rows. Fifteen monitoring wells were installed around an existing well (CSPDPW205) to monitor the study area. Of these, one well was installed 40 feet upgradient of the HRC injections to serve as a background control, one well was installed 5 feet downgradient of the injection array. The remaining 13 wells were installed in 5 rows progressively downgradient of the injection array at distances of approximately 20, 45, 80, 120, and 160 feet. The layout, depicted in Figure 1, was intended to allow groundwater to flow first through the area of the upgradient well (SWTMW29), through the zone of HRC injection, and then through each of the monitoring rows in sequence, such that the direction and rate of flow could be correlated with any observed treatment. During the study, however, meandering channelized flow was observed within the study area and groundwater flow proved to be non-perpendicular to the monitoring network.

HRC Injection. HRC was introduced at each of the thirty injection locations at a rate of 20 pounds per vertical foot of saturated thickness using direct-push tools and a high viscosity pump. Approximately 30% of the product injected consisted of HRC Primer; a less viscous, less polymerized form of HRC intended to increase solubility. The remainder of the product was standard HRC, used to provide extended delivery of the lactate substrate throughout the study.

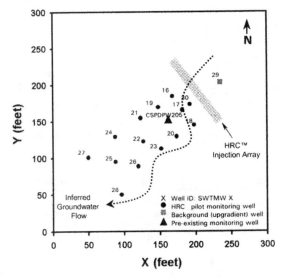

Figure 1: Study Area Layout.

Process Monitoring. An initial set of all monitored parameters was collected at the time of injection to establish baseline values for water chemistry and contaminant concentrations. Field parameters were collected and evaluated frequently to indicate treatment progress and to direct sampling for contaminants at appropriate times. In particular, Chemical Oxygen Demand (COD) and ORP measurements were used to evaluate distribution of the HRC product in the aquifer, and progress toward reducing conditions, respectively. Sampling for target contaminants was conducted at locations where COD and ORP verified that treatment progress would be likely. Full sampling events were conducted at days 0/1, 34/35, 57, 85/86, and 105/106 Additional sampling events are planned at day 180 and at undetermined intervals thereafter until active treatment has ceased.

RESULTS
Field Parameters. The primary parameters used as indicators of product delivery and treatment were COD and ORP. These showed dramatic changes *in situ* and are discussed here at length. Dissolved oxygen (DO), manganese, and ferrous iron also changed significantly. DO concentrations dropped from an average

background concentration of 5.0 mg/L to less than 2.0 mg/L in the affected areas. Manganese and ferrous iron concentrations, due to increased solubility of the reduced forms, increased from 0.32 mg/L and 0.04 mg/L, respectively to 2.2 mg/l and 5.5 mg/l, respectively in the most affected wells.

The remainder of the field parameter measurements showed much less change over time, but did serve to assess the overall impact on groundwater quality following treatment with HRC. For instance, groundwater pH averaged 7.41 prior to beginning the study. While high levels of activity resulted in a corresponding decrease in pH during bench-scale studies, very little change was observed *in situ*. Sulfate concentrations decreased by approximately 17% in affected areas and sulfide concentrations remained unchanged. Arsenic, barium, cadmium, chromium, lead, selenium, silver, and mercury were also monitored for changes associated with decreasing pH and redox, but no significant changes in concentration were noted.

The organic acids; acetic, lactic, propionic, and pyruvic were measured throughout the study to verify delivery and metabolism of the HRC product. Relatively high concentrations of lactic acid were quickly observed in wells near the injection area and moved progressively downgradient with time. Acetic, propionic and pyruvic acids were also detected in affected wells, providing good evidence of the microbial metabolism of the free lactate.

Table 1. Measured Parameters and Average Initial Conditions

Field Parameters	Initial Conditions	Contaminants & Transformation Products	Initial Concentration *(Risk-Based Standard)*
Temperature	15.9 C	Nitroaromatic explosives	a
PH	7.41	2,4-DNT	3.7 µg/L *(0.0885 µg/L)*
Dissolved oxygen (DO)	5.05 mg/L	RDX	9.3 µg/L *(0.55 µg/L)*
Oxidation-reduction potential (ORP)	138 mV	1,3,5-TNB	231 µg/L *(361 µg/L)*
Chemical oxygen demand (COD)	8 mg/L	Nitrate	6.1 mg/L *(10 mg/l)*
Ferrous Iron	0.04 mg/L	Hydrazine	0.939 µg/L *(N/A)*
Manganese	0.32 mg/L	n-nitrosodimethylamine (NDMA)	0.00177 µg/L *(N/A)*
Sulfate Sulfide	229 mg/L ND (5 mg/L)	Nitroso derivatives of RDX (MNX, DNX, and TNX)	MNX ND (0.20) µg/L DNX ND (0.50) µg/L TNX ND (0.20) µg/L *(All N/A)*

a – Analyses for the full list of USEPA Method 8330 explosives were conducted, however only analytes initially near or above the risk-based standard are discussed here.

COD. Since there was virtually no native organic matter in the groundwater COD was measured on-site to indicate total available oxidizable substrate resulting from the addition of HRC. Prior to injection of the HRC product, the average COD in the test area was 8 mg/L. COD measurements quickly indicated dissolution and transport of the product downgradient with elevated levels (>80 mg/L) detected as early as 16 days after injection in the first row of monitoring wells (SWTMW16). COD measurements showed movement of the HRC product downgradient through wells SWTMW17 and CSPDPW205 within 35 days and through wells SWTMW20 and 23 by day 85/86 while the upgradient location, well SWTMW29, remained at or near the background concentration throughout. After 105/106 days of treatment some decrease in the extent of elevated COD measurements indicated the possible dispersion and/or consumption of the more soluble primer component of the injected substrate. 30 days later, the area just downgradient of the injection zone continued to show elevated COD. The time series in Figure 2 depicts the appearance and spread of HRC, measured as COD, in the study area over time.

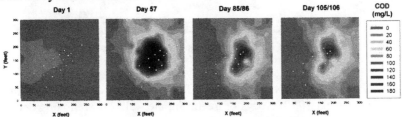

Figure 2: Time Series of COD Isopleths.

ORP. The average measured ORP in the study area prior to treatment with HRC was 138 mV, indicating aerobic, mildly oxidizing conditions. As can be seen in Figure 3, a rapid decrease in ORP was well correlated to the spread of substrate as evidenced by COD. The extent of reduced ORP actually exceeded the extent of elevated COD suggesting influence beyond the physical limit of substrate delivery, possibly due to the diffusion of hydrogen ion (H^+). Through 105/106 days of observation, negative ORP (reducing conditions) had been achieved in all wells except SWTMWs 26, 28, and 29. While the general direction of groundwater flow in the study area was initially believed to be North-Northeast to South-Southwest, COD and ORP measurements over time indicates non-linear channelized flow. Flow was nearly due South as groundwater entered the study area, turning South-Southwest as it passed the first full row of monitoring wells. Wells located in this primary flow path will be the focus of following discussions pertaining to contaminant removal.

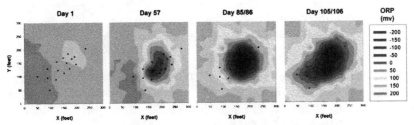

Figure 3: Time Series of ORP Isopleths.

Both COD and ORP proved to be predictive of the spatial extent of treatment as can bee seen in Figure 4 which depicts the aerial extent of measurable RDX removal at the same time points. The shape and extent of the area of greatest contaminant removal was very similar to the shapes and extents of COD and ORP influence. The relative magnitudes of the changes were also well correlated.

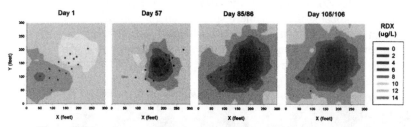

Figure 4: Time Series of RDX Isopleths Depicts Correlation to High-COD and Low-Redox Environments.

RDX. The average initial concentration of RDX in the study area was calculated to be 9.3 µg/L. After 105/106 days of treatment, the wells known to be affected by high COD and low ORP showed removal rates between 45 and 99 percent. The risk-based treatment standard of 0.55 µg/L was achieved by day 85/86 in SWTMW23. Both SWTMW17 and SWTMW23 were below the standard on day 105/106. RDX removal in these and other wells in the affected area is depicted in Figure 5. Removal rates as of day 105/106 at all locations are shown in Figure 6.

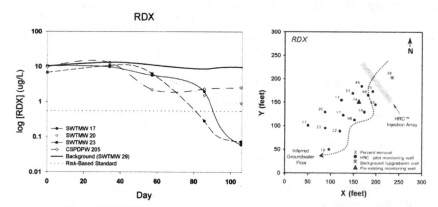

Figure 5: RDX Removal vs. Time.   Figure 6: Percent RDX Removal by Location.

2,4-DNT. The average concentration of 2,4-DNT observed in the study area prior to treatment was calculated to be 3.7 µg/L. After 105/106 days of treatment, removal rates in the affected wells ranged between 72 and 98 percent. The risk-based treatment standard of 0.0885 µg/L was achieved by day 85/86 in SWTMW17 and SWTMW23. Both SWTMW17 and SWTMW23 were below the analytical method detection limit of 0.0509 µg/L on day 105/106. 2,4-DNT removal in these and other wells in the affected area is depicted in Figure 7. Removal rates as of day 105/106 at all locations are shown in Figure 8.

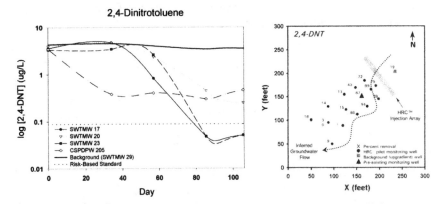

Figure 7: 2,4-DNT Removal vs. Time.

Figure 8: Percent 2,4-DNT Removal by Location.

1,3,5-TNB. The average 1,3,5-TNB concentration in the study area prior to treatment was calculated to be 231 µg/L. After 105/106 days of treatment, removal rates in the affected wells ranged between 86 and 100 percent and reached concentrations as low as the analytical method detection limit of 0.0811 µg/L. Although 1,3,5-TNB concentrations at the beginning of the study were below the risk-based treatment standard of 310 µg/L, reductions in concentration exceeding three orders of magnitude were observed (Figure 9). Percent removal at each well location is depicted in Figure 10.

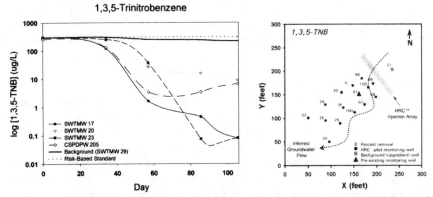

Figure 9: 2,4-DNT Removal vs. Time.

Figure 10: Percent 2,4-DNT Removal by Location.

Nitrate. Prior to treatment the average nitrate concentration in the study area was calculated to be 6.1 mg/L. After 105/106 days of treatment, removal rates in the affected wells ranged between 52 and 94 percent and reached concentrations as low as the analytical method detection limit of 0.2 mg/L. Figure 11 depicts nitrate removal in wells affected by the treatment. Percent removal at each well location is depicted in Figure 12. Although nitrate concentrations at the beginning of the study were below the risk-based treatment standard of 10 mg/L, nitrate removal proved to be widespread and well correlated with elevated COD and reduced ORP much like the explosives compounds.

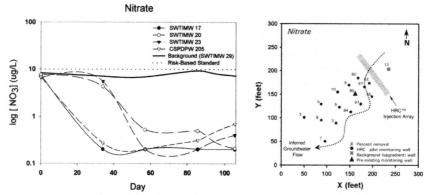

Figure 11: Nitrate Removal vs. Time.   Figure 10: Percent Nitrate Removal by Location.

Transformation Products. Samples were collected to verify that the postulated transformation products of RDX; the mono, di, and trinitroso derivatives of RDX (MNX, DNX, and TNX), the non-cyclic nitrosamine n-nitrosodimethylamine (NDMA), and hydrazine did not accumulate during reductive treatment. Initial samples showed low concentration detections of NDMA (0.0013 µg/L) and hydrazine (0.817 µg/L, estimated) only in SWTMW29, the upgradient background control well. After 106 days of treatment, additional samples were collected from three groups of wells; wells in the area of greatest RDX removal, wells in areas of partial RDX removal, and at SWTMW29. None of the post-treatment analyses returned detectable concentrations of any of the transformation products listed here.

CONCLUSIONS

Results through 105/106 days of treatment confirm that *in situ* reductive treatment using HRC can successfully, and simultaneously address RDX, 2,4-DNT, 1,3,5-TNB and nitrate contamination in the concentration ranges and under the hydrologic conditions described here. Site-specific risk-based action levels were achieved for all contaminants at a minimum of two monitoring locations within 85 to 106 days of treatment. COD and ORP proved to be effective surrogate measures of treatment progress with COD serving as a reliable indicator of product delivery and reduced ORP correlating very well to contaminant removal. Further, it appears that the use of soluble primer may facilitate the rapid development of reducing conditions and increase diffusion into areas not directly addressed by HRC product injection. While additional monitoring will allow the determination of HRC product life *in situ*, recent measurements at PCD suggest that low redox conditions are being maintained in the vicinity of the injections and that treatment is likely to be continuing.

REFERENCES

McCormick, N. G., J. H. Cornell, and A.M. Kaplan. 1981. Biodegradation of hexahydro-1,3,5-trinitro-1,3,5-triazine. *Appl. Environ. Microbiol.* 42:817-823.

Heaston, M. S., P. W. Barnes, and K. A. Alvestad. Reductive Biotransformation of Nitrate and Explosives Compounds in Groundwater." *In-Situ and On-Site Bioremediation. Proceedings of the Sixth International Symposium.* Battelle. June 4, 2001. San Diego, California.

Hawari, J., A. Halasz, T. Sheremata, S. Beaudet, C. Groom, L. Paquet, C. Rhofir, G. Ampleman, and S. Thiboutot. 2000. Characterization of Metabolites during Biodegradation of Hexahydro-1,3,5-Trinitro-1,3,5-Triazine (RDX) with Municipal Anaerobic Sludge. *Appl. Environ. Microbiol.* 66:2652-2657.

Hawari, J. 2000. Biodegradation of RDX and HMX: From Basic Research to Field Application. p. 277-310. In J. C. Spain, J. B. Hughes and H-J. Knackmuss (ed.), Biodegradation of Nitroaromatic Compounds and Explosives, CRC Press, Boca Raton, Florida.

ANAEROBIC BIOLOGICAL TREATMENT OF RDX IN GROUNDWATER

Douglas E. Jerger Ph.D. (IT Corporation, Knoxville, TN)
Todd Harris (Mason and Hanger Corporation, Amarillo, TX)
Amy H. Van Hout (IT Corporation, Knoxville, TN)
Daniel P. Leigh P.G. (IT Corporation, Concord, CA)

ABSTRACT: A Corrective Measures Study was conducted for the perched aquifer zone at the DOE Pantex Plant to evaluate potential alternatives for corrective actions of high explosives such as hexahydro-1,3,5-trinitro-1,3,5-triazine (RDX) in the groundwater (Purdy and Burton, 2000). In situ accelerated anaerobic biological treatment was evaluated and further developed as a promising alternative. Bench-scale, anaerobic serum bottle test results indicated that RDX was readily biodegradable under anaerobic conditions by indigenous microorganisms with the addition of molasses as a carbon source. Aqueous RDX concentrations of approximately 5,000 micrograms per liter ($\mu g/l$) were reduced to less than the detection limits of 5 $\mu g/l$ within 30 days. Soil concentrations of RDX were below the detection level of 0.5 milligram per kilogram (mg/kg). The appearance and disappearance of the mononitrosamine, dinitrosamine and trinitrosamine RDX transformation products were monitored throughout the tests. Octahydro-1,3,5,7-tetranitro-1,3,5,7-tetrazocine (HMX) was also degraded under these conditions. Concentrations of 2,4,6-trinitrotoluene (TNT) were below the detection limits. The results from the bench tests were used to perform aquifer biosimulations to establish a conceptual design for a field pilot test.

INTRODUCTION

RDX has been reported to be biologically degraded under anaerobic conditions when a supplemental carbon source was added to the culture medium (McCormick 1981, McCormick, 1985, Kaplan 1998, Awari, 2000). The initial transformation products, which were reported by these authors, were hexahydro-1-nitroso-3,5-dinitro-1,3,5-triazine (MNX), hexahydro-1,3-dinitroso-5-nitro-1,3,5-triazine (DNX), and hexahydro-1,3,5-trinitroso-1,3,5-traizine (TNX). Ring cleavage followed with the formation and subsequent degradation of 1,1-dimethylhydrazine, 1,2-dimethylhydrazine and hydrazine under batch culture conditions (McCormick, 1981). The production of formaldehyde and methanol was also observed in these tests. These intermediates were not observed during subsequent biodegradation tests under continuous culture conditions (McCormick 1985). In a recent review, research conducted to date suggests that biodegradation of cyclic nitramines RDX and HMX under ring cleavage following initial transformation by anaerobic microorganisms (Awari, 2000). Under these conditions formaldehyde, formic acid, carbon dioxide, nitrous oxide and ammonia are produced. Mineralization of RDX and HMX under these conditions typically exceeds 60%.

Indigenous bacteria in Pantex vadose zone soils were able to degrade RDX under microaerophilic and anoxic conditions (Schull et al 1999). Biodegradation was not observed under aerobic conditions. The addition of a readily biodegradable carbon source increased the rate of RDX biodegradation.

The primary objectives of the anaerobic bench-scale testing were to identify the conditions under which the microbial activity is stimulated, to determine whether the indigenous microflora can be stimulated to degrade the RDX, and to measure transformation products which have been reported to persist during the degradation process. These objectives need to be addressed to ensure successful process development toward pilot-scale design and testing.

MATERIALS AND METHODS

The laboratory testing included a two-part anaerobic serum bottle test to evaluate the test objectives. The anaerobic microcosm test 1 was conducted as a screening assay while test 2 was performed to further quantify the degradation of RDX. A baseline analysis of physical and chemical parameters in the groundwater and soils was done to determine the conditions for the anaerobic microcosm test.

Anaerobic Microcosm Test I. Four different treatments were prepared in duplicate in 125 milliliter (ml) sterile serum bottles (Table 1). Each anaerobic microcosm contained 25 grams (g) site soil, 75 ml groundwater (with or without amendments), and was spiked with RDX. Duplicate serum bottles were sampled at time zero, 20 days and 46 days from each treatment. Methane production, pH and the aqueous and soil fractions in the microcosms were analyzed for high explosives (HE).

TABLE 1. Anaerobic Microcosm I Experimental Design

Treatment	RDX Spike (mg/l)	Amendments
Abiotic Control	10	Groundwater, soil, rezasurin, 1000 mg/l Molasses, 750 mg/l Restore 375[b], autoclaved to inhibit microbial activity
Indigenous Microflora A	10	Groundwater, soil, rezasurin, 1000 mg/l Molasses, 750 mg/l Restore 375[b]
Indigenous Microflora B	--[a]	Groundwater, soil, rezasurin, 1000 mg/l Molasses, 750 mg/l Restore 375[b]
Bio-augmented	10	Groundwater, soil, rezasurin, 1000 mg/l Molasses, 750 mg/l Restore 375[b], and 10 ml WWTP[c]

[a]RDX was not spiked into sample.
[b]Restore 375 is composed of 50% Ammonia Chloride, 20% Disodium Phosphate, 12.5% Monosodium Phosphate, and 12.5% Sodium Tripolyphosphate.
[c]WWTP – Wastewater Treatment Plant anaerobic digester sludge.

Anaerobic Microcosm Test II. Anaerobic microcosm test II was performed to further quantify the degradation of RDX. The setup protocol was similar to anaerobic microcosm test I. RDX was added to the vessels at a final concentration of 2 milligrams per liter (mg/l). The experimental conditions are shown in Table 2. Duplicate serum bottles were sacrificed at day 0, day 11, and day 31 from each test condition. Methane production, pH, volatile fatty acid analyses and HE analyses were conducted.

TABLE 2. Anaerobic Microcosm II Experimental Design

Treatment	RDX Spike (mg/l)	Amendments
Abiotic Control	2	Groundwater, soil, rezasurin, sodium sulfide, refrigerated to inhibit microbial activity
Unamended Control	2	Groundwater, soil, rezasurin, sodium sulfide
Molasses	2	Groundwater, soil, rezasurin, 1000 mg/l Molasses, 750 mg/l Restore 375[a]
Molasses w/ Bioaugmentation	2	Groundwater, soil, rezasurin, 1000 mg/l Molasses, 750 mg/l Restore 375[a], 10 ml WWTP[b]

[a]Restore 375 is composed of 50% Ammonia Chloride, 20% Disodium Phosphate, 12.5% Monosodium Phosphate, and 12.5% Sodium Tripolyphosphate.
[b]WWTP – Wastewater Treatment Plant anaerobic digester sludge.

RESULTS AND DISCUSSION

Anaerobic Microcosm Test I. The results of the screening test indicated the production of RDX transformation products by the indigenous microflora under anaerobic conditions with molasses as a carbon source (Figure 1). The microcosm augmented with sludge from a WWTP anaerobic digester also showed degradation of the RDX, production and degradation of the MNX, DNX, and TNX (Figure 2). The microcosms that did not receive molasses showed no loss of RDX. The presence of the transformation products at time zero was due to the rapid degradation of RDX. The higher rates of RDX, MNX, DNX, and TNX degradation in the bioaugmented microcosms were probably due to the higher concentration of microbial biomass and the reducing environment created by the anaerobic sludge.

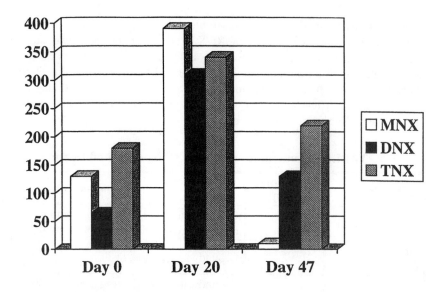

FIGURE 1. Anaerobic Biodegradation of RDX Transformation Products I – Indigenous Microflora A

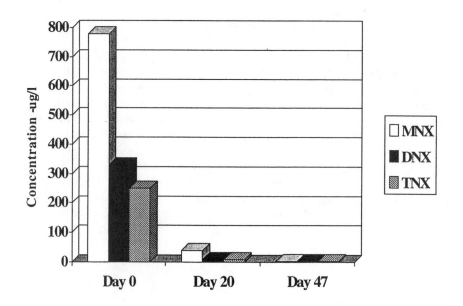

FIGURE 2 Anaerobic Biodegradation of RDX Transformation Products I – Bioaugmented

Anaerobic Microcosm Test II. The microcosms amended with molasses exhibited rapid and efficient degradation of RDX and HMX. The microcosms amended with only molasses reduced RDX concentrations from approximately 3000 µg/l to less than 5 µg/l in 30 days (Table 3, Figure 3). The production and degradation of the nitroso transformation products were also observed during the 30 day incubation period. HMX was biodegraded from an initial concentration of approximately 300 µg/l to less than (<) 4.7 µg/l. The microcosms amended with molasses and bioaugmented with WWTP anaerobic digester sludge exhibited more rapid degradation of RDX and HMX. Within 10 days RDX concentrations of approximately 500 µg/l were treated to <4.5 µg/l. The presence of the nitroso transformation products was observed during the initial sampling but they were below detection limits at 10 days. HMX was also treated to below the detection limits within 10 days. Losses of RDX and HMX were not observed in the abiotic controls or in the microcosms not receiving molasses. The concentrations of the explosive constituents in the microcosm soils were all <0.5 mg/kg after 30 days.

The chemical parameters monitored during the microcosm incubation period showed that the pH, ammonia-N and phosphate concentrations were maintained in an acceptable range for microbial activity during the incubation period. The high concentrations of acetic and butyric acid measured in the molasses amended microcosms in addition to the low concentrations of methane indicate that the methane fermentation is not necessary for RDX and HMX biodegradation. The TOC concentrations in these microcosms remained unchanged also indicating very little conversion of the molasses to carbon dioxide (CO_2). However under methanogenic conditions as shown with the bioaugmented microcosms, the degradation of RDX and HMX was more rapid with substantial production of methane and carbon dioxide. Methane concentrations in the bioaugmented microcosms exceeded 200 mg/l under the pressure developed in the microcosms from gas production in comparison to less than 0.001 mg/l in the other microcosms.

TABLE 3. Anaerobic Microcosm II: Explosives Analysis in Groundwater

Sample	Days	RDX (µg/l)	MNX (µg/l)	DNX (µg/l)	TNX (µg/l)	HMX (µg/l)
Abiotic Control 1	0	3400	<360	<360	<360	9.04
	11	3500	<200	<200	<200	340
	30	3100	200	160	300	510
Abiotic Control 2	0	3400	<360	<360	<360	<360
	11	2600	<210	<210	<210	7.26
	30	3900	< 45	80	300	680
Unamended Control 1	0	3700	<360	<360	<360	<360
	11	2700	<210	<210	<210	240
	30	3200	170	160	320	550
Unamended Control 2	0	3500	<360	<360	<360	<360
	11	2300	<210	<210	<210	210
	30	3600	130	150	340	640
Molasses 1	0	3200	<420	<420	<420	<420
	11	780	470	210	120	320
	30	<72	<160	< 37	17	<210
Molasses 2	0	2900	<210	<210	<210	270
	11	730	370	170	130	320
	30	<4.7	<24	190	<24	<4.7
Molasses Bioaugmented 1	0	480	100	91	170	120
	11	<14	<14	<14	<14	<14
	30	<4.7	<24	<24	<24	<4.7
Molasses Bioaugmented 2	0	540	110	95	180	6.57
	11	<12	<12	<12	<12	<12
	30	<4.5	<110	<90	<4.5	<4.5

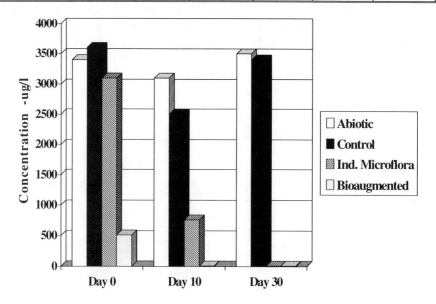

FIGURE 3. Anaerobic Degradation of RDX II

CONCLUSIONS

The microcosm tests performed with groundwater and soils from the Pantex site indicate that the biodegradation of RDX is feasible under anaerobic conditions. Anaerobic Microcosm II test results indicated degradation of RDX under both anaerobic non-methanogenic and methanogenic conditions with a supplemental carbon source (molasses). The degradation of the nitroso-metabolites, including MNX, DNX and TNX, under these conditions indicates further reduction to a hydroxyl amino group (Kaplan, 1998). McCormick suggested that production of the hydroxyamino triazines destabilizes the molecule and hydrolytic ring cleavage occurs (McCormick et al 1981). Following ring cleavage under reducing conditions mineralization of RDX is extensive (greater than 60%) and small organic molecules such as methanol, formic acid, CO_2, ammonia and nitrous oxide are produced (Awari, 2000). Awari (2000) also concluded that the anaerobic degradation pathway for cyclic nitramines remains controversial since McCormick (1981) proposed a very complicated pathway without extensive experimental evidence.

REFERENCES

Awari, J. 2000. "Biodegradation of RDX and HMX: From Basic Research to Field Application." in *Biodegradation of Nitroaromatic Compounds and Explosives*, J. Spain, J. Hughes, H. Knackmuss eds. Lewis Publishers, Boca Raton, FL. Chapt 11. pp 277-310.

Kaplan, D.L. 1998. "Biotransformation and Bioremediation of Munitions and Explosives". pp 549- 575. In Sikdar, S.K. and R.L. Irvine (eds.), Bioremediation: Principles and Practices Vol. II. Biodegradation Technology Developments. Technomic Publishing Inc., Lancaster-Basel.

McCormick, N.G., J.H. Cornell, and A.M. Kaplan. 1981. "Biodegradation of hexahydro-1,3,5-trinitro-1,3,5-triazine.".*Appl. Environ. Microbiol.* 42: 817-823.

McCormick, N.G., J.H. Cornell, and A.M. Kaplan. 1985. "The fate of hexahydro-1,3,5-trinitro-1,3,5-triazine (RDX) and related compounds in anaerobic, denitrifying continuous culture systems using simulated wastewater." Report TR85—008. U.S. Army Natick Research and Development Center, Natick, Massachusetts.

Purdy, C.B., and J. Burton. 2000. "Groundwater Contamination in the Perched Aquifer Zone at the DOE Pantex Plant: Successful Expedited Site Characterization". In Looney, G. and R. Falta (eds), Vadose Zone Science and Technology Solutions, Battelle Press, Columbus, OH>

Schull, T.L., G. Speitel, and D. McKinney. 1999. "Bioremediation of RDX in the Vadose Zone Beneath the Pantex Plant." Amarillo National Resource Center for Plutonium – 1999-1.

The Effects of Ubiquitous Electron Acceptors on the Initiation of RDX Biodegradation

Jeffrey Davis and Lance D. Hansen (ERDC, Vicksburg, MS)
Brenda O'Neal (ARA, Vicksburg, MS)

ABSTRACT: Remediation of hexahydro-1,3,5-trinitro-1,3,5-triazine (RDX) can be accomplished by facultative or obligately anaerobic microorganisms using RDX as a gratuitous terminal electron acceptor. Numerous studies have been performed that examine the effect of carbon source on the reductive transformation of RDX. This study examines the effects of important ubiquitous electron acceptors on the initiation of the reductive transformation of RDX.

A study was performed to examine the effects of the reductive transformation of naturally-occuring electron acceptors (nitrate and sulfate) on the initiation of biological degradation of RDX. This study found that the presence of nitrate and sulfate did not significantly affect the RDX biodegradation induction time of 45 days. The presence of nitrate did not significantly affect the first-order RDX removal rate of 0.00979 d^{-1}. The presence of sulfate did improve the removal rate of RDX by a factor of approximately 2. The in situ remediation of ground water contaminated with RDX can be significantly improved in the presence of sulfate.

INTRODUCTION

Currently the Department of Defense has a total of 21,425 contaminated sites on 1,769 installations contaminated by energetic compounds, solvents, and heavy metals. Remedial response has been completed on only 9,640 of those sites. Through the end of fiscal year 1994, funds spent on studies, interim actions, design and cleanup totaled $7 billion. The remaining cost to complete remedial response on all sites is currently estimated at $26.5 billion (DoD 1995).

The production and processing of military explosives has led to the contamination of soil and water by energetic compounds at approximately thirty-five Army Ammunition Plants and Depots in the United States. In a report prepared for the Executive Director of the Strategic Environmental Research and Development Program (SERDP 1993), it was estimated that 706,000 cubic yards of soil and 10 billion gallons of groundwater have been contaminated by energetic compounds at these sights. Due to the fact that most sites were in the preliminary assessment and remedial investigation stages at the time of this report, these estimates are rough and most likely low.

Sikka et al. (1980) provide evidence of microbial degradation in experiments where contaminated river water was combined with 1% sediment from the same contaminated stream. Significant degradation of RDX occurred after a 20 day lag period. Little or no loss of RDX occurred in the river water alone or with amendment of yeast extract. Approximately 80% of the RDX added was transformed within two weeks after degradation started. In radiolabled studies, 80% of the [^{14}C]RDX added was evolved as $^{14}CO_2$ when 1% river sediment was added to the flasks. Evolution of $^{14}CO_2$ was preceded by a 10 day lag phase. It is believed that the river sediment provides a large seed of microorganisms capable of degrading RDX and nutrients for the growth of these microorganisms.

RDX in nutrient broth cultures disappeared in approximately four days when inoculated with anaerobic activated sewage sludge. Transformation of RDX in nutrient broth was not observed when inoculated with aerobic activated sewage sludge and incubated aerobically. McCormick et al. (1981) proposed a pathway for the anaerobic biological degradation of RDX (FIGURE 1). This pathway suggests that the one or more nitro groups are reduced to the point where destabilization of the triazine ring occurs, and the ring is fragmented by hydrolytic cleavage. Fragments of the ring are further reduced ultimately resulting in a mixture hydrazines and methanol. Degradation intermediates identified were the mono- di- and trinitroso analogs of RDX, formaldehyde, methanol, hydrazine, and 1,1- and 1,2-dimethylhydrazine. It was suggested that an aerobic treatment may be used to mineralize these degradation intermediates (McCormick et al. 1981, Walker and Kaplan 1992).

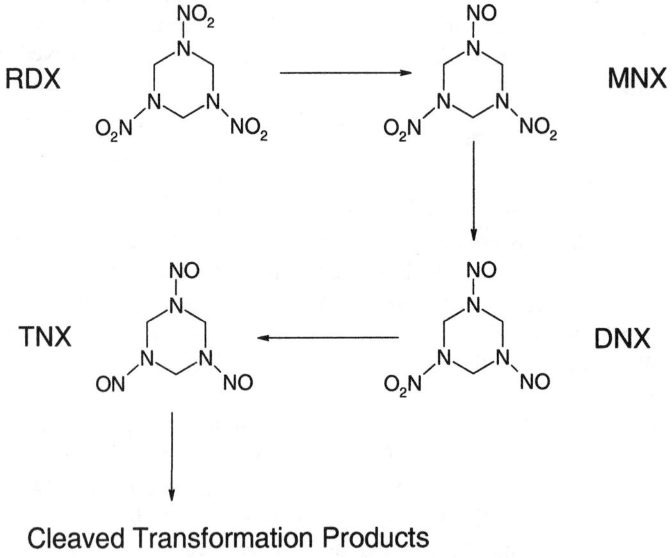

FIGURE 1. Anaerobic Pathway

Groundwater, typically, is co-contaminated with nitrate and/or sulfate. The pupose of the study was to examine the effects of these ubiquitous electron acceptors on the degradation of RDX.

MATERIALS AND METHODS

A triplicate set of batch experiments was conducted to examine the onset of RDX biodegradation in the presence of two common groundwater contaminants (nitrate and sulfate). Serum bottles (156 mL) were prepared according to TABLE 1, and then sealed using rubber stoppers and crimp-top caps. After sealing, serum bottles were inoculated (except sterile control) to 10% nitrate-reducing bacteria solution enriched from municipal anaerobic digester sludge.

TABLE 1 Experimental Design

	Acetate (ppm)	RDX (ppm)	Nitrate (ppm)	Sulfate (ppm)
Sterile Control	500	10	0	0
Donor Control	0	10	0	0
RDX Only	500	10	0	0
RDX + Nitrate	500	10	30	0
RDX + Sulfate	500	10	0	30
RDX + Nitrate + Sulfate	500	10	15	15

All replicates were prepared in a buffer containing 100 ppm K_2HPO_4 and 100 ppm KH_2PO_4 (pH 7), NH_4Cl added at 50 ppm(NH_4^+).

The serum bottles were shaken continuously on an orbital shaker at 150 rpm. A 2.5 mL aliquot was taken periodically from each serum bottle for explosives and ion analysis. Statistical analyses were performed using Jandel Scientific's SigmaStat© 2000. Analyses were performed at 95% confidence.

Explosives Analysis. Samples were filtered using a 0.45 μm syringe filter and diluted 1:1 (v/v) with HPLC grade acetonitrile, CH_3CN. Analysis was performed using a Waters 610 fluid unit pump, a Waters 717 plus autosampler including a 200ul loop injector and a Waters 486 tunable UV absorbance detector monitored at 245 nm. Millenium 2.1 chromatography software was used for data analysis. Chemical separation was achieved using a Supelco LC-18 reverse phase HPLC column (25 cm x 4.6 mm) with a Novapak C-18 pre-column for the primary column; and a Supelco LC-CN reverse phase HPLC column (25 cm x 4.6 mm) with a Novapak CN pre-column for the secondary column. The mobile phase (1:1 (v/v) methanol/organic-free reagent water) was run at a flow rate of 1.2 mL/min. A volume of 50 μL was injected. Daily calibration consisted of running standards at the beginning of the analytical run, every 8 hours of continuing analysis, and after the last sample of the day. A midpoint check standard was also run after every 10 samples. Standards were prepared from stock standard solutions, which were obtained from the Army Environmental Center at Aberdeen Proving Ground.

Ion Analysis. Samples were filtered using a 0.45 μm syringe filter. Analysis was performed on a DIONEX Ion Chromatograph. Chemical separation and detection were achieved using an Ionpac AS11 analytical column (4 x 250 mm) and a

Dionex conductivity detector (1.25 µL internal volume). The mobile phase (NaOH) was run at a flow rate of 1.5 mL/min. A volume of 25 µL was injected. Daily calibration consisted of running standards at the beginning of the analytical run and after every 10 samples. Standards were prepared from stocks, which were prepared on a monthly basis, except acetate, which was prepared weekly.

RESULTS AND DISCUSSIONS

The removal of RDX (Figure 2) and acetate was not detected in the sterile controls. This indicates that the removal of RDX in the other conditions was the result of microbial activity. The removal of RDX was not detected in sample incubated without additional acetate (donor control). This indicates that the inoculation process did not significantly (p=0.05) add a carbon source/ electron donor.

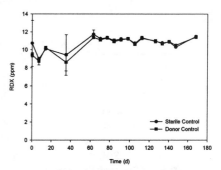

FIGURE 2. RDX results from sterile and donor controls.

The removal of RDX in the presence of nitrate and/or sulfate is shown in FIGURE 3. The sole electron acceptor in the serum bottles labeled as RDX only is RDX. This condition is the control. The removal of RDX was begun after approximately 50 days (TABLE 2). The induction times found for the other conditions are summarized in TABLE 2. A one-wat ANOVA statistical analysis (p=0.05) shows that no significant difference between the induction times was caused by the presence of sulfate or nitrate.

A first-order kinetic model was fitted to the time course of the removal of RDX and is shown in FIGURE 4. The results are summarized in TABLE 2. A one-way ANOVA with pair-wise t-test statistical analysis of the results reveals that sulfate significantly (p=0.05) increased the removal rate of RDX and nitrate did not significantly (p=0.05) increase the removal rate of RDX.

TABLE 2. First-order removal rates and induction times of RDX removal in the presence of nitrate and/or sulfate.

Condition	k^1 (d^{-1})*	Induction Time (d)*
RDX Only	-0.00979±0.00118	49.1±4.7
RDX + Nitrate	-0.00806±0.0086	50.4±7.3
RDX + Sulfate	-0.0177±0.0019	41.1±6.0
RDX + Nitrate + Sulfate	-0.0176±0.0030	47.8±2.7

* error represented is 95% confidence

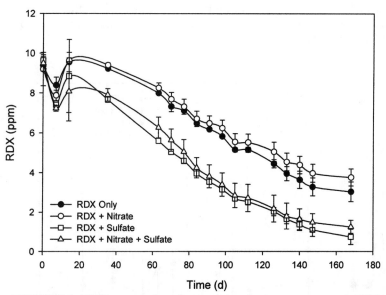

FIGURE 3. RDX removal under various electron acceptor conditions.

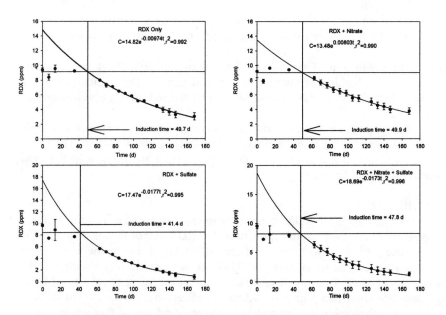

FIGURE 4. RDX removal in the presence of nitrate and/or sulfate.

It should be noted that the removal of RDX commenced after the removal of sulfate and nitrate. FIGURE 5 is a plot of the removal of RDX, nitrate and sulfate in the full test (RDX + Nitrate + Sulfate). The increased removal rates, as shown in TABLE 2, are initial removal rates. The competitive effects of electron acceptors was not determined in this study due to the long induction time required for RDX biodegradation.

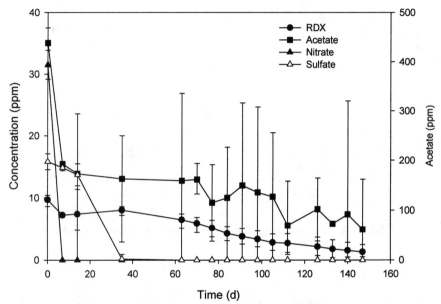

FIGURE 5. Time course of RDX, nitrate, and sulfate removal.

The production of the intermediates shown in FIGURE 1 was detected transiently. No buildup of intermediates was detected. This indicates that the complete degradation of RDX is rate limited by the reductive transformation of RDX.

All tests were conducted with excess acetate. Acetate removal was rapid in the initial week of the study then removal was negligible afterward. The initial removal was due to the presence of dissolved oxygen and oxygen in the head space at the start of the study.

CONCLUSIONS

The initiation of RDX removal was examined under various electron acceptor conditions. The induction time for RDX removal was found to be approximately 45 days with no significant effect found due to the presence of nitrate or sulfate. The removal rate of RDX was not significantly affected by the presence of nitrate. The presence of sulfate increased the removal rate of RDX significantly. These two primary conclusions indicate that the removal of RDX in situ may be enhanced in ground waters that contain sulfate, or that ground water may need to be amended with sulfate as well as an electron donor to improve the efficiency of RDX removal.

Future work will be performed to examine the competitive effects of electron acceptors. This work will be performed by insulting an active RDX degrading culture with nitrate and/or sulfate.

Acknowledgements

This work was supported by the United States Army Environmental Quality Technology Program. The authors wish to thank the Environmental Chemistry Branch of the Army Engineer Research and Development Center. The technical assistance provided by Lynn Escalon was greatly appreciated.

REFERENCES

DoD. 1995. *Defense environmental restoration program - annual report to congress for fiscal year 1994*. U.S. Department of Defense, Washington, DC.

McCormick, N. G., J. H. Cornell, and A. M. Kaplan. 1981. Biodegradation of hexahydro-1,3,5-trinitro-1,3,5-triazine. *Applied and Environmental Microbiology* **42 (5)**: 817-823.

SERDP. 1993. *An approach to estimation of volumes of contaminated soil and groundwater for selected army installations*. Labat-Anderson Inc., (May). Prepared for the Executive Director, Strategic Environmental Research and Development Program.

Sikka, H. C., S. Banerjee, E. J. Pack, and H. T. Appleton. 1980. *Environmental fate of RDX and TNT*. Technical Report 81-538, US Army Medical Research and Development Command, Fort Detrick, Frederick, MD.

Walker, J. E. and D. L. Kaplan. 1992. Biological degradation of explosives and chemical agents. *Biodegradation* **3**: 369-385.

REDUCTIVE TRANSFORMATION OF RDX IN A BENCH-SCALE SIMULATED AQUIFER

Lance D. Hansen, *Jeffrey L. Davis*, Lynn Escalon
US Army Engineer Research and Development Center, Vicksburg, MS 39180

ABSTRACT: The production and processing of military explosives has led to the contamination of soil and water by energetic compounds at more than 100 federal installations across the United States. Contaminants typically found at these sites include 2,4,6-trinitrotoluene (TNT), dinitrotoluene (DNT), hexahydro-1,3,5-trinitro-1,3,5-triazine (RDX), and octahydro-1,3,5,7-tetranitro-1,3,5,7-tetrazocine (HMX) and their transformation products. The use of in-situ biological degradation of these contaminants would provide significant cost savings over currently accepted treatment methods. Due to the mobility of RDX, it is currently the major contaminant of concern and the focus of this study. A packed bed soil column study was conducted to examine the biologically mediated reductive transformation of RDX using acetate as an electron donor/carbon source. The columns were inoculated with nitrate-reducing bacteria enriched from municipal anaerobic digester sludge. Several experimental runs were conducted to examine conditions necessary to 1) poise the system for reductive RDX transformation, 2) assess varying electron donor / electron acceptor ratios, and 3) develop preliminary in situ transformation kinetics for the established RDX⇒ MNX ⇒DNX ⇒TNX transformation pathway necessary for cost-effective field-scale application. Significant results include RDX concentration reduction from 8 mg/L to 56 µg/L (99.2% removal), with a first-order reaction coefficient of 3.0 hr^{-1}. A build-up of MNX and DNX was not detected. Current results show a build up of TNX followed by much slower TNX degradation to unidentified transformation products indicating potential rate limiting step for full-scale implementation. Based on current data a conservative first-order reaction coefficient for TNX of 0.017 d^{-1} was estimated. Additional studies are underway to examine effects of various electron donors on microbial community composition and biomass, as well as toxicity and mutagenicity of unidentified transformation products.

INTRODUCTION

Many active and formerly used federal facilities are plagued with a rapidly moving, relatively toxic, and expansive plumes of explosives contamination that threaten the available supply of potable water for surrounding communities. Currently there is no in situ alternative for remediation of explosives in groundwater. Available remediation alternatives are limited to long-term groundwater pumping and ex situ treatment followed by discharge or reinjection of treated water. The Best Available Technology (BAT) is sorption to granular activated carbon (GAC). Shortcomings of this approach include high capital and operation costs and expected long term duration of cleanup activities (often exceeding 30 years).

Many researchers have established that these explosives can be degraded through biological processes, but successful application of bioremediation to in-situ treatment of contaminated soils and waters has yet to be proven in the field. The purpose of this study was to determine the suitability of in-situ biological degradation of RDX for the remediation of groundwater.

RDX poses a threat to health from exposure by inhalation of dust particles and fumes, or ingestion. The major toxicological effects of exposure to RDX are nausea, irritability, convulsions, unconsciousness, and amnesia. Due to these effects shown in humans, the USEPA has established drinking water health advisories (HA) for exposure to RDX (McLellan et al. 1992).

The fate and transport of RDX in the environment can be influenced by many factors including photolysis by sunlight, hydrolysis, and biologically mediated degradation. Under normal environmental conditions, photodegradation of RDX occurs rapidly. The half-life of RDX when exposed to direct sunlight was found to be approximately 11 hours (Sikka et al. 1980). The transformation products identified from the photolysis of RDX include nitrate, nitrite, formaldehyde and nitrogen. Although significant hydrolysis of RDX has been shown to occur, it is unlikely that hydrolysis will be a significant factor in the fate of RDX under the conditions normally found in the environment (Sikka et al. 1980; Hoffsommer and Rosen 1973). Calculated half-lives for alkaline hydrolysis of RDX at pH 8 or higher range from approximately one to several years (Spanggord et al. 1980a).

Evidence of microbial degradation was shown in experiments where contaminated river water was combined with 1% sediment from the same contaminated stream. Significant degradation of RDX occurred after a 20 day lag period. Little or no loss of RDX occurred in the river water alone or with amendment of yeast extract. Approximately 80% of the added RDX was transformed within two weeks after degradation started. In radiolabeled studies, 80% of the [^{14}C]-RDX added was evolved as $^{14}CO_2$ when 1% river sediment was added to the flasks. Evolution of $^{14}CO_2$ was preceded by a 10-day lag phase. It is believed that the river sediment provides a large seed of microorganisms capable of degrading RDX and nutrients for the growth of these microorganisms (Sikka et al. 1980)

Results from anaerobic studies suggested that degradation of RDX was a co-metabolic process. A source of organic carbon and RDX had to be present at the same time to achieve RDX degradation. In flasks initially containing 10-mg/L RDX and 50-mg/L yeast extract, the RDX was completely transformed in three days. RDX has been found to be resistant to biodegradation under aerobic conditions (Spanggord et al. 1980b).

RDX in nutrient broth cultures disappeared in approximately four days when inoculated with anaerobic activated sewage sludge. Transformation of RDX in nutrient broth was not observed when inoculated with aerobic activated sewage sludge and incubated aerobically. A pathway was proposed for anaerobic biological degradation of RDX (Figure 1). This pathway suggests that one or more nitro groups are reduced to the point where destabilization of the triazine ring occurs, and the ring is fragmented by hydrolytic cleavage. Fragments of the ring are further reduced ultimately resulting in a mixture of hydrazines and methanol. Degradation intermediates identified were the mono- di- and trinitroso analogs of RDX, formaldehyde, methanol, hydrazine, and 1,1- and 1,2-dimethylhydrazine. It was suggested that an aerobic treatment might be used to mineralize these degradation intermediates (McCormick et al. 1981, Walker and Kaplan 1992).

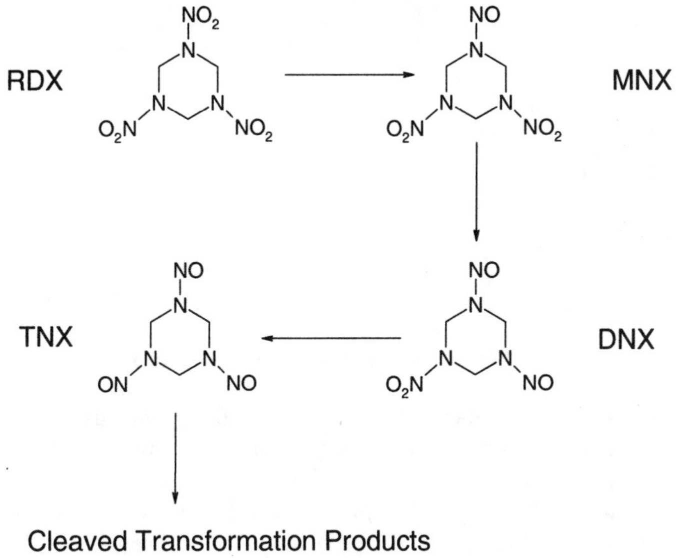

FIGURE 1. Anaerobic Pathway (adapted from McCormick et al., 1981)

This study used a column-based approach to examine the feasibility of using acetate as an amendment to enhance the degradation of RDX in synthetic aquifer soil. The studies were done under high and low concentrations of acetate and RDX.

METHODS AND MATERIALS

Column Setup. A bank of eight columns was prepared using 1.5 inch, (3.8 cm) schedule 40 clear PVC pipe with schedule 80 PVC end caps tapped with 1/8 inch (0.32 cm) NPT threads. Columns were wet packed with washed and sieved homogeneous sand (size range 1.0 – 0.1 mm). The wetting solution consisted of nitrate-reducing bacteria enriched from municipal anaerobic digester sludge. Figure 2 is a schematic of the column bank configuration. Columns were connected in series and operated in an upflow mode. All influent material was autoclaved and degassed prior to being connected to the system.

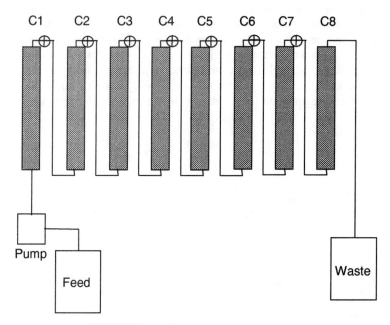

FIGURE 2. Schematic of column setup

The columns were operated at a steady flow rate using a positive displacement pump. Four studies were performed to examine the effects of RDX and acetate concentrations on the removal of RDX. Samples were taken to determine the effluent history and bed profile. Samples were divided for analysis of acetate (ion chomatography) and RDX (HPLC).

Explosives Analyses. Samples were filtered with a 0.45 µm syringe filter and split for analysis of acetate and explosives. Samples for explosives analysis were diluted 1:1 (v/v) with HPLC-grade acetonitrile, CH_3CN. Analysis was performed using a Waters HPLC with a tunable UV absorbance detector monitored at 245 nm. Chemical separation was achieved using a Supelco LC-18 reverse phase HPLC column (25cm X 4.6mm (5um)) with a Novapak C-18 pre-column for the primary column; and a Supelco LC-CN reverse phase HPLC column (25cm x 4.6mm (5um)) with a Novapak CN pre-column for the secondary column. The mobile phase (1:1 (v/v) methanol/organic-free reagent water) was run at a flow rate of 1.2 ml/min. A volume of 50 µl was injected. Standards were prepared from stock primary source standard solutions, which were obtained from the Army Environmental Center at Aberdeen Proving Ground, and secondary source standard solutions, which were obtained from Ultra Scientific.

Ion Analysis. Acetate analysis was performed on a DIONEX Ion Chromatograph. Chemical separation and detection were achieved using a Ionpac AS11 analytical column (4 x 250 mm) and a Dionex conductivity detector (1.25

µl internal volume). The mobile phase (NaOH) was run at a flow rate of 1.5 ml/min using the gradient method. A sample volume of 25 µl was injected.

RESULTS AND DISCUSSION

Four conditions tested are outlined as experimental runs 1 through 4 in Table 1. The first test was performed at a relatively high flow rate (0.478 L/d). The second and subsequent tests were conducted at a low flow rates. The third and fourth tests were performed at high acetate concentrations (~100 ppm). The fourth test was performed at low RDX concentration, similar to those reported on-site.

TABLE 1. Experimental conditions

Experiment Run	Flow Rate (L/d) Hydraulic Retention Time (d)	Acetate (ppm)	RDX (ppm)
1 - High Flow Rate	0.478±0.012 3.82	33.03±2.85	18.43±2.69
2 - Low Flow Rate	0.153±0.0008 11.9	20.46±2.61	20.11±4.68
3 - High Acetate	0.152±0.004 12.0	99.25±3.73	18.87±1.06
4 - Low RDX	0.149±0.008 12.2	107.11±10.24	7.48±1.46

Table 2 reports detailed effluent concentrations observed during each experimental run. During Run 1 the system was allowed to come to equilibrium with respect to the acetate and RDX effluent concentrations.

TABLE 2. Summary of Results

Run #	RDX-In (mg/L)	RDX Out (mg/L)	MNX (mg/L)	DNX (mg/L)	TNX (mg/L)	% RDX Removal
1	18.43±2.69	16.59	ND	ND	ND	9.98
2	20.11±4.68	5.71	8.24	3.82	0.25	71.61
3	18.87±1.06	2.39	7.32	8.25	0.65	87.33
4	7.48±1.46	0.059	0.034	ND	6.08	99.21

Figure 3 shows the complete break through of RDX in Run 1. RDX equilibrium across the column shows no degradation. Further, there is no evidence of biologically mediated reductive transformation products indicating that losses in RDX in follow-on Runs 2 through 4 was not due to physical removal. RDX breakthrough occurred during Run 1 validating the selection of silica sand for the test media. All loss of RDX following Run 1 is therefore demonstrated to be due to biological activity.

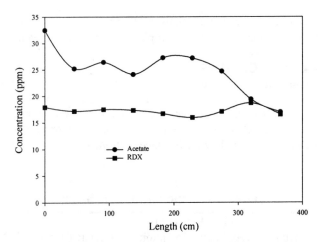

FIGURE 3. RDX Run 1 (High Flow/High RDX/Low Acetate)

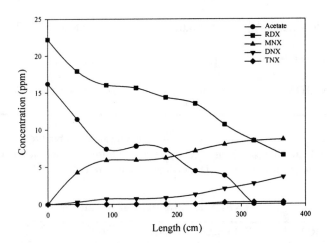

FIGURE 4. RDX Run 2 (Low Flow/High RDX/Low Acetate)

Run 2 (Figure 4) demonstrates successful RDX removal over the course of the column corresponding to a sequential increase in known biologically mediated reductive transformation products. RDX degradation progressed to 6.42 mg/L corresponding to 65.8% removal. The near stoichiometric increase in reductive transformation products supports the conclusion of biologically mediated reductive transformation. Acetate analysis at column 7 indicates complete removal of acetate from the column suggesting an electron donor limitation during this run.

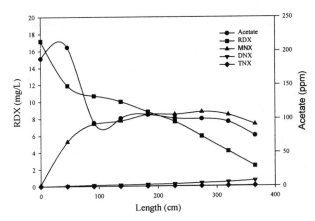

FIGURE 5. RDX Run 3 (Low Flow/High RDX/High Acetate)

To overcome the electron donor limitation experienced in Run 2, the influent acetate was increased to ~10 times during Run 3 (Figure 5) the theoretical stoichiometric requirement for complete reductive transformation of RDX→MNX→DNX→TNX. This change resulted in a greater removal of RDX (~87%) over the course of the bed profile and maintained the near stoichiometric increase in reductive transformation products. It was concluded that the observed rate and extent of transformation of RDX was optimized for the column study as configured. To improve the extent of transformation, it was proposed that the flow rate be reduced to increase the residence time in the biologically active zone. However, the ~12 day retention time was the longest achievable with available equipment.

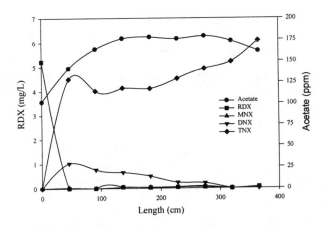

FIGURE 6. RDX Run 4 (Low Flow/Low RDX/High Acetate)

It was identified that concentrations used were not representative of likely site contamination. Therefore, RDX concentration was reduced to concentrations representative of the highest likely concentrations observed in DoD facilities.

Run 4 was conducted at the same low flow rate, and the influent retained acetate in excess. RDX was reduced to a concentration of ~7.5 mg/L. This correlated with representative RDX concentrations at contaminated sites and correlated with RDX effluent from the column bank.

Figure 6 shows near complete removal of RDX by the end of column 1 correlating with stoichiometric conversion of RDX to known reductive transformation products. In contrast to Runs 1, 2, or 3 a dramatic increase in TNX was identified. TNX did not appear to degrade over the course of the column bank. A black film was observed in the Teflon tubing toward the end of the column bank correlating with an observable odor of sulfide during sample collection indicating sulfate-reducing conditions in the columns.

REFERENCES:

DoD. 1995. *Defense environmental restoration program - annual report to congress for fiscal year 1994.* U.S. Department of Defense, Washington, DC.

Hoffsommer, J. C. and J. M. Rosen. 1973. Hydrolysis of explosives in sea water. Bulletin of Environmental Contamination and Toxicology 10 (2): 78-79.

McCormick, N. G., J. H. Cornell, and A. M. Kaplan. 1981. Biodegradation of hexahydro-1,3,5-trinitro-1,3,5-triazine. *Applied and Environmental Microbiol.* 42 (5): 817-823.

McLellan, W. L., W. R. Hartley, and M. E. Brower. 1992. *Hexahydro-1,3,5-trinitro-1,3,5-triazine. In Drinking water health advisory: munitions (USEPA Office of Drinking Water Health Advisories)*, ed. W. C. Roberts and W. R. Hartley. Boca Raton: Lewis Publishers.

Sikka, H. C., S. Banerjee, E. J. Pack, and H. T. Appleton. 1980. *Environmental fate of RDX and TNT. Technical Report 81-538,* US Army Medical Research and Development Command, Fort Detrick, Frederick, MD.

Spanggord, R. J., T. Mill, T. -W. Chou, W. R. Mabey, J. H. Smith, and S. Lee. 1980a. *Environmental fate studies on certain munition wastewater constituents, final report, phase I – literature review.* SRI Project No. LSU-7934, SRI International, Menlo Park, CA.

Spanggord, R. J., T. Mill, T. -W. Chou, W. R. Mabey, J. H. Smith, and S. Lee. 1980b. *Environmental fate studies on certain munition wastewater constituents, final report, phase II – laboratory studies.* SRI Project No. LSU-7934, SRI International, Menlo Park, CA.

Walker, J. E. and D. L. Kaplan. 1992. Biological degradation of explosives and chemical agents. *Biodegradation* **3**: 369-385.

IDENTIFICATION OF BOTTLENECKS TO THE *IN SITU* BIOREMEDIATION OF DINITROTOLUENE

Shirley F. Nishino and *Jim C. Spain*
(Air Force Research Laboratory, Tyndall AFB, FL)

ABSTRACT: The vadose zones beneath waste pits at Badger Army Ammunition Plant (BAAP) are heavily contaminated with 2,4-dinitrotoluene (2,4-DNT) and 2,6-dinitrotoluene (2,6-DNT) in a ratio of about 25 to 1. We conducted bench scale studies in columns and in shake flasks to determine the feasibility of *in situ* bioremediation at BAAP. Indigenous bacteria readily acclimated to degrade 2,4-DNT, but 2,6-DNT degradation was slow and did not occur until 2,4-DNT was removed. The onset of degradation was 6-8 weeks at 13°C, the *in situ* soil temperature. The lag period was shortened to 1-2 weeks when the experiments were conducted at 20°C. There was a broad pH optimum around pH 8.and degradation stopped below pH 7. The addition of 10 mM phosphate stimulated the degradation considerably, however much lower concentrations of phosphate were required in soil-free systems. In column studies conducted with contaminated soil from BAAP, initial concentrations of 14.03 g/kg and 0.55 g/kg of 2,4- and 2,6-DNT were reduced to 2-14 mg/kg and 4-12 mg/kg in inoculated columns at room temperature, and to 5-9 mg/kg and 11-98 mg/kg in inoculated columns at 13°C. The bench scale experiments clearly established the biological aspects of the feasibility of bioremediation of dinitrotoluene (DNT) at BAAP and a subsequent pilot scale field study based on the laboratory results has validated the findings and clearly shown that DNT can be biodegraded *in situ*.

INTRODUCTION

DNT is widely used in the production of explosives and polyurethane foams. Release of DNT into the environment has led to the evolution of bacteria able to use the compound as a sole source of carbon, nitrogen and energy during aerobic growth. Such bacteria have been isolated from a variety of DNT contaminated sites (Nishino et al., 2000), yet DNT persists at the sites. Laboratory studies conducted to determine the bottlenecks to *in situ* degradation reveal that the key requirement is the presence of DNT degrading bacteria. Strains able to grow on 2,4-DNT seem to be ubiquitous at contaminated sites, strains able to degrade 2,6-DNT are less common, and neither type is found at uncontaminated sites. Simple laboratory experiments with batch cultures revealed that under appropriate conditions of pH, moisture, oxygen, and temperature both isomers of DNT can be mineralized (Nishino et al., 1999). In some instances the isomers are degraded sequentially and 2,6-DNT is degraded only after 2,4-DNT is depleted (Zhang et al., 2000).

2,4-DNT and 2,6-DNT are listed compounds of concern (Stone & Webster Environmental Technology & Services, 1998) at the BAAP Propellant Burning Ground (PBG). Waste pits at the PBG contain 2,4-DNT in concentrations up to

28% by weight (Stone & Webster Environmental Technology & Services, 1998). Similar waste pits at the Deterrent Burning Ground (DBG) are less severely contaminated. The approved treatment remedy for the PBG was excavation to 110 feet and incineration of the soil at an estimated cost of $75M. *In situ* bioremediation was proposed as an alternative to the approved remedy. Bench scale studies conducted with contaminated soil from BAAP evaluated the potential for bioremediation of DNT in the subsurface. Shake flask studies and column studies were conducted to determine the ability of bacteria enriched from BAAP soils to degrade DNT at ambient soil temperatures and to determine the physical parameters necessary to optimize DNT degradation. The studies were designed to investigate the critical parameters that would affect *in situ* DNT degradation in a field scale demonstration in which groundwater was to be recirculated through the PBG waste pits accompanied by air injection.

MATERIALS AND METHODS

Soils freshly collected from the PBG and DBG, composited, dried at room temperature and sieved (to pass 20 mesh), were stored at 4°C until used. The composite soils were analyzed for DNT content and organic content (Table 1) and were used for all bench scale studies. Clean groundwater from BAAP was stored at 4°C in the shipping drum until used in experiments. All materials and vessels used to handle and transfer the soils and groundwater were autoclaved prior to use to avoid cross contamination of the microbial populations.

TABLE 1. Initial concentrations of DNT in test soils from BAAP.

Soil	Fraction		2,4-DNT (g/kg)	2,6-DNT (g/kg)	% Organic Matter
	Mesh	%*			
PBG 9901 UZ†	<20	68	14.03	0.55	2.7
	>20	31	9.63	0.27	
PBG 9901 SZ‡	<20	68	7.2E-3	<1E-3	0.5
	>20	32	1.9E-3	<1E-3	
PBG Clean UZ	<20	62	<1E-3	<1E-3	1.5
	>20	38	<1E-3	<1E-3	
DBG 9901 UZ	<20	57	1.40	0.42	1.3
	>20	43	0.39	0.13	

*by weight, † unsaturated zone, ‡ saturated zone

Column studies. Dried soil (75 g) was placed in autoclaved glass columns (25 mm I.D. x 30 cm) on top of a 3 cm layer of washed, autoclaved sand, then topped with additional sand (total sand approximately 70 g). Groundwater (pH 7.7) was recirculated upwards through the columns at approximately 5 ml/min with a peristaltic pump. Filtered air was pumped through the reservoir to provide oxygen. Samples were withdrawn for analysis of pH, DNT (Nishino et al., 2000), nitrite (Smibert and Krieg, 1994), and nitrate (Parsons et al., 1984).

DNT-degrading bacteria. DNT-degrading strains were maintained on agar plates containing the appropriate DNT isomer (Nishino et al., 1999). A mixed culture

(MI) inoculated with *Burkholderia* sp. DNT (Spanggord et al., 1991), *Burkholderia cepacia* JS872, *B. cepacia* JS922, *Hydrogenophaga palleronii* JS863 (Nishino et al., 1999), and *B. cepacia* JS850 (Nishino et al., 2000) was grown in 250 ml shake flasks containing nitrogen-free minimal medium (Bruhn et al., 1987). 2,4-DNT (1 mM) and 2,6-DNT (250 µM) were provided as needed. The culture was incubated with shaking at 30°. A DNT-degrading culture (IB) that was enriched from shake flasks inoculated with soil from the PBG was maintained under identical conditions. Both cultures were transferred periodically and cells were harvested as needed for inoculation of shake flask and column studies. A portion of the IB culture was subcultured at room temperature. When growth on the added DNT was complete, the bacteria were harvested by centrifugation and the entire pellet was used to inoculate a new culture which was incubated at a temperature 2 degrees lower than the previous temperature. Five similar transfers were made over a 3-week period to obtain a DNT-degrading culture adapted to growth at 13°C (LTIB).

RESULTS AND DISCUSSION

DNT serves as the sole carbon, nitrogen and energy source for bacteria able to use 2,4- or 2,6-DNT as growth substrates (Nishino et al., 2000). Because the C/N ratio in DNT is 7/2, the excess N (about 80%) accumulates as nitrite in the growth medium. The release of nitrite is a sensitive measure of DNT degradation, provided that conversion to nitrate or ammonium does not change the nitrogen pools. Preliminary studies demonstrated that 2,4-DNT degrades, accompanied by stoichiometric release of nitrite, in both inoculated and uninoculated soil from the PBG when 50 mM phosphate buffer is recirculated through the columns at room temperature. Other preliminary work demonstrated that constant moisture allows continuous degradation of DNT. Up to a 2-week lag period resulted when soil columns were allowed to dry. Additional studies determined specific effects of phosphate amendment, pH, and ambient soil temperature on DNT degradation.

Soil and groundwater effects. Columns were constructed to determine whether DNT degradation would occur in the various BAAP soils when groundwater, rather than phosphate buffer was the recirculated fluid. For soils that contained little or no DNT (Table 1), 2,4-DNT was dissolved aseptically in groundwater to approximate the total amount of 2,4-DNT in the contaminated PBG UZ soil.

DNT degradation in the inoculated columns was much slower when groundwater was recirculated through the columns than when phosphate buffer was recirculated (Figure 1). The obvious difference between the phosphate buffer and groundwater systems is the possible nutritive benefits of the phosphate, if the soil-water system is phosphate limited. A collateral effect of the switch from phosphate buffer to groundwater was the change in pH of the systems. The system with 50 mM phosphate buffer maintained pH in a narrow range between 7.2 and 7.0 throughout the experiment, while the pH in the systems with groundwater increased to 8.2-8.5 from the initial groundwater value of 7.7. Shake flask studies indicated that both phosphate and pH were limiting. Sterile phosphate buffer was added to the recirculating groundwater and the pH was lowered to 7.5.

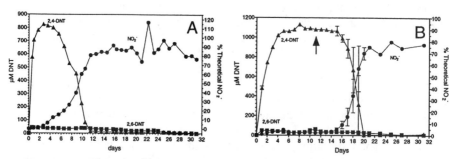

FIGURE 1. DNT degradation in columns with (A) 50 mM phosphate buffer, and (B) BAAP groundwater. Arrow indicates phosphate addition.

2,4-DNT degradation began in the inoculated PBG UZ columns 5 days following phosphate amendment and pH adjustment. DNT degradation did not begin in the uninoculated PBG UZ columns before the termination of the experiment at 32 days. The total incubation time was selected to allow a 2-week lag period after inoculation for columns to respond to the phosphate addition. 2,4-DNT began to degrade in uninoculated columns constructed with soil from the DBG within 11 days of phosphate amendment. The response of the native bacteria in the DBG soil and not in the PBG soil might be due to the 10 fold lower concentrations of DNT in the DBG soil. There was no detectable DNT degradation in columns constructed with clean soil or soil from the saturated zone. The last two results might be expected from the lack of DNT in the soils sufficient to support DNT-degrading population.

Phosphate amendment and temperature effects. The effects of phosphate amendment and temperature were examined in columns constructed with PBG UZ soil. Based on shake flask experiment results the filter sterilized groundwater was amended with phosphate buffer (pH 7.7) at final concentrations ranging from 0 to 10 mM. Columns were incubated at room temperature (19-25°C) or at the soil temperature at BAAP (13°C).

Shake flask experiments indicated that 10 mM orthophosphate is the minimal concentration necessary for sustained, rapid degradation of DNT in the presence of soil from BAAP, but that DNT would degrade completely in the presence of soil without added phosphate after an extended incubation period. The result was used to define the range of phosphate amendment treatments in the column experiments. Addition of orthophosphate to the groundwater reservoirs caused the immediate formation of a white precipitate, the amount of which appeared to be proportional to the phosphate buffer concentration. Although the precipitate did not appear to impede water flow through the columns, concerns about possible clogging led to further tests with alternate forms of phosphate.

2,4-DNT degradation began in room temperature columns 3 days after inoculation and the start of groundwater circulation as indicated by nitrite release and the transient accumulation of 4-methyl-5-nitrocatechol, the first intermediate in the 2,4-DNT degradation pathway (Spanggord et al., 1991). Initial degradation rates were similar regardless of the phosphate concentration (Figure 2), however; after 2 weeks, the degradation rates slowed dramatically in all but the columns with 10 mM phosphate. The slowdown in the degradation rates suggests that phosphate became limiting after consumption of roughly half of the original DNT in the soil. The result also suggests, in combination with shake flask experiments that phosphate is in equilibrium in the soil-water columns and becomes biologically available, albeit slowly, during an extended incubation period.

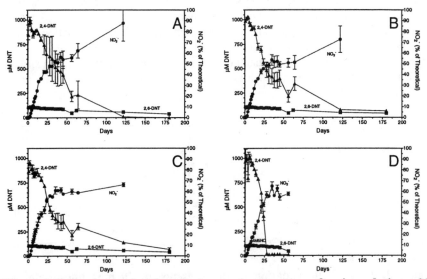

Figure 2. Soil columns incubated at room temperature after inoculation with ITIB enrichment. Orthophosphate was added to the groundwater at the following concentrations: A) 0 mM, B) 0.5 mM, C) 2 mM, and D) 10 mM. Error bars represent 1 standard deviation.

Columns incubated at 13°C that were inoculated with active LTIB culture began to degrade 2,4-DNT after a lag period that was similar to the lag period for room temperature columns (Figure 3A). However, the time to completion of degradation was unpredictable for the low temperature columns. For uninoculated columns (Figure 3B), the lag periods were longer, and time to completion even less predictable. Under field conditions, where the bacteria have been subjected to less disturbance from drying, mixing and temperature changes, and where the inoculum source is potentially much larger than 75 g of soil, the initiation of DNT degradation by the indigenous bacteria might be expected to occur sooner.

Because degradation of 2,4-DNT in room temperature columns showed a positive effect from phosphate amendment, a set of shake flasks was started at 13°C to determine whether enhancement of 2,4-DNT degradation by phosphate amendment also occurs at 13°C. The shake flasks with and without inoculation were compared to the columns presented in Figure 3B. The shake flasks were constructed so that the quantities of soil and groundwater used would be the same as in the columns. The LTIB culture was used to inoculate 2 of the 4 shake flasks. The shaking used to provide aeration also caused higher concentrations of DNT to be dissolved in the aqueous phase than in the comparable columns. Nitrite release showed that DNT degradation began almost immediately in the inoculated shake flasks, but at a rate too low to be detectable by a decrease in the aqueous DNT concentration. The prolonged but steady release of nitrite, and concomitant degradation of 2,4-DNT was similar to that in the room temperature columns with no added phosphate (Figure 2A). Nitrite release in the uninoculated flasks showed that DNT degradation began after 3 months of incubation. The result is similar to that for uninoculated columns with 10 mM phosphate at 13°C (Figure 3B). The 13° flask and column experiments clearly demonstrate that inoculation is not necessary, but can greatly hasten both the initiation and completion of 2,4-DNT degradation in the PBG soil at ambient soil temperature. They also show a similar result for phosphate amendment. It is not necessary, but can increase the rate of DNT degradation.

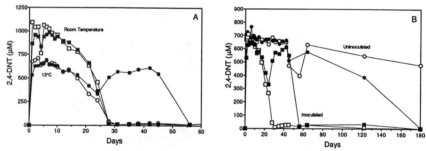

Figure 3. 2,4-DNT degradation in the presence of 10 mM phosphate. A) Inoculated columns at room temperature and 13°C, B) columns incubated at 13°C with and without inoculation.

The concentrations of 2,6- and 2,3-DNT gradually declined in the groundwater in all the columns, with the disappearance in the room temperature columns being slightly faster. 2,6-DNT remained constant in the 13° shake flasks. Inoculation had no effect on the rate of either isomer's disappearance.

At the termination of the column studies, both the organic content and the residual acetonitrile extractable DNT were dramatically reduced in all treatments except the 13°, no phosphate, uninoculated shake flasks (Table 2). Most columns reached residual 2,4-DNT levels that are 10 to 100 times lower than in the preliminary experiments where initial concentrations of 10.92 g/kg of 2,4-DNT were reduced to 0.22 g/kg, and 2,6-DNT levels that are 2 to 6 times lower than in preliminary experiments where initial concentrations of 2,6-DNT were reduced from 0.22 g/kg to 0.02. The extended incubation periods (up to 180 days) that were up to 5 times longer than in the preliminary experiments might have been responsible for the difference.

Table 2. Residual DNT and % organic matter in soil at termination of column study Values are averages for 2 replicates.

PO_4 (mM)	°C	LTIB	2,4-DNT (g/kg)	2,6-DNT (g/kg)	2,3-DNT (g/kg)	% Organic Matter
0	19-25	+	2.3E-3	1.2E-2	3.0E-4	0.04
0.5	19-25	+	1.3E-2	1.1E-2	4.8E-4	0.05
2	19-25	+	1.4E-2	1.1E-2	2.3E-4	0.05
10	19-25	+	6.6E-3	3.8E-3	<5E-5	0.11
10	13-15	+	9.4E-3	2.0E-2	8.3E-4	0.04
10	13-15	-	4.7E-1	1.1E-2	1.5E-3	0.05
0	13	+	4.7E-3	9.8E-2		0.46
0	13	-	6.58	0.19	3.2E-3	0.71

pH and Macronutrients. When the MI culture was inoculated into slurries of groundwater and soil, the optimum pH for 2,4-DNT degradation was between 7.0 and 7.5. In a similar test, the IB culture enriched from BAAP soil, showed maximal 2,4-DNT degradation in the presence of phosphate at pH 8.25. The pH of BAAP soil at field capacity ranges from 8.0 to 8.5, thus the indigenous DNT-degrading are adapted to the pH conditions at BAAP. Phosphate addition to soil-water slurries enhanced DNT degradation, but no effects were detected upon amendment with sulfate (and magnesium). In soil-free systems, 100 µM phosphate was required to degrade 500 µM 2,4-DNT.

DNT supplies more nitrogen than bacteria need for growth and nitrite accumulation is both a regulatory problem and inhibits 2,4-DNT degradation when nitrite exceeds 20 mM (Nishino et al., 2000). One method of removal is to take advantage of denitrifying bacteria that reduce nitrite to nitrogen gas under anaerobic conditions. Addition of methanol to nitrogen sparged fluid from column reservoirs, was sufficient to establish conditions that lead to the removal of nitrite (data not shown). Based on this result, a reduction zone to establish denitrifying conditions was incorporated into the final field demonstration.

Conclusions. Likely bottlenecks to *in situ* DNT degradation at BAAP were investigated, including nutrients, temperature, moisture, pH and nitrite. The studies demonstrated that 2,4-DNT-degrading bacteria are present in the soil and groundwater at the BAAP PBG as well as at the DBG, but are not present in uncontaminated soils. Even when present, DNT-degrading bacteria are inactive when the soil is dry. The indigenous bacteria are capable of degrading 2,4-DNT in the Waste Pit soils at the ambient soil temperature, but the onset of degradation at the low temperature is highly variable. Phosphate amendment enhances 2,4-DNT degradation, particularly in the PBG soil. Phosphate may have to be added in a form that will not cause a precipitate in the soil. The indigenous bacteria degrade DNT most rapidly at normal BAAP soil pH. The findings were used as the basis for a pilot scale field study that took place at BAAP from April-August, 2000 (Cuffin et al., 2001). The initial demonstration was successful and the treatment is being extended to two other waste pits at the site.

REFERENCES

Bruhn, C., H. Lenke, and H.-J. Knackmuss 1987 "Nitrosubstituted aromatic compounds as nitrogen source for bacteria." *Appl. Environ. Microbiol.* 53: 208-210.

Cuffin, S. M., P. M. Lafferty, P. N. Taylor, J. C. Spain, S. F. Nishino, and K. A. Williams. 2001. "Bioremediation of dinitrotoluene isomers in the unsaturated/saturated zone," In *abstr. Poster Session B1, Sixth International In Situ and On-Site Bioremediation Symposium,* pp.

Nishino, S. F., G. Paoli, and J. C. Spain 2000 "Aerobic degradation of dinitrotoluenes and pathway for bacterial degradation of 2,6-dinitrotoluene." *Appl. Environ. Microbiol.* 66: 2139-2147.

Nishino, S. F., J. C. Spain, and Z. He. 2000. "Strategies for aerobic degradation of nitroaromatic compounds by bacteria: process discovery to field application," In Spain, J. C., J. B. Hughes and H.-J. Knackmuss (Eds.). *Biodegradation of nitroaromatic compounds and explosives,* pp. 7-61. Lewis Publishers, Boca Raton.

Nishino, S. F., J. C. Spain, H. Lenke, and H.-J. Knackmuss 1999 "Mineralization of 2,4- and 2,6-dinitrotoluene in soil slurries." *Environ. Sci. Technol.* 33: 1060-1064.

Parsons, T. R., Y. Maita, and C. M. Lalli. 1984. "A manual of chemical and biological methods for seawater analysis," In pp. 14-17. Pergamon Press, Oxford.

Smibert, R. M. and N. R. Krieg. 1994. "Phenotypic characterization," In Gerhardt, P., R. G. E. Murray, W. A. Wood and N. R. Krieg (Eds.). *Methods for general and molecular bacteriology,* pp. 607-654. American Society for Microbiology, Washington, D. C.

Spanggord, R. J., J. C. Spain, S. F. Nishino, and K. E. Mortelmans 1991 "Biodegradation of 2,4-dinitrotoluene by a *Pseudomonas* sp." *Appl. Environ. Microbiol.* 57: 3200-3205.

Stone & Webster Environmental Technology & Services. 1998. *Draft alternative feasibility study Propellant Burning Ground and Deterrent Burning Ground, waste pits, subsurface soil, Badger Army Ammunition Plant, Baraboo, Wisconsin.* Draft, U.S. Army Corps of Engineers, Omaha District.

Zhang, C., S. F. Nishino, J. C. Spain, and J. B. Hughes 2000 "Slurry-phase biological treatment of 2,4- and 2,6-dinitrotoluene: role of bioaugmentation and effects of high dinitrotoluene concentrations." *Environ. Sci. Technol.* 34: 2810-2816.

IN SITU BIOREMEDIATION OF TNT IN SOIL

André Gerth and Anja Böhler (BioPlanta GmbH, Delitzsch, Germany)
Hartmut Thomas (WASAG DECON GmbH, Haltern, Germany)

Abstract: WASAG DECON GmbH (Haltern, Germany) and BioPlanta GmbH (Delitzsch, Germany) have carried out extensive trials of an *in situ* (pilot-scale) method of remediation of TNT-contaminated soils at the site of WASAG-CHEMIE, Sythen (Germany). Tests were designed to selectively stimulate the microorganisms existing on the site, assumption that TNT-tolerant and TNT-degrading species will have selectively accumulated in the soil as a result of the many years of contamination. In soils having low-to-medium nitroaromatic contamination, the remediation strategy provides for the biotransformation of the compounds concerned based on the interaction of white rot fungi with rhizosphere microorganisms. For this study, pot experiments were carried out to assess the impact of plants, microorganisms both autochthonous and allochthonous to the site, and the addition of supplementary C substrates on the bioremediation performance under field conditions. At Sythen, the technological sequence of operations for this microorganism-based process and the remediation performance achievable by it were tested *in situ* on a reference area of 250 m^2. The results demonstrate the fundamental efficacy of the *in situ* process described.

INTRODUCTION

Materials contaminated with TNT and other nitroaromatics as a result of the production, processing, and storage of explosives are both toxic and mutagenic. Despite the low water solubility of TNT, soil contamination presents hazards to the groundwater - a protected natural resource.

State of the art in Germany.
As regards the treatment of explosive-contaminated soils, the state of the art in Germany consists of landfilling or incineration. For soils containing medium-to-low concentrations of nitroaromatics, biological ex situ processes are currently being tested.

At sites with large areas of medium-to-low contamination, *in situ* remediation processes are of interest both economically and ecologically. Such processes have many advantages:
- Suitable for use over large areas
- No excavation of the contaminated soil necessary
- Little interference with the ecosystem
- Cost-effective.

Site description. Since 1895, explosives have been produced for civilian use on the company grounds of WASAGCHEMIE Sythen GmbH. During the two world wars, plants were operated for the fabrication of military explosive devices, and after World War I for the purpose of dismantling such devices. In the areas of the site in which, up to 1945, the production and disassembly of military explosive devices was carried out, the topsoil exhibits levels of TNT contamination that fluctuate greatly depending on the location. While the "hot spots" were excavated in 1999, relatively large areas are still contaminated with mainly low-level (below 10 mg TNT per kg), in isolated cases moderately high (10-50 mg TNT per kg) concentrations of nitroaromatics. The contaminated material is restricted to the top 50 cm of the soil profile.

Groundwater analyses have not revealed any overall contamination by compounds typical of explosives. The impact on the groundwater produced by the areas of contaminated soil is restricted to the works site. In the immediate vicinity of each inlet point, the groundwater exhibits a considerable increase in nitroaromatic concentrations. Currently, only traces of these compounds are detectable in the groundwater current downstream of these points and outside the company grounds.

Objectives. The aim of extensive investigations was to develop a process for the biological *in situ* remediation of TNT-contaminated soils.

As a result of the inhomogeneous distribution of the TNT crystals, which had had a negative impact on previous tests involving real soils, the TNT degradation process was investigated in model systems based on artificially-contaminated quartz sand. The aim was to record the stimulation of biological TNT transformation in quantitative terms by adding additional sources of carbon and to compare the degradation performance of autochthonous microorganisms with that of admixed white rot fungi. Microbiological tests of the contaminated soil at the Sythen site in 1999 had already confirmed the existence there of various species of fungus tolerant to TNT.

Parallel to this, a field trial concerning *in situ* remediation was started on the company grounds of WASAGCHEMIE Sythen GmbH. This trial was designed to estimate the time spans necessary for remediation and to optimize the technical sequence of remediation operations.

Transformation of TNT. TNT is only prone to a limited degree to attack by oxygenases formed by bacteria in an aerobic environment. Preferably, a biological transformation process is carried out under reducing conditions. The transformation of TNT starts with the reduction of a nitro group - via a hydroxylamine - to the amino group.

This reduction process, the products of which are 2-amino-4,6-dinitrotoluene (2-ADNT) and 4-amino-2,6-dinitrotoluene (4-ADNT), and also, after a second reduction step, the isomeric diaminonitrotoluenes (DANTs), can also be initiated abiotically by means of reducing agents. The reduction of all three nitro groups to the highly-reactive triaminotoluene (TAT) is only possible under extreme reduction conditions.

In the presence of a readily degradable carbon source, TNT is also transformed under oxidizing conditions. Reductive conditions accelerate the transformation process and reduce the accumulation of transformation products. The reduced transformation products are reactive and react with functional groups in the organic soil material. The products of the reaction with humic substances are complex, with various types of covalent bonds being formed.

MATERIALS AND METHODS

Tests in the model system. Quartz sand was used as the substrate. This was supplemented with TNT dissolved in acetone sufficient to provide 800 mg TNT per kg, and then intensively homogenized.

Soil from the Sythen site containing autochthonous microorganisms, or white rot fungi (*Stropharia rugosoannulata*) precultivated on straw, were used to inoculate the substrate. Molasses and rotting birchwood were used as the carbon sources. *Medicago sativa* (lucerne) was planted in the substrate as an alternative treatment for the purpose of recording the possible effects of plants on TNT attenuation in the soil.

The tests were carried out as pot experiments under field conditions using vessels with a volume of 30 L. Soil samples were taken at fortnightly intervals by means of a rod-type probe, with eight individual samples being combined into one composite sample for each vessel.

In response to treatment employing autochthonous microorganisms and molasses as the carbon source, the TNT concentration dropped very rapidly, thus this treatment was reproduced in a second study, consisting of a substrate to which molasses had been added without any additional inoculation with microorganisms. The sampling interval was reduced to 2 days for this treatment.

Ecotoxicity test. Determination of the germination capacity and the root length growth of cress (*Lepidum sativum*) was the method used for ecotoxicological evaluation of the substrate. For this purpose, soil filtrates were prepared to establish the germination capacity of the plant seeds and the lengths of the roots.

Microbiological tests. Enumeration of the bacteria and fungi in the soil was ascertained quantitatively after the trial had finished by means of various culture media.

Field trial *in situ*. A birch-covered area of about 250 m^2, exhibiting a mean TNT concentration of 6,000 mg per kg, was selected as the test area. The birch trees present in the area were rooted out and crushed using a shredder. The shredded tree material was ploughed into the ground using a rotary hoe together with rotting birchwood chips some of which were overgrown with fungi (Figure 1). This mechanical hoeing also served to homogenize the soil.

The area was irrigated at the beginning of the trial and during relatively long dry periods. In the course of the trial, the ground was hoed over once again and shredded birchwood was ploughed in. Samples were taken at weekly intervals from the upper soil horizon (0–30 cm). The individual samples (21 samples) were composited to give representative mixed samples.

FIGURE 1. Mechanical treatment of the soil

Analysis. The air-dried, screened samples of soil were analyzed by GC/ECD following ultrasonic extraction with acetone.

RESULTS

Tests in the model system. In the control preparation (uninoculated and containing no C source or plants) the TNT concentration remained constant for the duration of the trial.

In response to the treatment containing *Stropharia* but no carbon, no reduction (average concentration 620 mg/kg DW) in the TNT concentration was detectable; in the treatment containing molasses and a scattering of birchwood, on the other hand, a reduction to 40-45% was recorded. The addition of molasses alone sufficed to bring about a reduction in the TNT concentration. In the first sample, which was collected after 1 week of treatment, the TNT concentration was only < 20 mg/kg. The TNT concentration was further reduced to < 10 mg/kg after 8 weeks.

The most rapid attenuation of TNT was established in the treatment containing autochthonous microorganisms and molasses (Figure 2). Within as little as 2 days the TNT concentrations dropped to < 50 mg/kg. In the treatments containing wood chips as the carbon source, a comparable reduction in concentrations was observable, these falling to a residual TNT content of 350 mg/kg in about 100 days. The addition of autochthonous microorganisms and diluted molasses enabled a reduction in TNT concentrations to 20 mg/kg to be attained within 14 days.

Even in the treatment to which, apart from molasses, no microorganisms had been added, a clear reduction in TNT contamination was observable. This reduction was, however, not detectable until 12-14 days had elapsed.

In the control sample, total ADNT concentrations of 25-45 mg/kg DW were detectable. Considerably higher concentrations of primary metabolites were observed in the treatments in which a sizeable reduction in TNT was detectable. While 2-ADNT was predominant in the test preparations not containing molasses, in those containing molasses the metabolite concentrations to be found were nearly identical or ones with an elevated 4-ADNT content. In some treatments, it was observable that the metabolite concentration initially increased before subsequently falling again.

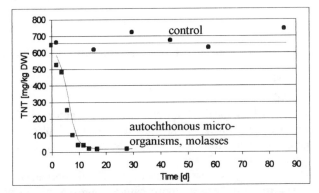

FIGURE 2. TNT concentration treatment autochthonous microorganisms and molasses, control

Ecotoxicity test. On treating TNT contaminated soil with autochthonous mircoorganisms and molasses the seeds of cress developed without visual impairments. The roots reached a length of about 6.5 till 7.0 cm (control 9.5 cm). In the other variants the seeds were partially yellowed and the root growth was reduced.

Microbiological tests. Tests on the contaminated soil confirmed the existence of various species of microorganisms tolerant to TNT. Three isolates grew at a initial TNT-concentration of 1.000 ppm.

Field trial *in situ*. A roughly 80% reduction in the TNT contamination of the soil was detectable over a 100-day period. The TNT concentration in the soil samples analyzed exhibited great fluctuations (Figure 3).

ADNT concentrations were in the 25-50 mg per kg DW. At the same time, 4-ADNT content was somewhat higher - a fact that, based on the results of the model tests, is indicative of active microbiological transformation.

As a result of natural succession, a variety of grasses and herbs that are typical of the site were able to establish themselves.

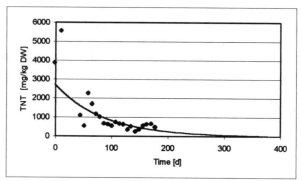

FIGURE 3: TNT concentration in the soil subjected to the field trial

DISCUSSION

The rate of biological TNT transformation depends on the availability of readily-usable carbon sources. A transformation also took place in the presence of poorly-degradable cosubstrates (e.g., wood chips); however, the degradation rates recorded were considerably lower.

Our results suggest that no "specialized" microorganisms were required for TNT transformation: because those naturally present in soil were capable of mediating TNT removal. As had already been the case in the pilot tests of 1999, the addition of the fungus *Stropharia rugosoannulata*, which has been very successfully tested in laboratory experiments, exhibited no measurable impact on TNT degradation.

In the microbiological tests, this fungus was no longer detectable in the substrate at the end of the trial; other species, such as fungi and bacteria, were presumably more competitive.

The toxicity of the substrates from the pot experiments on cress seedlings diminished parallel to the TNT concentration. As expected, filtrates from treatments in which no TNT degradation was observable exhibited the highest ecotoxicity. Although the seedlings in the treatments with wood chips as the carbon source grew better, the seedlings' root growth was impaired. At the end of the trial, all the soil filtrates from treatments in which the TNT concentration had dropped considerably exhibited only very slight phytotoxicity. This would suggest that no phytotoxicity emanated from metabolites and/or humus adducts that were not examined.

A comparison of the test results showed that degradation rates of a similarly high level are attainable in the field trial. An extrapolation of the degradation curve permits the assumption to be made that the remediation period for the Sythen site will be 2 years. The remediation period required, which depends on the aim that is set for the remediation process, can be reduced by using suitable carbon sources. This will be the subject of additional trials in 2001.

Specially-precultivated fungi produced no measurable increase in microbiological degradation performance. The latter were not able to establish themselves in a viable fashion in competition with the autochthonous microflora. The material introduced in the form of appropriate carbon sources and soil containing autochthonous microflora is able to bring about transformation of the TNT.

CONCLUSIONS

1) In the trials, the white rot fungus *Stropharia rugosoannulata* did not bring about any detectable TNT transformation.
2) TNT attenuation in the soil was successfully accomplished via stimulation by means of autochthonous microorganisms.
3) The rate of microbiological TNT transformation in the soil was dependent on the type and quantity of the carbon source available.
4) In the pot experiment, the planting of *Medicago sativa* in the substrate had no measurable impact on TNT attenuation.
5) It was established that the toxicity recorded in the soil diminished as the TNT concentration was reduced.
6) Enhanced biotransformation is a practicable option for TNT-contaminated soils. In the case of the process tested, remediation of the Sythen site can be expected to take 2 years.

REFERENCES

Michels, J.: Forschungsverbund "Biologische Verfahren zur Bodensanierung". Eine Zusammenstellung der laufenden Projekte. Umweltbundesamt, Berlin 1999

McFarlan, S. "2,4,6-Trinitrotoluene Pathway Map." The University of Minnesota Biocatalysis/Biodegradation Database.

Pennington, J. C., Bowen, R., Brannon, J. M., Zakikhani, M., Harrelson, D. W., Gunnison, D., Mahannah, J., Clarke, J., Jenkins, T. F., and Gnewuch, S. 1999. "Draft protocol for evaluating, selecting, and implementing monitored natural attenuation at explosives-contaminated sites," Technical Report EL-99-10, U.S. Army Engineer Research and Development Center, Vicksburg, MS.

Price, C. B., Brannon, J. M., Yost, S. L., and Hayes, C. A. 2000. "Adsorption and transformation of explosives in low-carbon aquifer soils," ERDC/EL TR-00-11, Engineer Research and Development Center, Vicksburg, MS.

Bayman, P., Radkar, G.V. 1997. "Transformation and Tolerance of TNT (2,4,6-trinitrotoluene) by Fungi." *International Biodeterioration And Biodegredation.* 39 (1), 45-53.

Meharg, A.A., Dennis, G.R., Cairney, J.W.G. 1997. "Biotransformation of 2,4,6-trinitrotoluene (TNT) by ectomycorrhizal basidiomycetes." *Chemosphere.* 35 (3).

Brannon, J. M., and Myers, T. E. 1997. "Review of fate and transport processes of explosives." Technical Report IRRP-97-2, U.S. Army Engineer Waterways Experiment Station, Vicksburg, MS.

Pennington, J. C., Thorn, K. A., Inouye, L. S., McFarland, V. A., Jarvis, A. S., Lutz, C. H., Hayes, C. A., and Porter, B. E. 1999. "Explosives conjugation products in remediation matrices: Final report," Technical Report SERDP-99-4, U.S. Army Engineer Research and Development Center, Vicksburg, MS.

Thorne, P.G., and Leggett, D.C. 1999. "Investigations of Explosives and Their Conjugated Transformation Products in Biotreatment Matrices." U.S. Army Cold Regions Research and Engineering Laboratory Special Report 99-3.

APPLICATION OF LIGNINOLYTIC FUNGI FOR BIOTRANSFORMATION OF CHLOROPHENOLS

Alexey A. Leontievsky, Nina M. Myasoedova, Boris. B. Baskunov and Ludmila A. Golovleva (Institute of Biochemistry and Physiology of Microorganisms, Russian Academy of Sciences, Pushchino Moscow region, Russia) Chris Bucke and Christine S. Evans (University of Westminster, London, UK)

ABSTRACT: The toxicity of thirteen isomers of mono-, di-, tri- and pentachlorophenols was tested in potato-dextrose agar cultures of the white rot fungi *Panus tigrinus* and *Coriolus versicolor*. 2,4,6-Trichlorophenol was chosen for further study of its transformation in soil, sawdust, and aqueous solution by cultures of these fungi and immobilized laccase from *C. versicolor*. The ligninolytic enzyme systems of both fungi were found to be responsible for 2,4,6-TCP transformation. 2,6-Dichlorohydroquinone, 2,6-dichlorobenzoquinone and 3,5-dichlorocatechol were found as products of primary oxidation of 2,4,6-TCP by intact fungal cultures and purified ligninolytic enzymes, including immobilized laccase from *C. versicolor*.

INTRODUCTION

Chlorophenols are known as one of the most toxic and widely spread environmental pollutants. Degradation of these compounds by bacterial species has been relatively well studied. Fungal degradation of chlorophenols is much less understood despite the well documented involvement of lignin-degrading white rot fungi in biodegradation of different types of hazardous xenobiotics. The ability of these fungi to attack persistent chemical structures, including chlorophenols, in many cases depends on the expression of the ligninolytic enzyme system. Ligninolytic enzymes of white rot fungi are characterised by their wide substrate specificity which is the basis for numerous attempts of application of these enzymes to solve biotechnological problems.

Different approaches have been used to study fungal biodegradation of chlorinated phenols: intact cultures of fungi (Reddy et al., 1998), separation of mycelium and cultural liquid (Armenante et al., 1994), application of free and immobilized enzymes (Shuttleworth & Bollag, 1986).

The aim of the present work was to investigate transformation of 2,4,6-trichlorophenol (2,4,6-TCP) in process of model remediation of soil, sawdust, and aqueous solutions by intact cultures of white rot fungi *Panus tigrinus* and *Coriolus versicolor* and immobilized laccase from *C. versicolor*.

MATERIALS AND METHODS

Microorganisms and methods of cultivation. The white-rot fungi *Panus tigrinus* 8/18 (IBPM RAS) and *Coriolus versicolor* VKM F-116 were grown at 29^0C on potato-dextrose agar (PDA) plates under stationary conditions or in liquid media with agitation. Inoculum for liquid media for both fungi were grown for seven days on soya-glycerol medium (g/L): NH_4NO_3, 0.2; KH_2PO_4, 0.2; K_2HPO_4, 0.02; $MnSO_4 \cdot 7H_2O$, 0.1; peptone, 0.5; soya powder, 0.5; glycerol, 2.0 ml/L. Cultivation was carried out in 750-ml flasks with 100 ml of mineral medium, containing "low" or "high" concentration of nitrogen sources, equal to 2.4 and 24.0 mM of ammonium nitrogen (Kirk et al., 1978). As a carbon source, 1% glucose (*C. versicolor*) or 1% maltose (*P. tigrinus*) were used. Submerged cultures grown for purification of extracellular enzymes contained additionally 2.0 mM 2,4-dimethoxyphenol (*P. tigrinus*) or 0.2 mM tannic acid (*C. versicolor*) as inducers.

2-, 3-, 4-chlorophenols, 2,3-, 2,4-, 2,5-, 2,6-, 3,4-, 3,5-dichlorophenols (DCP), 2,3,4-, 2,4,5-, 2,4,6-trichlorophenols (TCP) and penthachlorophenol (PCP) were added to PDA cultures (to a final concentration 5-200 mg/L). 2,4,6-TCP was added to liquid media as a dimethylformamide solution (100 µl) on the 5-th day of cultivation to a final concentration of 50 mg/L.

The toxicity of chlorophenols in PDA cultures was evaluated by measurements of diameters of colonies in comparison with control plates with no chlorophenols added. Control cultures of *P. tigrinus* colonized the surface of the plates in 84 h, and control cultures of *C. versicolor* in 60 h. In liquid cultures the effect of 2,4,6-TCP was estimated by comparison of dry biomass weight and extracellular enzyme activities compared with control variants without of 2,4,6-TCP.

Immobilization of laccase. Laccase was covalently bound to R-637 Celite (Manville, USA) with glutaraldehyde. The mixture of 25 µl of 1 mg/ml aqueous solution of laccase with 400 µl of 1 mg/ml aqueous solution of gelatine was added to 0.5 g of Celite. Then 900 µl of cold (-20^0 C) acetone was added and the preparation was held on ice for 10 min. Finally, 300 µl of 25% glutaraldehyde was added and the mixture was incubated at room temperature for two hours. Afterwards, Celite with immobilized laccase was washed three times with 20 mM Na-acetate buffer, pH 5.0 and stored in the same buffer at 4^0C.

Enzyme reactions with 2,4,6-TCP. Reaction mixtures (3 ml) contained 25 µg of purified laccase and 50 or 400 mg/L of 2,4,6-TCP in 20 mM Na-acetate buffer, pH 5.0, or 50 mM Na-borat-acetate-phosphate buffer, pH 5.0 or 7.0 were used. Reaction vials were incubated for 10 min -24 h at 29^0 C or 50^0 C with agitation. Reaction mixtures with Mn-peroxidases (MnP) were held in 25 mM Na-lactate-succinate buffer, pH 4.0 with 0.1 mM $MnSO_4$ and 50 mM H_2O_2. A 5-ml plastic syringe filled with 1.0 g of Celite with 25µg of immobilised laccase in appropriate buffer solution was used as a flow-through column. Flow rate of 4 ml/h was created with peristaltic pump, 2 ml fractions were collected. 1 mM 2,2,6,6-Tetramethylpiperidine-N-oxyl (TEMPO) or 4-OH-TEMPO were added to the reaction mixture as redox mediators when necessary.

Mn-peroxidases and laccases of *P. tigrinus* and *C. versicolor* were purified as described earlier (Leontievsky et al., 1990) with some modifications.

Isolation and identification of 2,4,6-TCP oxidation products. Reaction mixtures or culture liquid, were extracted by ethylacetate evaporated on a vacuum at 40^0 C. The residues were dissolved in methanol and analysed by HPLC, thin-layer chromatography (TLC) and mass-spectrometry (MS).

HPLC analysis of 2,4,6-TCP was conducted with a ODS-2 column (4.6 x 250 mm) at 50^0 C. Elution was carried out at a flow rate of 0.8 ml/min with 65% of methanol in 5 mM KH_2PO_4, pH 2.0.

Qualitative analysis of 2,4,6-TCP and its metabolites was performed by TLC on Kieselgel 60 F_{254} plates developed with benzene-dioxane-acetic acid (90:10:2). After processing of the plates, metabolites were detected under UV light, with diazotized benzidine (as a reagent for aromatic hydroxy-groups) and silver nitrate solution in acetone (as a reagent for chlorine substituents). The spots with definite Rf values were eluted and used for MS-analysis (Finnigan Mat 8430, Germany).

RESULTS AND DISCUSSION

Toxicity of chlorophenols. The toxicity of thirteen chlorophenols was tested in PDA cultures of the white rot fungi *P. tigrinus* and *C. versicolor*. In static PDA plate cultures, the diameter of hyphal growth of the fungi was measured in the presence of increasing concentrations of chlorophenols (Table 1). The toxic effect of chlorophenols increased with the number of chlorine substituents. In general, both of the fungi tested were affected by the chlorophenols to a similar degree with a few exceptions, where *C. versicolor* was more resistant to 2-CP and PCP and much more sensitive to 3,5-DCP, than *P. tigrinus*. The chlorophenols that produced the most inhibition effect on growth rate were 2,3,4-TCP, 2,4,5-TCP, 2,4,6-TCP and PCP at the lowest concentrations of 15-50 mg/L. Comparison of the most toxic chlorophenols showed that PCP completely inhibited growth of *P. tigrinus* and *C. versicolor* at different concentrations, of 25 and 75 mg/L respectively. All TCP molecules inhibited hyphal growth completely at 50 mg/L, but 2,4,6-TCP was marginally less toxic at 15-25 mg/L than 2,3,4- and 2,4,5-substituted compounds. The results found for PDA cultures with 2,4,6-TCP correlated with the data of liquid cultures of the fungi.

2,4,6-TCP and PCP were chosen for the further experiments as a representatives of the relatively toxic group of chlorophenols.

TABLE 1. Concentrations of chlorophenols (mg/L) providing 100% inhibition of fungal growth in agar cultures of *P. tigrinus* and *C. versicolor*.

	2CP	3CP	4CP	2,3DCP	2,4DCP	2,5DCP	2,6DCP
P. tigrinus	150	150	150	75	75	100	75
C. versicolor	200	150	200	50	75	75	75

	3,4DCP	3,5DCP	2,3,4TCP	2,4,5TCP	2,4,6TCP	PCP
P. tigrinus	75	50	25	15	50	25
C. versicolor	50	50	15	15	50	75

Removal of chlorophenols from different media. The results on toxicity of chlorophenols were used for development of approaches for remediation of natural media with corresponding contamination. Hardwood and softwood sawdust, sandy loam soil and liquid mineral medium, contaminated with 2,4,6-TCP or PCP, were taken as a model. Solid-state fermentation of hardwood and softwood sawdust contaminated with 2,4,6-TCP by *P. tigrinus* resulted in 80-98 % disappearance of toxicant for 2 months at 15-20^{0}C (Figure 1). No products other than polymer were found.

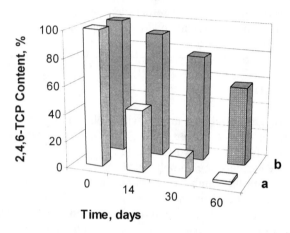

FIGURE 1. 2,4,6-TCP removal from pine sawdust by *P. tigrinus*. The initial concentration of toxicant was 200 mg/kg of straw. (a) Solid-state fermentation of sawdust by *P. tigrinus*. (b) Control variant, no culture of *P. tigrinus*.

Bioremediation of 2,4,6-TCP or PCP contaminated soil by *P. tigrinus* was studied in 50 x 50 x 60 cm lysimeters in field experiments at May-August 1999 and 2000. The soil in the lysimeters was mixed with sterilized wheat straw or with wheat straw fermented by *P. tigrinus* for two weeks. The optimal ratio soil: straw was found to be 1:0.5 kg. The initial concentration of toxicants was 340 (PCP) and 500 (2,4,6-TCP) mg/kg of soil. For four months the extractable concentration of chlorophenols in control variants without of fungus decreased to 70 % in contrast to 35-40% in the experiments with fungus (Figure 2). In liquid cultures both *P. tigrinus* and *C. versicolor* showed a greater rate of 2,4,6-TCP oxidation in low nitrogen media than in high nitrogen media (Figure 3), suggesting that the lignin-degrading enzyme systems participated in this process (Kirk et al., 1978).

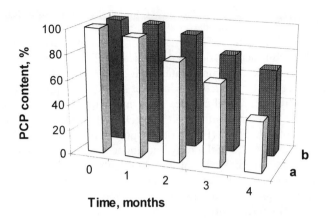

FIGURE 2. PCP removal from sandy loam soil in the field lyzimeters by *P. tigrinus* (May-August 1999). The initial concentration of toxicant was 340 mg/kg of soil. (a) Lyzimeter with introduction of *P. tigrinus* into soil. (b) Control lyzimeter, no culture of *P. tigrinus*.

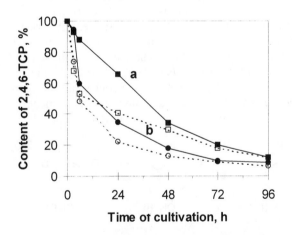

FIGURE 3. 2,4,6-TCP removal from submerged liquid medium by *P. tigrinus* (a) and *C. versicolor* (b). Solid lines – "high" nitrogen, suppressed ligninolytic system, dashed lines – "low" nitrogen, activated ligninolytic system. 50 mg/L of 2,4,6-TCP were added to 5-th day cultures of fungi.

Transformation of 2,4,6-TCP by fungal cultures in liquid media. During submerged cultivation of *P. tigrinus* and *C. versicolor* in low nitrogen liquid media without addition of 2,4,6-TCP, both fungi produced Mn-peroxidase and laccase. No lignin peroxidase activity was found in culture liquid of these fungi. The addition of 2,4,6-TCP on the 5-th day of growth of *P. tigrinus* cultures almost completely suppressed extracellular laccase activity but stimulated a reciprocal increase of MnP activity. When 2,4,6-TCP was added under the same conditions to culture of *C. versicolor*, its effect on the enzyme composition in the culture liquid was opposite to that of *P. tigrinus*. The dynamics of laccase activity was not significantly changed, but MnP activity declined drastically. This permits us to suggest that MnP was responsible for primary attack of 2,4,6-TCP in *P. tigrinus* cultures, when in the cultures of *C. versicolor* – laccase. As the main products of 2,4,6-TCP (I) degradation by both fungal cultures a polymer with MW over 1000 kDa (VI) and oligomeric compounds with MW 360 kDa (V) were found. The products of bioconversion of 2,4,6-TCP (I), besides of polymers, were identified as 2,6-dichlorobenzoquinone (II), 2,6-dichlorohydroquinone (III) (Figures 4). A trace amount of 2,6-dichloro-4-methoxyphenol and 2,6-dichloro-4-methoxy-1,3-dihydroxybenzene were also detected (data not shown). The formation of two chlorine-containing derivatives demonstrated that the fungi effected dechlorination of 2,4,6-TCP in *para*-position.

Purified MnPs and laccases from *P. tigrinus* and *C.* versicolor incubated with 0.5-6.0 mM 2,4,6-TCP for 2-24 h at 30^0C oxidized 70-100% of 2,4,6-TCP. 2,6-dichlorobenzoquinone (II) and 2,6-dichlorohydroquinol (III), as well as condensed oligomeric compound (V) and polymer (VI) were found as the only products of this reaction (Figures 4). Application of immobilized laccase from *C. versicolor* (see below) resulted in additional identification of 3,5-dichlorocatechol (IV) as a product of this reaction. To our knowledge this is the first description of this intermediate in a subsequent reaction of 2,4,6-TCP with a fungal enzyme system. The existence of this intermediate suggests the possibility of direct splitting of the aromatic ring of 2,4,6-TCP by a corresponding dioxygenase in the next reaction. It means, that at least in the case of *C. versicolor*, the 2,4,6-TCP degradation pathway may be similar to the bacterial modified *ortho*-pathway and different from the fungal *P. chrysosporium* pathway which suggested the involvement of a reductive dehalogenation stage during 2,4,6-TCP degradation (Reddy et al. 1998).

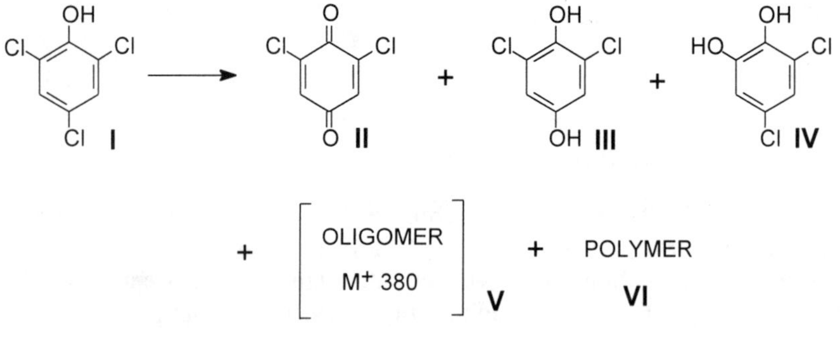

FIGURE 4. Scheme of 2,4,6-TCP transformation by fungal cultures. I, 2,4,6-TCP; II, 2,6-Dichloro-1,4-benzoquinone; III, 2,6-Dichloro-1,4-hydroquinone; IV, 3,5-Dichlorocatechol; V, Oligomer with M+ 380; VI, Polymer.

Immobilization of laccase from *C. versicolor*. Purified laccase from *C. versicolor* was immobilized on Celite R-637 by covalent binding with glutaraldehyde with effectivity of 98%. However, subsequent measurements of immobilized enzyme activity during the first 4-6 h of storage at 4^0 C revealed a decrease up to 50%. No soluble laccase activity or protein were found in the buffer above the carrier. Further storage at this temperature (in acetate buffer) showed stability of immobilized laccase was maintained for at least one month.

In comparison with soluble enzyme, immobilized laccase had pH optima shifted to 0.4-0.9 unit to alkaline region for syringaldazine (from 5.1 to 5.5), 2,4,6-TCP (from 4.7 to 5.2) and 2,6-dimethoxyphenol (from 4.3 to 5.2). Also immobilized enzyme had increased temperature optimum (from 55 to 60^0C), temperature stability ("half-life" time at 60^0C from 10 min to 1 h) and persistence to inhibitor ("half-life" time from 1.6 to 16.0 µM NaN_3).

Incubation of 25 µg of soluble laccase from *C. versicolor* with 400 mg/L of 2,4,6-TCP at pH 5.0 and 25^0C resulted in 100% removal of TCP during 3 h. The immobilized laccase gave the same result for 30 min at pH 7.0 and 50^0C. The addition of redox mediators (1 mM TEMPO or 4-OH-TEMPO) to reaction medium with immobilized laccase diminished formation of polymeric products with concomitant enrichment of the products set with compounds II and III (Figures 4). When Celite with immobilized laccase was placed in the column with 4 ml/h flow rate of 2,4,6-TCP solution, 3,5-dichlorocatechol (IV) was found as a predominant primary product. This means that formation of compound IV and formation of compound II and III might be the result of opposite and successive processes. Firstly, 2,4,6-TCP-oxidation resulted in 3,5-dichlorocatechol formation with subsequent polymerization. Then, after saturation of the reaction mixture with polymeric materials (late fractions from the flow through column or batch system with polymerization blocked by mediators), the conditions occurred for accumulation of compounds II and III, which are probably more stable than 3,5-dichlorocatechol.

ACKNOWLEDGEMENTS

This work was supported by NATO Science for Peace Program, grant SfP 972294 and Russian Foundation for Basic Research, grant RFBR 99-04-48185.

REFERENCES

Armenante, P. M., N. Pal, and G. Lewandowski. 1994. "Role of mycelium and extracellular protein in the biodegradation of 2,4,6-trichlorophenol by *Phanerochaete chrysosporium*." Appl. Environ. Microbiol. *60*(6): 1711-1718.

Kirk, T. K., E. Schultz, W. J. Connors, L. F. Lorens, and J. G. Zeicus. 1978. "Influence of culture parameters on lignin metabolism by *Phanerochaete chrysosporium*." Arch. Microbiol. *117*(3): 277-285.

Leontievsky, A. A., N. M. Myasoedova, O. V. Maltseva, N. G. Termkhitarova, V. I. Krupyanko, and L. A. Golovleva. 1990. "Mn-dependent peroxidase and oxidase of *Panus tigrinus* 8/18: purification and properties." Biochemistry (Engl. Transl. Biokhimiya) 55(10): 1375-1380.

Reddy, G. V. B., M. D. S. Gelpke, and M. H. Gold. 1998. "Degradation of 2,4,6-trichlorophenol by *Phanerochaete chrysosporium*: involvement of reductive dechlorination." J. Bacteriology. *180*(19): 5159-5164.

Shuttleworth, K. L., and J. M. Bollag. 1986. "Soluble and immobilized laccase as catalysts for the transformation of substituted phenols." Enzyme Microb. Technol. *8*:171-177.

PROCEDURES FOR COMPOSTING PENTACHLOROPHENOL CONTAMINATED SOILS

Michael K. Foget, P.E.
(SHN Consulting Engineers & Geologists, Inc., Eureka, California)
John Andrews R.G.
(SHN Consulting Engineers & Geologists, Inc., Redding, California)
Bob Vogt
(Louisiana-Pacific Corporation Samoa, California)
Neil Sherman
(Louisiana-Pacific Corporation Samoa, California)

Abstract: Pentachlorophenol-based fungicides were used at many primary wood product facilities prior to the mid 1980's. Drips and spills from fungicide application systems commonly resulted in pentachlorophenol (PCP) contaminated soil and groundwater. Because of these practices, Louisiana-Pacific Corporation (L-P) has been addressing PCP contaminated soil at its wood product facilities. A pilot study was initiated in 1997 to evaluate the feasibility of composting contaminated soil to reduce PCP levels to regulatory levels. Five different composting mixtures were evaluated using 1,000 yd^3 (765 m^3) of PCP contaminated soil excavated from a former sawmill facility. Each mixture consisted of approximately 200 yd^3 (153 m^3) of contaminated soil and predetermined amounts of wood fines, steer manure, fertilizer, and water. Once mixed, the contaminated soil and soil additives were placed in 3-ft. (0.9 m^3) high windrows and covered with plastic. PCP levels in three out of five mixtures decreased from 160 mg/kg to less than 10 mg/kg in 16 weeks. The procedures developed during this pilot study have been used to cost-effectively compost more than 10,000 yd^3 (7,650 m^3) of PCP-contaminated soil excavated from a former sawmill site. PCP concentrations at the full-scale project have decreased from 60 mg/kg to less than 15 mg/kg in one-year.

INTRODUCTION

A recent study completed in Finland demonstrated that bioremediation or composting was effective at reducing PCP levels in contaminated soil obtained from a sawmill facility (Laine and Jorgensen, 1997). Based on this study, L-P excavated 1,000 yd^3 (765 m^3) of contaminated soil from a former sawmill site, and initiated a pilot study to develop cost-effective composting procedures for remediating PCP-contaminated soil. The primary goals of the pilot study were to develop procedures that: 1) were suitable for addressing large volumes of contaminated soil and 2) would not require routine maintenance after the soil was placed in the compost cell.

Five different composting mixtures were evaluated during the pilot study, the results of which were used to develop a compost mix to treat 10,000 yd^3 (7,650 m^3) of PCP-contaminated soil.

BACKGROUND

Pilot Scale. The former sawmill site operated from 1942 to 1990. The site is located in the unincorporated area of Tehama County, California, approximately 3 miles (4.8 km) south of the City of Red Bluff.

Full Scale. The former Vertical Grain and Jamb Plant operated from 1945 to 1990. The site is located in the southwest section of Red Bluff, California.

MATERIALS & METHODS

During composting, microorganisms break down organic matter to produce carbon dioxide, water, heat, and humus. As such, compost piles are designed to provide an environment conducive to microbial growth. Design elements that can be controlled to enhance microbial growth include: 1) carbon to nitrogen ratio, 2) oxygen content, 3) moisture content, and 4) nutrient balance (Richards, 1998; Von Fahnestock, 1998).

Pilot Scale. Materials used to construct the composting cells are described in this section. Table 1 summarizes the chemical and physical characteristics of the materials used to develop the proper mixing ratios.

TABLE 1. Pilot Scale Initial Analytical Results

	PCP (mg/kg)	pH	Total N (mg/kg)	TOC (%)	C:N Ratio	Bulk Density (g/cc)	Moisture (%)
Source Material	160	6.8	300	0.9	30	1.8	14.7
Wood Fines	---	5.7	4,000	32.5	81	---	46.0
Fertilizer	---	---	190,000	---	---	0.8	5.0

PCP = pentachlorophenol TOC = total organic carbon
Total N = total nitrogen C:N = carbon to nitrogen ratio.

Approximately 1,000 yd^3 (765 m^3) of PCP-contaminated soil was excavated from a former dip and spray system in February 1997. The excavated soil was comprised of silty clay. The soils were stockpiled and covered on site from the time they were excavated in February until they were placed in the composting cells in May 1997.

Wood fines (pine and fir bark) were obtained from a nearby wood sorting facility, and were 3/8 inch (1 cm) or less in size.

Granular ammonium phosphate sulfate was used to supply nitrogen. The fertilizer contained 19% total nitrogen, 9% available phosphoric acid, 0% potassium, and 19% sulfur.

Fresh steer manure was obtained from a local feed lot, and water was from an on-site water supply well. The manure was added as a bulking agent and to introduce a variety of microorganisms into one of the composting mixtures.

Full Scale. During construction of the compost cells, wood fines were reclaimed from an abandoned log deck. The full-scale composting project included 10,000 yd^3 (7,650 m^3) of source material, 3,500 yd^3 (2,675 m^3) of wood fines, and 2,700 gallons (10,220 L) of fertilizer. The soils consist primarily of clayey silty sands. The initial contaminant conditions were: average PCP concentration of 58 mg/kg, maximum PCP concentration of 110 mg/kg, average petroleum hydrocarbon as Stoddard solvent (Stoddard) concentration of 350 mg/kg, and maximum Stoddard concentration of 870 mg/kg. A liquid fertilizer mix with 30% available nitrogen was used. The wood fines were 1/4 inch (0.6 cm) or less in size. Well water was supplied to the mixing area by water trucks. The target moisture content was 22% by weight.

PROCEDURES

Pilot Scale. Five different compost mixtures were evaluated during the pilot study. The mixtures included:
- Cell 1--130 yd^3 control pile (H$_2$O)
- Cell 2--200 yd^3 pile of soil mixed with fertilizer (H$_2$O+N)
- Cell 3--250 yd^3 pile of soil mixed with fertilizer and fines (H$_2$O+N+fines)
- Cell 4--250 yd^3 pile of soil mixed with fertilizer, fines and manure (H$_2$O+N+fines+manure)
- Cell 5--200 yd^3 pile of soil mixed with commercially available fines and nutrients (Commercial)

The carbon to nitrogen ratio and moisture content of each mixture was adjusted to enhance microbial activity. The target carbon to nitrogen ratio and moisture content were 30:1 and 26%, respectively. In addition, wood fines were added to enhance porosity. The mixing ratios used to achieve the target values were generated by modifying a spreadsheet developed by Tom Richards at Cornell University (Richards, 1998). These ratios are summarized in Table 2.

The mixing procedures were selected because mill personnel could implement them, they could be modified to compost large volumes of contaminated soil, and it would not be necessary to rework or add nutrients to the piles once they were covered.

The soil was mixed in 33 yd^3 (25 m^3) batches. In general, 11, 3 yd^3 (2.3 m^3) loader scoops of contaminated soil were obtained from the stockpile and spread to a depth of one foot in the mixing area. As the soil was spread, the bucket of the loader was used to break up soil clumps. Then the appropriate amendments were added to the contaminated soil. At this point, two loaders worked in tandem to mix the soil, wood fines, and manure into a homogeneous mixture. Water, with the appropriate amount of dissolved fertilizer, was added during the mixing process. The soil mixture was placed into a composting cell once sufficient water was added.

The composting unit consisted of five individual cells. Each cell was approximately 3 feet (0.9 m) high, 20 feet (6 m) wide, and 100 feet (30 m) long. The cells were constructed by placing the soil mixture on 6-mil plastic. When full, the cells were covered with 12-mil plastic to inhibit moisture loss and infiltration. The perimeter of each cell was sealed using rejected roofing shakes to secure the plastic from a nearby roofing shake manufacturing plant. Approximately two cells were constructed per day.

TABLE 2. Pilot Scale Final Mix Ratios

Cell	Source Material (yd^3)	Wood Fines (yd^3)	Manure (yd^3)	Fertilizer (lbs.)	Water (gallons)	Final Volume (yd^3)
1	1	0	0	0	3	130
2	1	0	0	0.9	3	200
3	1	0.5	0	4.2	3.8	250
4	1	0.25	0.21	0.6	1.5	250
5	1	Commercially available fines & nutrients			1.5	200

Cell number 5 was prepared under the direct supervision of a commercial contractor. In general, the mixing procedures were identical to the procedures used to construct cells 1 through 4. The ingredients and mix ratios were not disclosed.

Full Scale. The soil was treated using the compost mixture developed for Cell 3 of the pilot study, which was soil, wood fines, and nutrients. The compost mix from cell 3 was selected for the full-scale application because it was the easiest to implement and the most cost effective mix studied. The target mixing ratios for the full-scale study is presented in Table 3. Soil was mixed in 40 yd^3 batches using the same method that as was used in the pilot study. Each cell was approximately 3 feet (0.9 m) high, 100 feet (30 m) wide and 100 feet (30 m) long. A total of seventeen treatment cells were constructed at a rate of approximately one cell per day.

During construction of each cell, approximately 4 yd^3 (1.2 m^3) of soil (one loader scoop) was removed from each batch and stockpiled in a separate location. Following the completion of mixing activities, this stockpile contained approximately 1,000 yd^3 (765 m^3) of material. After the material was remixed to increase its moisture content, it was placed into cells 16 and 17 as the composite sample piles.

TABLE 3. Full Scale Target Mixing Ratios

	Stockpile Ratio	Batch Ratio	Field Mix
Soil	1 yd^3	40 yd^3	10 scoops (4 yd^3 bucket)
Wood Fines	0.32 yd^3	13 yd^3	3 scoops (4 yd^3 bucket)
Fertilizer*	0.20 gallons	8 gallons	2.5 gpm for 3 minutes
Water	32 gallons	1,280 gallons	Add until soil begins to clump

* = Liquid fertilizer at 30 percent available nitrogen.

RESULTS

Pilot Scale. Following the completion of the construction activities, the cells were monitored periodically to evaluate PCP degradation in the different mixtures. Six soil samples were collected from each cell and were combined into one or two composite samples for chemical analysis. In general, the samples were analyzed for total bacteria, PCP, tetrachlorophenol (TCP), total organic carbon (TOC), total nitrogen, nitrogen as ammonia, nitrogen as nitrate (NO$_3$), total phosphate, soluble phosphate, and moisture content. Selected analytical results are summarized in Table 4, and PCP results are plotted in Figure 1.

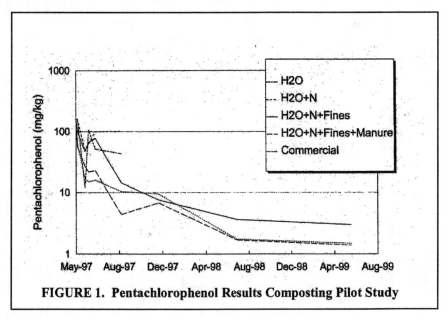

FIGURE 1. Pentachlorophenol Results Composting Pilot Study

TABLE 4. Pilot Scale Representative Sample Results

Date	Cell	Plate Count (MCFU/g)	Temp. °F	pH	PCP (mg/kg)	TCP (mg/kg)	Total N (mg/kg)	Nitrate N (mg/kg)	Ammonia N (mg/kg)	Total P (mg/kg)	Soluble P (mg/kg)	Moisture (%)	TOC (%)	C:N (ratio)	Density (g/cc)
05/06/97	1	1.3	70	5.3	160	30	350	6	12	377	42	10	1.3	37	—
	2	5.8	71	6.6	160	34	350	1.1	70	302	50	15	1.3	37	—
	3	66	76	5.7	100	21	950	4.4	230	478	124	23	11	116	—
	4	1,200	80	8	57	6.4	1,100	130	47	457	202	22	6.7	61	—
	5	600	78	7.8	62	5.6	2,100	12	10	521	188	18	6.6	31	—
05/27/97	1	1.71	84	6.9	12	1	150	1	10	155	32	—	0.68	45	1.6
	2	2.93	89	6.8	49	4.2	710	1	66	176	56	—	0.91	13	1.7
	3	4.59	94	5.9	47	5.8	850	2.9	250	202	56	—	8.6	101	1.2
	4	4.93	100	7.5	27	2	1,100	3.9	35	420	89	—	9.7	88	1.2
	5	3.44	95	7.8	17	1.8	2,200	25	10	400	274	—	5.6	25	1.3
06/25/97	1	0.3	90	6.9	52	4	460	4.2	<10	197	41	9.7	1.1	24	2
	2	0.38	90	6.8	100	9.6	600	1.2	58	225	42	8.9	1.3	22	1.8
	3	1.4	95	5.4	78	11	1,100	<1.0	220	249	129	18	8.4	76	1.6
	4	1.2	105	7.8	23	1.7	1,500	14	55	247	156	18	4.1	27	1.7
	5	0.4	100	7.8	16	2.1	2,600	35	<10	316	204	14	6.4	25	1.5
12/16/97	3	—	—	4.9	7.6	<1.0	1,400	12	72	86	613	18	5	36	1.3
	4	—	—	7.4	6.8	<1.0	2,000	17	<10	136	710	17	4.9	25	1.3
	5	—	—	7.6	9.5	<1.0	4,300	120	<10	352	652	16	3.7	9	1.6
7/2/98	3	—	—	—	2.6/4.7	—	—	—	—	—	—	—	—	—	—
	4	—	—	—	1.8/1.6	—	—	—	—	—	—	—	—	—	—
	5	—	—	—	1.4/2.1	—	—	—	—	—	—	—	—	—	—
6/7/99	3	—	—	—	3	—	—	—	—	—	—	—	—	—	—
	4	—	—	—	1.4	—	—	—	—	—	—	—	—	—	—
	5	—	—	—	1.5	—	—	—	—	—	—	—	—	—	—

Note: Values represent average concentrations if more than one sample from each cell was submitted for analysis. Cells were remixed between 06/25/97 and 12/16/97 to increase moisture and decrease bulk density.
— indicates not analyzed.

PCP levels decreased from 160 mg/kg to less than 10 mg/kg in approximately 16 weeks in Cells 3, 4, and 5, for a 94% reduction. After 14 months, PCP levels in Cells 3, 4, and 5 were reduced by approximately 97%.

Although decreasing PCP concentrations were also observed in Cells 1 and 2, the concentrations generally remained above 50 mg/kg. The results in Cells 1 and 2 showed significant variability, possibly due to incomplete mixing. Additionally, wood fines were not added to these mixtures and clods formed when water was added to these mixtures. The clods made it difficult to collect representative samples. In contrast, the texture of the material placed in Cells 3, 4, and 5 was more homogeneous, and the sample results were more consistent over time.

Full Scale. Five months after the treatment cells were constructed, PCP concentrations were reduced in the composite cell by 94% while Stoddard concentrations were reduced by 99%. Two years after cells were constructed, several of the individual treatment cells were sampled and PCP concentrations were still greater than the target 1 mg/kg. However, PCP concentrations were reduced by 95% from the initial premix concentrations and Stoddard solvents concentrations were reduced by 99%. Analytical results for PCP and Stoddard are summarized in Table 5.

TABLE 5. Full Scale Average PCP & Stoddard Solvent Results (mg/kg)

Location	Date	PCP	Stoddard
Pre-Mix	5/17/99	58	350
Post Mix	8/2/99	28	37
Composite Sample Pile	12/8/99	3.3	3.7
Pile #3	5/4/00	14.6	9.3
Pile #13	5/4/00	14.4	18.6
Composite Sample Pile	5/10/01	1.0	<1.0
Pile #3	5/10/01	2.4	2.4
Pile #13	5/10/01	3.0	6.0

CONCLUSIONS

Five different composting mixtures were evaluated using 1,000 yd^3 (765 m^3) of PCP-contaminated soil excavated from a former sawmill facility. Each mixture consisted of approximately 200 yd^3 (153 m^3) of contaminated soil and predetermined amounts of wood fines, steer manure, fertilizer, and water. Once mixed, the contaminated soil and additives were placed in 3-foot (0.9 m) high windrows and covered with plastic. PCP levels in the three mixtures that included the addition of a bulking agent decreased from 160 mg/kg to less than 10 mg/kg in 16 weeks. After 25 months, the PCP levels in these three mixtures were 3.0, 1.4, and 1.5 mg/kg, respectively.

Based on the final PCP results in Cells 3, 4, and 5, the treated material was transported to L-P's Red Bluff wood waste landfill with approval from the California Regional Water Quality Control Board, Central Valley Region. This material will be incorporated into the foundation layer of the final landfill cover.

The procedures developed during the pilot study were used to cost-effectively compost more than 10,000 yd^3 (7,650 m^3) of PCP-contaminated soil excavated from three former sawmill sites in Northern California. After two years, PCP concentrations were reduced by 98% in the composite sample pile, and by 95% in select individual piles. Once PCP levels are below 1 mg/kg, this material will also be incorporated into the foundation layer of the final landfill cover of the Red Bluff woodwaste landfill.

Criteria for selecting the composting procedures included: 1) space availability, 2) time availability, 3) using readily available material such as fertilizer and wood fines, 4) procedures could be implemented by mill personnel, 5) procedures could be modified to compost large volumes of contaminated soil, 6) procedures would not be necessary to rework the piles after they were covered, and 7) treated soil could be disposed of on site.

The pilot scale and full scale soils were treated at a cost of $20/yd^3 and $23/yd^3, respectively. The unit costs for the pilot scale were less than the full scale costs because wood fines were provided at no cost. These results demonstrate that large volumes of PCP-contaminated soil can be successfully composted with locally available materials to reduce PCP levels.

ACKNOWLEDGMENTS

The authors would like to acknowledge California Regional Water Quality Control Board personnel for their efforts in approving the pilot study. Recognition is also given to Mr. Jeff Pluim and Mr. Max Thomas, mill personnel responsible for implementing the procedures and collecting the soil samples.

REFERENCES

Laine, M. and K. Jorgensen. 1997. "Effective Composting of Chlorophenol-Contaminated Soil in Pilot Scale," *Environmental Science and Technology*, 31(2) pp. 371-378.

Richards, T. 1998. *Moisture and Carbon/Nitrogen Ratio Calculation Spreadsheet*. Department of Agricultural and Biological Engineering, Cornell University, www.cfe.cornell.edu/compost.

Von Fahnestock, F. et al. 1998. *Biopile Design, Operation, and Maintenance Handbook for Treating Hydrocarbon-Contaminated Soils*, Battelle Press.

BIODEGRADATION OF PENTACHLOROPHENOL IN SIMULATED SAND SEDIMENTS CO-CONTMINATED WITH CADMIUM

S. R. *Kamashwaran* and Don L. Crawford (University of Idaho, Moscow, Idaho)

ABSTRACT: Pentachlorophenol (PCP), a carcinogen and inhibitor of oxidative phosphorylation, is a contaminant of soil and groundwater. In the environment, PCP is often present with co-contaminants such as heavy metals. Little is known of its fate under these conditions. Here, degradation of PCP (50 ppm) in the presence and absence of soluble Cadmium (50 ppm) was studied in 20 cm deep simulated sand sediments, either uninoculated, or inoculated with sewage sludge, and established under conditions promoting enrichment of sulfate-reducing (SRB) or methanogenic (MET) consortia. Two sand sediments were used, one low (0.35 mg/g) and one high (4.5 mg/g) in iron. Oxidation-reduction potential (Redox) and pH were monitored at different depths. While PCP was removed in the sediments in the presence and absence of Cd, redox conditions, the presence and absence of Cd, and the sand type strongly influenced the rate and pathways of PCP biodegradation.

INTRODUCTION

Polychlorinated phenols have been used in large quantities, as pesticides, herbicides and wood preservatives. Now environmental contamination with polychlorinated phenols has become a major global concern due to their widespread distribution and toxicity (Escher et al., 1996). Contamination of subsurface environments with polychlorinated phenols has become a serious threat to drinking water sources. Many urban centers depend on groundwater as their only source of drinking water. Chlorophenols that reach nontarget wetlands, and upland and aquatic environments, associate with colloidal and particulate matter and eventually settle into surface soils. Depending on the presence of various degrading microorganisms and the appropriate conditions for their activity, the chlorophenols may be biodegraded.

Pentachlorophenol (PCP) is one such polychlorophenol that is found in significant concentrations in soils, natural water courses, and the atmosphere (Larsen et al., 1991). Inadequate handling, accidental spills and leaching from dumpsites have resulted in its contamination into soil and groundwater (Kitunen et al., 1987). PCP is an inhibitor of oxidative phosphorylation, and it is carcinogenic (Apajalahti et al., 1984). The US EPA has listed PCP as a priority pollutant (Guthrie et al., 1984).

PCP is highly persistent in nature, and its biodegradation is dependent on various factors, including temperature, nutrients, availability of electron acceptors and donors, and the presence of toxic metals (Chang et al., 1996; Haggblom et al., 1993; Kohring et al., 1989; Kuo et al., 1996). In natural habitats these factors are often present as gradients that result in a continuum of microbial activities (Elisa et al., 2000). Sediments and wetland soils are usually characterized by the presence of an oxic surface layer and a redox stratification of the oxygen depleted zone (Zehnder et al., 1988). This stratification results in the creation of distinct niches in which different redox processes are active (Kuivila et al., 1989). An important factor that determines the stratification and development of natural redox gradients is the presence or absence of oxygen. The steepness of this gradient is dependent on the amount of bioavailable organic matter in the surface layer, and thus on the biological oxygen demand (Revsbech et al., 1980). The production of reduced compounds, such as ammonia, ferrous iron, sulfide, and methane, within anoxic zones also influence the steepness of the oxygen gradient (Heiner et al., 2000). These reduced compounds may be reoxidized at the oxic/anoxic transition zone, which may in turn influence the depth of oxygen penetration (Kuivila et al., 1989). Collectively all these factors in contaminated sediments change in time and space, and are mediated by a diversity of organisms. These factors also influence the biodegradation of polychlorinated phenols.

We have previously shown that anaerobic degradation of PCP occurs in the presence and absence of Cadmium (Cd), in anaerobic cultures under two different physiological conditions, sulfidogenic (SRB) and methanogenic (MET) (Kamashwaran et al., 2000). We monitored the degradation and transformation of PCP and the co-precipitation of Cd from its initial soluble phase by the two distinct consortia. We also monitored the ability of these anaerobic bacteria to mineralize PCP and incorporate the carbon from PCP into biomass, by supplementing cultures with either ^{14}C-PCP or ^{13}C-enriched PCP. We then followed the appearance of $^{14}CO_2$ and/or $^{14}CH_4$, and incorporation of ^{13}C into the phospholipid fatty acids of the microorganisms. Both mineralization and incorporation were observed under both physiological conditions. The objective of the present study was to compare the degradation of PCP in the presence and absence of Cd in contaminated sediments with established redox gradients under these two enrichment conditions. Our goal was to (i) determine the influence of the redox gradients on the biodegradation of PCP, (ii) understand the influence of Cd on the degradation of PCP under oxic and anoxic conditions, (iii) determine the influence of the added electron donors such as lactate and acetate (enriching for sulfate-reducing or methanogens, respectively) on the degradation of PCP in sediments, (iv) determine the PCP intermediary degradation products produced under oxic and anoxic conditions, and (v) determine the rate of PCP removal in the presence and absence of Cd under oxic and anoxic conditions.

MATERIALS AND METHODS

Chemicals. Pentachlorophenol (99%) and anhydrous Cadmium Chloride (99.2%) were purchased from Sigma Chemical Company, St. Louis, MO and J.T. Baker Chemical Company, Phillipsburg, NJ, respectively. Other chemicals were of analytical and HPLC grade and purchased from commercial sources.

Enrichment preparation. Sewage sludge obtained from the anaerobic digester of the Moscow, Idaho municipal wastewater treatment plant was used as the primary inoculum. The sludge was centrifuged at 7000 rpm, and the solid pellets were resuspended in water. Resuspended sludge was mixed with sand and contaminant to make a slurry.

Sediments. Sands were obtained from the US Department of Energy, Natural and Accelerated Bioremediation (NABIR) reference material collection (US DOE, Washington, DC). The sands came from Abbott Pitt, Mappsville, VA. Two types of sands were used, one low (0.35 mg/g) and one high (4.5 mg/g) in iron. The total organic carbon in the sediments varied from 0.01% to 0.03% (w/w). When air-dried, they consisted of 95-98% sand. These sands had no known previous exposure to PCP. They were mixed with the inoculum with or without PCP or PCP/Cd solution to give approximately the final concentration of 50 ppm PCP or 50 ppm each of PCP and Cd.

Column studies. Degradation of PCP (50 ppm) in the absence and presence of soluble Cd (50 ppm) was studied in simulated sand aquifers, either uninoculated, or inoculated with sewage sludge, and established under conditions promoting enrichment of sulfate-reducing (SRB; lactate as electron donor) or methanogenic (MET; acetate as electron donor) consortia. The standing columns consisted of glass cylinders (25 X 5 cm) with sampling ports situated at 1cm intervals along their length. The columns were filled with 20 cm of sand slurry. The columns established were stagnant and one cm of water was maintained above the top surface of the sand throughout the course of the experiment. Controls consisted of uninoculated sediment uniformly contaminated with approximately 50 ppm of PCP or PCP/Cd, with no electron donor added (Column 1). To another, only sewage sludge inoculum was added (10% w/v), with no electron donor (Column 2). To the SRB enrichment (Column 3), inoculum and lactate (5.8 mg/ml) were added; to the MET enrichment (Column 4), inoculum and acetate (2.5 mg/ml) were added. The sediment columns were incubated unstirred standing at room temperature (25°C), and periodically over 75 days, 1-ml samples of the sediment slurry from each column were removed at depths of 1cm, 4cm, 7cm, 11cm, 15cm and 20cm. The aqueous and solid phases of the samples were then analyzed for PCP and PCP degradation products. Oxidation-reduction potential (Redox) and pH were monitored at each sampling depth during sampling.

Microsensor measurements. Redox and pH were measured with Clark-type microelectrodes purchased from Microelectrodes, Inc., Bedford, NH. The electrodes were 20 cm in length with an outer body diameter of 2 mm and with a tip diameter between 50 and 100 μm.

PCP analysis. Samples were periodically removed via syringe, from each column at depths of 1cm, 4cm, 7cm, 11cm, 15cm and 20cm. The samples were centrifuged at 14,000 rpm for 5 minutes. The supernatant was collected and passed through a 0.45-μm filter (Acrodisk GHP) into HPLC vials. The solid pellet obtained after centrifugation was extracted by vortexing with ethanol (95%) and then centrifuged at 14,000 rpm for 5 minutes. The supernatant was collected and passed through

0.45-μm filter into HPLC vials. Each filtered sample was chromatographed using a Hewlett Packard 1090 HPLC equipped with a diode array detector set at 217 nm to determine the amount of PCP present. The column was a reverse phase Envirosep PP column (125 X 3.2 mm) (Phenomenex, Torrance, CA). The column temperature was 41°C, and the flow rate was 0.8-ml/minute. The injection volume was 10-μl, and the mobile phase was an isocratic mixture of 40% phosphoric acid (100 mM) and 60% acetonitrile. The combined values from the respective aqueous and solid phases represented the total PCP left in the medium.

Cadmium precipitation. For estimating soluble versus insoluble Cd, 1-ml sample was diluted 5-fold, and extracted with 1M HNO_3. Each filtered extract was assayed for Cd by Inductively Coupled Plasma-Atomic Emission Spectroscopy (ICP-AES). Assays were run by the EPA certified analytical laboratory of the College of Agriculture, University of Idaho.

PCP dechlorination. To qualitatively determine the presence of intermediary PCP degradation products, 2-ml samples from each column, taken at 7 cm and 15 cm depths, were removed on day 40 and mixed with 1-ml of 50% aqueous acetonitrile. They were then acidified with 0.5-ml 1N HCl. This mixture was extracted three times with 8-ml of ethyl acetate. The combined ethyl acetate layers were concentrated to 50-μl using a rotatory evaporator. The concentrated ethyl acetate sample was subjected to GC-MS analysis (Schenk et al., 1989). GC-MS analysis was performed with a gas chromatograph (5890 series II; Hewlett-Packard, Wilmington, Del), which was connected to a mass spectrometer (Hewlett-Packard, model 5989). The injector temperature was 250°C, the oven temperature was 105°C and the detector temperature was 300°C. The carrier gas used was H_2 with a flow rate of 2-ml/minute.

RESULTS AND DISCUSSION

Redox depth profiles were measured with a microelectrode. During the course of the experiment, redox fell from an average value of +350mV at the sediment surface, to an approximate average of −150mV at a depth of 15 cm or below. Figures 1(a) and 1(b) show the change in redox profile at 7 cm depth in the iron poor sediment columns in the presence and absence of Cd respectively. At ≤4 cm depth the redox always remained positive in all columns. At ≥7 cm depth in all sediments and in the presence and absence of Cd, except the control, redox became negative within 15 to 30 days, depending upon the treatment. Additions of lactate and acetate accelerated the process, especially in the iron poor sand sediments, where the redox turned negative within 15 days. The presence of Cd slowed the process in the iron poor sediments, where the redox in the inoculated column in the absence of additional electron donor, turned negative by day 80, whereas in the absence of Cd, the time required was only 22 days.

The pH of the columns at different depths was monitored using a microelectrode. The aqueous phase pHs varied slightly, depending on the enrichment conditions. The control columns were acidic (pH ≅ 5.5) at all the depths

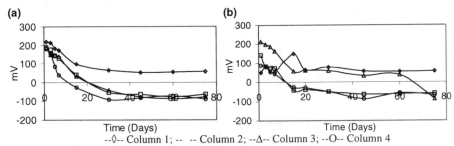

--◊-- Column 1; -- -- Column 2; --Δ-- Column 3; --O-- Column 4

FIGURE 1. Change in redox profile at 7 cm depth in the iron poor sediment columns in (a) the absence and (b) presence of Cd.

through out the course of the experiment. The pH at the surface sediment of the other columns averaged 7.8, and at the bottom, 7.0. The pH at the different depths in all the columns showed a similar profile pattern, where the pH at the surface sediment was higher, but decreased with the increasing depth. The pH at similar depths in each column varied little during the experiment, fluctuating within a range of ±1. Others have found that at pH lower than 6.7, PCP degradation is inhibited (Juteau et al., 1995).

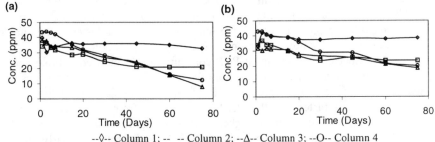

--◊-- Column 1; -- -- Column 2; --Δ-- Column 3; --O-- Column 4

FIGURE 2. PCP removal from the iron poor sediment columns in the presence of Cd at the depth of (a) 7 cm and (b) 15 cm.

After incubation for 10 days at 25°C, gas bubbles due to microbial activity became visible within columns 2 and 4, in the presence and absence of Cd, in both the low and high iron treatments. The bottom layer of column 3 turned black, an indication of the production of sulfide by SRBs. Concentrations of 50-mg/L PCP were used in the experiment, as waste groundwater below wood treatment plants have been reported to have PCP concentrations typically ranging from 25-mg/L to 100-mg/L or more (Thompson et al., 1971). Except for the control, PCP was removed from the sediments in the presence and absence of Cd. As shown in Figures 2 and 3, PCP degradation rates in the presence of Cd at depths of 7 and 15 cm were faster in high iron sediments than in those with low iron. In the presence of Cd, PCP removal was more rapid in the oxic zones (7 cm depth), but in the absence of Cd, faster PCP removal was observed in the anoxic zone (15 cm depth) (data not shown).

Cd at an initial soluble concentration of 50 ppm did not prove highly toxic, although it transiently slowed the rate of PCP degradation. The presence of Cd more strongly inhibited the removal of PCP when the column was not enriched with an electron donor. It was observed that by day 80, in the presence of Cd, the SRB-enriched column (Column 3) showed the best rate of PCP removal which was followed by the MET-enriched column (Column 4).

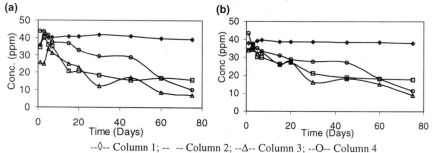

--◊-- Column 1; -- -- Column 2; --Δ-- Column 3; --O-- Column 4

FIGURE 3. PCP removal from the iron rich sediment columns in the presence of Cd at the depth of (a) 7 cm and (b) 15 cm.

In the absence of Cd, faster PCP removal was observed irrespective of the presence or absence of an additional carbon source. Figure 4 shows the removal of PCP from the iron poor sediments at the depths of 7 and 15 cm, in the absence of Cd stress. In the presence of Cd, the removal of PCP was faster in the anoxic zones of the iron rich column in comparison to the iron poor column.

We expected that sediment PCP concentrations would equilibrate across all of the depths due to its diffusion from the anaerobic to the aerobic zones and vice versa. However, in the presence of Cd the rate of PCP degradation in the aerobic zones was faster than the PCP diffusion rates from the anaerobic zones; therefore, over time we observed higher PCP concentrations within the anaerobic phases than within the aerobic phases of these sediments. Similarly, in the absence of Cd the rate of PCP degradation in the anaerobic zones was faster than the PCP diffusion rates from the aerobic zones.

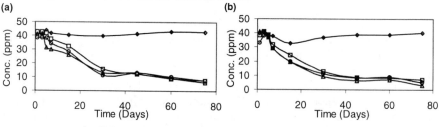

--◊-- Column 1; -- -- Column 2; --Δ-- Column 3; --O-- Column 4

FIGURE 4. PCP removal from the iron poor sediments in the absence of Cd at the depth of (a) 7 cm and (b) 15 cm.

The presence of PCP dechlorination products in the iron rich and iron poor sediments, for cultures grown in the presence of Cd, was determined qualitatively on day 40. Various tetrachlorophenol (TeCP), trichlorophenol (TriCP) and dichlorophenol (DiCP) isomers were detected in the anoxic zones of both iron rich and poor sediments. Tetra and trichlorophenol isomers detected included 2,3,4,5-TeCP, 2,3,4,6-TeCP, 2,3,5,6-TeCP, 2,3,6-TriCP, 2,4,5-TriCP, 2,3,5-TriCP, 2,4-DiCP, 3,5-DiCP. The oxic zones also contained various chloro-hydroquinones, including 2,3,5,6-Tetrachloro-p-hydroquinone, p-chlorophenol, and 2,6 Dichlorohydro- quinone. The appearance of these less chlorinated phenols in the anoxic regions of the columns demonstrated that reductive dechlorination was occurring. The appearance of the chloro-hydroquinones in the oxic zones showed that the PCP was also being hydroxylated under aerobic conditions. Thus, the oxic and anoxic microbial populations were probably using different pathways for PCP biodegradation. The inoculated control sediments produced no dechlorination products during the same time, indicating no loss of PCP.

This study demonstrated the capability of microbial populations to degrade PCP in sediments in the presence of a co-contaminant heavy metal, Cd. The rates and pathways by which PCP degradation occurs are regulated by such factors, as redox, pH, the availability of electron donors and acceptors, and the presence and absence of Cd and iron. These results may be useful in predicting the environmental persistence of PCP in groundwater and sediments as a function of site-specific conditions. Future studies should focus on identifying and comparing the microbial species involved in these transformations as influenced by these various factors.

REFERENCES

Apajalahti, J.H.A., and M.S. Sakinoja-Salonen. 1984. "Absorption of pentachlorophenol (PCP) by bark chips and its role in microbial PCP degradation". *Microbiol. Ecol. 10*: 359-367.

Chang, B. V., J.X. Zheng, and S.Y. Yuan. 1996. "Effects of alternate electron donors, acceptors, and inhibitors in pentachlorophenol dechlorination in soil. *Chemosphere 33*: 313-320.

Elisa, M. D' Angelo and K.R. Reddy. 2000. "Aerobic and anaerobic transformations of pentachlorophenol in wetland soils" *Soil. Sci. Soc. Am. J. 64*: 933-943.

Escher, B.I., M. Snozzi, and R.P. Schwarzenbach. 1996. "Uptake, speciation, and uncoupling activity of substituted phenols in energy transducing membranes". *Environ. Sci. Technol. 30*: 3071-3079.

Guthrie, M.A., E.J. Kirsch, R.F. Wukasch, and C.P.L. Grady, Jr. 1984. "Pentachlorophenol biodegradation-II, Anaerobic". *Water Res. 18*(4): 451-461.

Haggblom, M.M., M.D. Rivera, and L.Y. Young. 1993. "Influence of alternate electron acceptors on the anaerobic biodegradability of chlorinated phenols and benzoic acids. *Appl. Environ. Mocrobiol. 59*: 1162-1167.

Heiner Ludemann, Inko Arth and Werner Liesack. 2000. "Spatial changes in the bacterial community structure along a vertical oxygen gradient in flooded paddy soil cores". *Appl. Environ. Microbiol* 66(2): 754-762.

Juteau, P., R. Beaudet, G. McSween, F. Lepine, S. Milot, and G. G. Bisaillon. 1995. "Anaerobic biodegradation of pentachlorophenol by a methanogenic consortium". *Appl. Microbiol. Biotechnol.* 44: 218-224.

Kamashwaran, S.R., and Don L. Crawford. 2000. "Anaerobic Biodegradation of Pentachlorophenol in Mixtures with Cadmium by Two Physiologically Distinct Microbial Consortia". *Ind. Microbiol. & Biotechnol. (Submitted).*

Kitunen, V.H., R.J. Valo, and M.S. Salkinoja-Solenen. 1987. "Contamination of soil around wood preserving facilities by polychlorinated aromatic compounds". *Environ. Sci. Technol.* 21: 96-101.

Kohring, G.W., J.E. Rogers, and J. Wiegel. 1989. "Anaerobic biotransformation of 2,4-dichlorophenol in freshwater lake sediments at different temperatures". *Appl. Environ. Microbiol.* 55: 348-353.

Kuivila, K. M., J. W. Murray, and A. H. Devol. 1989. "Methane production, sulfate reduction and competition for substrates in the sediments of Lake Washington". *Geochim. Cosmochim.* 53: 409-416.

Kuo, C., and B.R.S. Genther. 1996. "Effect of added heavy metal ions on biotransformations of 2-chlorophenol and 3-chlorobenzoate in anaerobic bacterial consortia. *Appl. Environ. Microbiol.* 62: 2317-2323.

Larsen, S., H.V. Hendriksen, and B.K. Ahring. 1991. "Potential for thermophilic (50°C) anaerobic dechlorinators of pentachlorophenol in different ecosystems". *Appl. Environ. Micobiol.* 57: 2085-2090.

Revsbech, N. P., B. B. Jorgensen, and T. H. Blackburn. 1980. "Oxygen in the sea bottom measured with a microelectrode". *Science 207*: 1355-1356.

Schenk, T., R. Muller, F. Morsberger, M.K. Otto, and F. Lingens. 1989. "Enzymatic Dehalogenation of Pentachlorophenol by Extracts from *Arthrobacter sp.* strain ATCC 33790". *J. Bacteriol. 171*(10): 5487-5491.

Thompson, W.S., and J.V. Dust. 1971. "Pollution control in the wood preserving industry. I. Nature and scope of the problem". *Forest Products J.* 21:70-75.

Zehnder, A. J. B., and W. Stumm. 1988. "Geochemistry and biogeochemistry of anaerobic habitats", p. 1-38. *In* A. J. B. Zehnder (ed.), *Biology of anaerobic microorganisms.* John Wiley & Sons, Inc., New York, N. Y.

NEW BASIDIOMYCETES ON BIOREMEDIATION OF ORGANOCHLORINE CONTAMINATED SOIL

Dácio Roberto Matheus (Instituto de Botânica, SMA, São Paulo, Brazil)
Vera L.R. Bononi (Instituto de Botânica, SMA, São Paulo, Brazil)
Kátia M. G. Machado (Fund.Centro Tecnológico de Minas Gerais, MG, Brazil)
e-mail:dmatheus@smtp-gw.ibot.sp.gov.br

ABSTRACT: The possibilities of bioremediation of organochlorine compounds of contaminated soils at Baixada Santista, São Paulo State, Brazil and the biodegradation of hexachlorobenzene (HCB) and pentachlorophenol (PCP) by Brazilian basidiomycetes fungi for application in the bioremediation processes were evaluated. Fungi's tolerance to the soil compounds and their capacity to reduce PCP and HCB levels and mineralize ^{14}C-PCP and ^{14}C-HCB were evaluated. Out of a total amount of 125 isolated basidiomycetes, *Psilocybe* cf. *castanella* CCB 444, *Trametes villosa* CCB 176, *Peniophora cinerea* CCB 204 and *Lentinus* cf. *zeyheri* CCB274 showed reasonable characteristics to be applied in soil remediation systems: quick growth, tolerance to high pollutant levels and capacity to remove pollutant from the soil. *P. castanella* removed 64.50% of PCP and 24.61% of HCB of the soil and mineralized 4.95% of ^{14}C-PCP and 0.68% of ^{14}C-HCB. *T. villosa* and *P. cinerea* removed 58.10 % and 78.00% of PCP and mineralized 8.15 and 7.11% of ^{14}C-PCP of the soil, respectively. *L. zeyheri* removed 18.61% of HCB and mineralized 0.65% of ^{14}C-HCB of soil. Current efforts have been focused on experiments to optimize the biodegradation rates and to evaluate a pilot scale.

INTRODUCTION

The capacity of basidiomycetes to degrade organic pollutants, which until recently were considered recalcitrant to biological degradation, has been an object of research as to the possibility of using these organisms in processes of reclamation of contaminated soils. The enzymatic system that is involved in the degradation of the pollutants by white-rot wood decay basidiomycetes has a non-specific nature (Barr & Aust 1994).

An enormous diversity of basidiomycetes exists: 22,244 species are known worldwide (Hawksworth et al. 1995). Thus it is clear that a selection must be made of those species with greater potential to degrade xenobiotics, and at the same time able to reproduce and grow in contaminated environments or materials that must be treated. Moreover, in many countries, including Brazil, there are restrictions placed on the use of certain allochthonous microorganisms in systems of treatment open to the environment. Thus it is desirable to screen indigenous species.

In the State of São Paulo, the most industrialized region of Brazil, there is widespread concern over locating, monitoring and treating soil areas that - particularly in the sixties and seventies - were contaminated with organic pollutants, among them residuals containing organochlorine compounds such as hexachlorobenzene (HCB) and pentachlorophenol (PCP).

HCB is an extremely stable compound, chemically inert and slightly solubile in water. The contamination of several areas of the world, including Australia, Europe, Japan and North America, occurred mainly through its use as a fungicide in the treatment of seeds. Its long-term persistence in the environment and the detection of residuals in animal and human tissues has justified restricting the use of HCB (Verchuerem, 1983). The United Nations Environmental Program (UNEP) has classified the molecule as one of twelve persistent organic pollutants (POP). A worldwide attempt is being made to ban the production of, and the residuals containing HCB, because of the high risks it represents for the environment and for human health.

PCP is listed as one of the priority pollutants by the United States Environmental Protection Agency and European Community. The highest reported usage of PCP is in the wood preserving and treatment industry, particulary for utility poles, fences and railway ties. Various of these treatments and process can cause pollution of soil and water in the area, mainly if them occur in open-air basins (McAllister et al., 1996).

Objective. In the present work the potential use of basidiomycetes from Brazilian ecosystems in the bioremediation of contaminated soils with organochlorine compounds has been evaluated, as to the: (i) tolerance to organochlorine compounds in the soil, and (ii) capacity to biodegrade HCB and PCP.

MATERIAL AND METHODS
Fungi: the 36 fungal strains studied were previously selected for their capacity to decolorize the dye Remazol brilliant blue - R in a solid medium and growth rate superior to 1.0 cm day^{-1} in a medium of malt extract and agar (MEA 2%) (Machado 1998). All the cultures were from Brazilian ecosystems and were maintained in MEA 2% at 4°C. They have been deposited in the Culture Collection of Basidiomycetes (CCB) of the Instituto de Botânica of the Secretary for the Environment, São Paulo State, Brazil.

Soil: the contaminated soils (PCP-soil, containing high level of PCP and HCB-soil, containing high level of HCB) were collected from an area situated in the municipality of São Vicente, São Paulo, Brazil, in a region where organochlorine industrial residuals with high concentrations of organochlorine compounds from a factory producing sodium pentachlorophenate and carbon tetrachloride, had been inadequately disposed of. The material was taken off the top of piles of soil mixed with residuals, which had been removed from a depth of up to 1 m. The soils contained 98.0% sand, pH 3.65, cation exchange capacity (CEC) 5.5 mEq $100g^{-1}$ soil, 2.3% organic matter, 0.06% nitrogen, 1.0 µg g^{-1} phosphorus and 0.01 mEq 100 ml^{-1} potassium. The native microbiota of the HCB-soil consisted of $4.98.10^7$ u.f.c. of bacteria and $1.28.10^7$ u.f.c. of fungi. No bacteria or fungus were isolated from PCP-soil where the dilution plating was used.

Tolerance of the fungi to the organochlorine compounds in soil: Plugs of 5-mm taken from the actively growing margin of fungal colonies from master plates were inoculated in non-inclined test tubes containing 10 mL of MEA 2% and were incubated at 30°C for about one week. Both HCB-soil and PCP-soil described above, with 51,166 mg of HCB and 46,300 mg of PCP kg^{-1} of soil, respectively, were air-dried, sieved through a mesh of 2 mm and was diluted with 50% and 90% of non-contaminated soil, of the same physical-chemical characteristics. After dilution, the concentrations corresponding to 25,000 mg and 5,000 mg of HCB and 23,000 mg and 4,600 mg of PCP kg^{-1} of soil, respectively. Non-contaminated soil was used as control. Each portion was sterilized in fractions, at 100°C, for three consecutive days and analysed by gas chromatography as to content of organochlorines. When growth of the fungi commenced, a column of 15 g soil was placed over each colony, in quadruplicate, for each one of the concentrations of HCB and PCP which were to be evaluated. Those fungi that managed to colonize more than 25% of the soil column were considered tolerant to organochlorine compounds. Thus only four strains were selected for the experiments of degradation of HCB or PCP in soil.

Fungal inoculum: for the experiments of HCB and PCP degradation and mineralization, the inoculum of each fungus was prepared as described for the production of edible mushrooms using wheat grains as substrate (Bononi et al. 1995).

Degradation of HCB and PCP: both HCB-soil and PCP-soil were air dried, sieved through a mesh of 2 mm and autoclaved as described above. A 25 g (dry weight) mixture of contaminated soil-$CaSO_4$-sugar cane bagasse (95:2.5:2.5 per dry weight), was put into each 800 mL bottle. The moisture content was adjusted to 75% of the mixture's water retention capacity and corrected weekly by gravimetry. Under aseptic conditions, 2.5 g of inoculum of each fungus was inoculated into the mixture of soil-gypsum-bagasse. As control treatment, 2.5 g sterilized wheat grains was added to the soil mixture. All treatments were replicated four times. The degradation of the HCB and PCP were inferred from the decrease of the concentration of the compound in the treated soil, as compared to the initial concentration, by CG, in duplicate. Both the formation of inorganic chloride and the formation of pentachloroanisole were also measured.

Chromatographic analyses: The extraction from the total content of each bottle was done with a soxhlet apparatus and carried out for 4 hours, with 100 mL toluene with 2 mL 5% phosphoric acid added. The concentrations of organochlorines were determined by high-resolution gas chromatography (HP5890, integrator HP3396, capillary column of 5% phenyl, 95% of polymethyl siloxane and detection by flame ionization (FID), with nonadecane as internal standard. The temperatures for operation were 250°C in the injector and 300°C in the detector. Hydrogen (1 mL min^{-1}) and nitrogen (30 mL min^{-1}) were the carrier gases and hydrogen (29 mL min^{-1}) and oxygen (80 mL min^{-1}) the flame gases. The analyses were standardized with: PCP, pentachloranisol, hexachlorethane, hexachlorobutadiene, tetrachlorobenzene, pentachlorobenzene and HCB. The limit of detection was 10 mg kg^{-1} of soil.

Determination of inorganic chloride: 60 mL of Milli-Q water was added to two of the replicates from each treatment. The aqueous extract was manually agitated and filtered. Free chlorides were determined by potentiometric titration with silver nitrate 0.01M, making use of a silver electrode. The concentration of chloride was based on 25 g of dry soil.

Mineralization of ^{14}C-HCB and ^{14}C-PCP: Both HCB-soil and PCP-soil were diluted in non-contaminated soil in order to obtain approximately 1,300 mg of HCB or PCP kg^{-1} of soil, respectively. In 300 mL Erlenmeyer bottles, 100 g of the same mixture of soil-gypsum-bagasse as described above was placed. This mixture was autoclaved previously as priorly described. An amount of 3.0×10^6 d.p.m. of HCB-UL-^{14}C or PCP-UL-^{14}C (Sigma, St. Louis) was added for each 100 g of HCB-soil or PCP-soil, respectively. The moisture content was adjusted and mantained to 75%. In each bottle, 10 g of fungal inoculum of each studied species was inoculated. Sterilized wheat grains were inoculated in the control treatment. The incubation was done in a dark chamber at 23±2°C. The applied initial radioactivity was determined by combustion in Biological Material Oxidizer (Harvey Instruments, OX500) followed by liquid scintillation counting (Packard Tri-Carb 1600TR) of the $^{14}CO_2$ produced. During the period of incubation (120 days) the $^{14}CO_2$ produced by the mineralization of the ^{14}C-HCB or ^{14}C-PCP was periodically captured in a trap of soda-lime. The acid extraction of the CO_2 was done in a hermetically closed system where the gas formed was collected in solution of methanol/monoethanolamine (7:3 v/v) as described by Anderson (1990). The mineralization of ^{14}C-HCB or ^{14}C-PCP was calculated from the percentage of the produced $^{14}CO_2$.

Assay's conditions: In the Table 1 the conditions employed on HCB and PCP degradation and mineralization assays are presented.

TABLE 1: Conditions employed on HCB and PCP degradation and mineralization assays in soil.

Assay's conditions	HCB-soil		PCP-soil	
	Degradation	Mineralization	Degradation	Mineralization
Organochlorine concentration (mg kg^{-1}soil)	29,180	1,300	1,278	1,280
Temperature incubation (°C)	28	23	28	23
Time incubation (days)	65	90	128	128
Fungi inoculated in soil	CCB274, CCB176, CCB444, CCB204		CCB161, CCB213, CCB176, CCB444, CCB204	

CCB 161= *Agrocybe* perfecta, CCB213 and CCB176= *Trametes villosa*, CCB444= *Psilocybe* cf. *castanella*, CCB204=*Peniophora* cf. *cinerea*, CCB274= *Lentinus* cf. *zeyheri*.

Statistical analysis: the percentages of degradation and mineralization of HCB or PCP and the production of inorganic chloride by the selected basidiomycetes were statistically analysed. Analyses of variance (ANOVA) were conducted ($\alpha=0.05$)

using the ANOVA program of Excel 5.0 for Windows®. Whenever a significant effect of the treatments was observed, differences among treatment means were detected by comparation media test (P≤0.05) (Vieira & Hoffmann 1989).

RESULTS AND DISCUSSION

Tolerance to the organochlorine compounds in the soil: of the 36 strains of basidiomycetes studied the strains described below were capable to colonize HCB-soil or PCP-soil with high concentration of organochlorine compounds, and were selected to biodegradation tests. *Peniophora* cf. *cinerea* CCB204 and *Lentinus* cf. *zeyheri* CCB274 (up to 25,000 mg of HCB.kg^{-1} of soil), *Agrocybe perfecta* CCB161, *Trametes villosa* CCB213, *Trichaptum bisogenum* CCB203 and *Lentinus villosus* CCB271 (up to 4,600 mg of PCP kg^{-1} of soil), *Psilocybe* cf. *castanella* CCB444 and *Trametes villosa* CCB176 (up to 50,000 mg of HCB.kg^{-1} and 4,600 mg of PCP kg^{-1} of soil), The saprotrophic fungus *Psilocybe* cf. *castanella* CCB444 and the lignicolous fungus *Trametes villosa* CCB176 were among the strains most tolerant to organochlorines in both HCB-soil and PCP-soil.

Biodegradation of HCB and PCP in soil: the extent of HCB and PCP degradation was evaluated at 65 and 90 days after the inoculation of the fungi in HCB and PCP soils, respectively, from the residual concentrations of the compound in the soil in relation to the initial concentration (29,180 mg HCB kg^{-1} of HCB-soil and 1,278 mg PCP kg^{-1} of PCP-soil). In the HCB-soil, *Psilocybe* cf. *castanella* produced the greastest decrease of HCB (24.61%), statistically differing (P≤0.05) from the non-inoculated control treatment (13.81%).

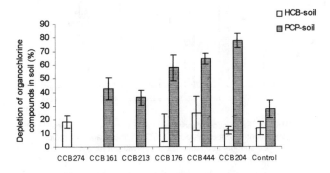

FIGURE 1. Biodegradation of HCB and PCP by basidiomycete in organochlorine-contaminated soil. CCB number and incubation conditions=see table 1, control=without fungus.

L. zeyheri (18.61%) did not differ statistically from *P. castanella*, but also did not differ from the control treatment at the significance level of 0.05. In its turn, the decrease of HCB by the strains *T. villosa* CCB176 (13.98%) and *P. cinerea* (12.10%) did not differ statistically from the control treatment. In the PCP-soil *P. cinerea* produced the greatest decrease of PCP (77.97%) followed by P. castanella (64.46%) and T. villosa CCB176 (58.07%).These percent decreases were significantly greater (P<0.05) than that which occurred in the non-inoculated control treatment (27.74%). The other fungi did not differ statistically from the control treatment (Figure 1). In PCP-soil incubated with *A. perfecta*, *Peniophora cinerea* e *T. villosa* CCB 213 the PCA production was detected (35, 25,5 e 16 mg kg^{-1} de PCA). PCA was not detected in soil incubated with *P. castanella* and *T. villosa* CCB 176, nor control treatment.

Production of inorganic chloride ions: the initial concentration of HCB (29,180 mg kg^{-1} of HCB-soil) theoretically could produce 21,240 mg de Cl^- kg^{-1} of HCB-soil. The largest Cl^- concentration was encountered in the HCB-soil incubated with *T. villosa* CCB176 (152.5 mg Cl^- kg^{-1} of soil). This concentration was significantly greater than residual Cl^- concentration from the other treatments wich, in this turn, differed from the control treatment (P ≤ 0.05) (Figure 2). In the PCP-soil, the concentration of PCP (1,278 mg kg^{-1} of PCP-soil) could produce 851 mg Cl^- kg^{-1} of PCP-soil. The largest Cl^- concentration was encountered in the PCP-soil incubated with *P. cinerea* and *A. perfecta* (440 mg and 418 mg Cl^- kg^{-1} of soil, corresponding to 51,7% and 51.3% of the initial concentration, respectively) (Figure 2).The other treatments did not differ among themselves, but did differ statistically from the control treatment (P ≤ 0.05). The results indicate the transformation of the PCP. During the degradation of chlorophenols by *Phanerochaete chrysosporium*, the withdrawal of chlorine from the aromatic ring

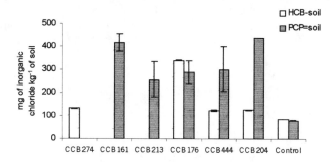

FIGURE 2. Production of inorganic chloride by basidiomycete in organochlorine-contaminated soil. CCB number and incubation conditions=see table 1, control=without fungus.

occurs before the breaking of the ring, through several reactions of oxidation, reduction and methylation of quinones. Evidence exists for the participation of peroxidases - such as lignin peroxidase and manganese peroxidase - in the process of dehalogenation of chlorophenols by basidiomycetes (Valli & Gold 1991). The effect of high concentrations of chlorides in the metabolism of basidiomycetes - including those species presented here - is still unknown. This will have to be the subject of later studies.

Mineralization of ^{14}C-HCB and ^{14}C-PCP: In HCB-soil, although the rates of mineralization were very low (>1%), by comparing averages, it was possible to statistically distinguish the strains CCB444 and CCB274 from CCB176 and CCB204 which, in their turn, differed from the control (P≤0.05) at 128 days of incubation. The extent of PCP and HCB mineralization is shown in Figure 3. PCP mineralization was much greater than HCB mineralization (HCB mineralization was only marginally greater than the control). The extent of PCP mineralization in the various treatments was 3.07, 3.17, 4.95, 7.11 e 8.15%, when the soil was incubated with *T. villosa* CCB 213, *A. perfecta, P. castanella, P. cinerea*, e *T. villosa* CCB 176.

Conditions employed in this experiment were compatible with the good colonization of the fungus in the soil. In this study, sugar cane bagasse was used because of its easy availability and its ability to support growth of the basidiomycetes studied in soil. The differences in behaviour among the isolates may have been a consequence of the decision to establish just a single condition for selection and evaluation of all isolates, in both contaminated-soils (HCB and PCP).

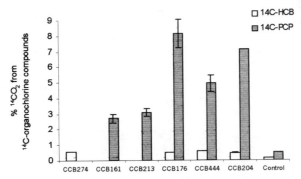

FIGURE 3. Mineralization of ^{14}C-HCB and ^{14}C-PCP by basidiomycete in organochlorine-contaminated soil, CCB number and incubation conditions=see table 1, control=without fungus.

However, the optimization of manageable parameters (e.g. aeration, soil moisture content, C/N) was not evaluated. Studies designed to evaluate the adequacy of the culture conditions, as well as the metabolites that probably are generated in the degradation of the organochlorines in soil are currently being carried out.

Until this moment we have observed that the manipulation of C/N ratio and the addition of unsaturated fatty acid-containing vegetable oil have made possible the mineralization over than 10% of 14C-HCB, during 56 days of incubation, in soil with up to 30,000 mg HCB kg^{-1} of soil, by selected fungi.

ACKNOWLEDGEMENTS

This project is a result of an agreement between Rhodia Brazil Ltda and the Universidade Estadual Paulista, together with the Instituto de Botânica of the Secretary for the Environment of São Paulo State, Brazil. We are grateful for the support received from Dr. Regina T. R. Monteiro of the Centro de Energia Nuclear na Agricultura - USP, Piracicaba, Dr. Mara M. D'Andrea of the Instituto

Biológico of the Secretaria da Agricultura e Abastecimento do Estado de São Paulo, and from Dr. Marina Capelari of the Instituto de Botânica of the Secretaria do Meio Ambiente do Estado de São Paulo, Brazil. Help of Fundação para o Desenvolvimento da Universidade Estadual Paulista, Fundação para o Desenvolvimento da Pesquisa Agropecuária, Conselho Nacional de Desenvolvimento Científico e Tecnológico, Fundação de Amparo a Pesquisa do Estado de Minas Gerais-FAPEMIG and Fundação de Amparo a Pesquisa do Estado de São Paulo-FAPESP is also gratefully acknowledged.

REFERENCES

Anderson, J.P.E. 1990 Principles of and assay systems for biodegradation. *Advances in Applied Biotechnology* series 4, 129-145.

Barr, D.P. and Aust, S.D. 1994 Mechanisms white rot fungi use to degrade pollutants. *Environmental Science and Technology* 28, 78-87.

Bononi, V.L.R., Capelari, M., Maziero, R. & Trufem, S.F.B. 1995 *Cultivo de cogumelos comestíveis.* p.p.37-39. São Paulo: Ícone. ISBN: 85-274-0339-0

Hawksworth, D.L., Kirk, P.M., Sutton, B.C., Pegler, D.N. 1995 *Ainsworth & Bisby's Dictionary of the fungi.* p.48. Wallingford: C.A.B. International. ISBN: 0-8519-8885-7.

Lamar, R.T. and Dietrich, D.M. 1992. Use of lignin-degradin fungi in the disposal os pentachlorophenol treated wood. *Journal of Industrial Microbiology* 9(3-4): 181-191.

Lamar, R.T., Glaser, J., Evans, J.W. 1993 Solid-phase treatment of a pentachlorophenol-contaminated soil using lignin-degrading fungi. *Environmental Science and Technology* 27, 2566-2571.

Machado, K.M.G. 1998 *Biodegradação de pentaclorofenol por fungos basidiomicetos ligninolíticos em solos contaminados por resíduos industriais.* PhD Thesis, Rio Claro: Universidade Estadual Paulista, Brasil.

McAllister, K.A., Lee, H., Trevors, J.T. 1996. Microbial degradation of pentachlorophenol. *Biodegradation,* 7: 1-40.

Valli, K. and Gold, M.H. 1991 Degradation of 2,4-dichlorophenol by the lignin-degrading fungus *Phanerochaete chrysosporium. Journal of Bacteriology* 173: 345-352.

Verschueren, K. (ed.) 1983 *Handbook of environmental data on organic chemicals.* New York: Academic Press. ISBN: 0-4422-8802-6.

Vieira, S. and Hoffmann, R. 1989 *Estatística Experimental.* São Paulo: Atlas. ISBN: 8-5224-0449-6.

EVALUATION OF AEROBIC AND ANAEROBIC DEGRADATION OF PENTACHLOROPHENOL IN GROUNDWATER

Scott J. MacEwen, P.E., CH2M HILL, Herndon, VA
Frances Fadullon, P.E., CH2M HILL, Herndon, VA
Dawn Hayes, U.S. Navy, LANTDIV NFEC, Norfolk, VA

ABSTRACT: Bench-scale laboratory tests were conducted to evaluate and compare the enhanced bioremediation of pentachlorophenol (PCP) in groundwater under anaerobic and aerobic conditions. While tests under both conditions showed a significant reduction of PCP over the 3-month test period (greater than 90 percent reduction compared to 40 percent in controls), removal to concentrations below the 10 µg/L method detection limit was only observed under aerobic conditions in the time allowed. The results of the laboratory tests were used to select an approach for a follow-up field test to evaluate application at a Navy-owned facility in Virginia Beach, Virginia that has been placed on the National Priorities List (NPL) by EPA Region III. The field test, which involved the injection of an Oxygen Release Compound® (ORC) slurry into the aquifer, was designed based on historic site data. The monitoring system was installed in August 2000 and a baseline round of data was collected. ORC was injected in October 2000. The test is scheduled to run for thirty-six weeks (July 2001).

INTRODUCTION

Naval Amphibious Base (NAB) Little Creek is located in Virginia Beach, Virginia. The facility was listed on the National Priorities List (NPL) in May 1999. Site characterization and remediation activities are being performed by the Navy under the review of EPA Region III and the Virginia Department of Environmental Quality (VDEQ). A former pentachlorophenol (PCP) wood treatment dip tank (Site 13) at a NAB Little Creek released a PCP and diesel fuel mixture to the surrounding soil and water table aquifer. Soil removal addressed the soil contamination at the site, however, groundwater contains residual PCP contamination.

Site-specific bench-scale laboratory tests were conducted to evaluate methods for enhancing the biological remediation of PCP in groundwater in order to design and implement a field test and ultimately aid in the selection of a full-scale remedial action.

PCP has been shown to degrade both aerobically and anaerobically under the right conditions. Side-by-side laboratory-scale test-tube trials were conducted using site soils spiked with PCP to evaluate which approach (if any) would be most effective for the site. Aerobic tests were prepared using Oxygen Release Compound® (ORC), a proprietary magnesium peroxide-based product. The anaerobic test-tube trials were prepared using Hydrogen Release Compound (HRC), a proprietary lactate-based product. Both products are manufactured by Regenesis, San Clemente, California.

Site Description. The Public Works PCP Dip Tank was an in-ground tank that had been used to treat wood with a 1-to-10 mixture of PCP and diesel fuel. The dip tank was operated from the early 1960s until 1974.

The tank had been cleaned and removed several years after wood treatment operations had ceased. The site is currently paved with asphalt and used as a storage area. The upper 23 ft (7 m) of geology in the vicinity of the former dip tank is primarily composed of fine to medium sands with some silt and clay and occasional gravels. Interbedded, discontinuous layers of clay, ranging in thickness from approximately 2 to 5 ft (0.6 to 1.5 m) occur within the upper 10 ft (3.0 m). The water table aquifer (Columbia Aquifer) is present at a depth of 5 to 7 ft (1.5 to 2.1 m) below ground surface (bgs). The top of a clay confining layer, estimated to be 35 to 40 ft (11 to 12 m) thick, is present at approximately 23 ft (7 m) bgs. This clay layer effectively isolates the Columbia Aquifer from deeper confined aquifers. The flow direction of the Columbia Aquifer at the site is predominantly to the southwest. The hydraulic conductivity has been estimated at 110 ft/day (33 m/day). The average gradient at the site is 0.0010 ft/ft. Using an effective porosity of 0.35, the groundwater flow velocity is estimated to be approximately 115 ft/yr (35 m/yr). Based on the current estimated extent of the PCP plume and the likely release dates, the net average migration rate of PCP is between 10 and 15 ft/yr (3 and 5 m/yr).

Soil and groundwater sampling at the site between 1986 and 1999 indicated that PCP was present in both media. The maximum concentration of PCP in the groundwater was 2,000 µg/L, compared to the drinking water maximum contaminant level (MCL) of 5 µg/L and an EPA Region III risk-based concentration (RBC) of 0.56 µg/L. The maximum concentration of PCP in the soil above the water table was approximately 900 mg/kg. Both of these maximum concentrations were found directly below the location of the former dip tank.

In April 1999, the PCP-contaminated soil in a 47-by-17 ft (14.3-by-5.2 m) area beneath and around the former dip tank was excavated to depths ranging from 6 to 8 ft (1.8 to 2.4 m). The maximum concentration of PCP remaining in the soil after the removal action was 36 mg/kg, however, most post-removal samples were found to contain less than 10 mg/kg PCP.

Anaerobic and Aerobic Degradation Pathways for PCP. PCP has been observed to degrade anaerobically by reductive dechlorination. To complete anaerobic degradation, each chlorine molecule acts as an electron acceptor and is replaced by hydrogen, producing first tetrachlorophenol (TeCP), then trichlorophenol (TCP), dichlorophenol (DCP), chlorophenol (CP), and finally phenol before the ring is broken relatively late in the process. Possible intermediate breakdown products include three isomers of TeCP, five isomers of TCP, six isomers of DCP, and three isomers of CP. The pathway that is followed at a specific site appears to depend on the type of microorganism present in the system (Mahaffey, 1997).

While PCP is considered to be more toxic than its potential breakdown isomers, several of them are still considered to represent risks to human health.

PCP and 6 of its 18 possible breakdown intermediates have been assigned human health RBCs for drinking water by EPA Region III.

In order for anaerobic degradation of PCP to occur, highly anaerobic conditions must exist in the aquifer; dissolved oxygen levels below 0.5 mg/L are necessary in the groundwater.

In aerobic degradation, the phenol ring is broken during an early stage of the process and complete mineralization to carbon dioxide, water and chloride occurs much more quickly than through the anaerobic pathway. Initial intermediate products that form prior to breaking the phenol ring may include tetrachloroatechol, tetrachlorohydroquinone (TeCHQ), tetrachlorobenzoquinone (TeCBQ), trichlorohydroxylbenzoquinone (TCBHQ), TCHQ, DCHQ, and CHQ (Mahaffey 1997). For aerobic oxidation to occur, a dissolved oxygen concentration in the groundwater of at least 2 mg/L is typically required.

The aerobic intermediate products of PCP have been found to be relatively innocuous. None of these intermediate products are listed in EPA Region III RBC tables. Also, none are listed in EPA's Integrated Risk Information System (IRIS) and Health Effects Assessment Summary Table (HEAST) databases. The intermediate breakdown products have fewer chloride atoms than does PCP and they degrade quickly by cleavage of the aromatic ring. They are labeled as short-lived, as they have not been found to accumulate in the environment at similar sites. It is assumed that the risk posed by these intermediate products is significantly less than PCP.

LABORATORY TESTS

Methods. A series of test-tube microcosm tests was conducted to determine if reagent-enhanced aerobic or anaerobic conditions could enhance the biodegradation of PCP at Site 13. The tests were conducted by Applied Power Concepts (APC) of Anaheim, CA (APC 1999A, APC 1999B). In these bench-scale tests, the Regenesis products HRC and ORC were used to determine which, if either, of these two products would be most appropriate to be applied in the proposed field treatability test. HRC, a polylactate ester made from reduced sugars and lactic acid tetramer, releases hydrogen in the environment via fermentation of the lactic acid. This hydrogen is available to promote the conversion of chlorinated hydrocarbons to dechlorinated hydrocarbons. ORC is a patented formulation of magnesium peroxide, MgO_2, intercalated with phosphate ions, which slowly releases oxygen when moist.

The tests were conducted over a 3-month period under idealized conditions and were designed to provide a relatively quick positive or negative response. Each test used 150 ml of a 10 mg/L solution of PCP mixed with 10 grams of soil collected from the upper portion of the Columbia Aquifer beneath the former tank location. Standard microbial counts were performed at the completion of each test.

Two types of tests were run with addition of ORC: tests with a low ORC dose (0.25 grams) and tests with a high ORC dose (0.75 grams). The tests were run for three months with samples analyzed once a month. The test included one control sample containing no ORC. Finally, an attempt was made to run a "sterile" sample using sterile sand as opposed to site soil and no ORC addition. However, at the end of the test, plate counts indicated that microorganisms were present in the sterile sample.

A parallel series of tests using the same procedures and contaminant concentrations were run to test the effect of HRC addition. As with the ORC tests, two levels of HRC addition were tested (0.25 grams and 0.75 grams).

Microbial counts using standard plate pour techniques were done to measure aerobic populations (using a glucose nutrient agar), anaerobic populations (using glucose nutrient agar under nitrogen), and sulfate reducing populations (SRB) using the standard AWWA test. Plate counts were run in triplicate.

Results. The results of the ORC tests were as follows (Table 1):

TABLE 1. Results of side-by-side laboratory tests using ORC and HRC to degrade PCP.

Test	PCP (mg/L) at elapsed time				Microbial Counts at Completion (cfu/ml)[E]		
	0 days[A]	30 days	60 days	90 days	Aerobic[B]	Anaerobic[C]	SRB[D]
ORC Tests							
Low ORC	10.00	4.18	1.59	0.93	9	4	41
					Spreader	Spreader	TNTC
					0	0	TNTC
High ORC	10.00	0.20	<0.01	<0.01	416	Spreader	TNTC
					82	44	896
					260	224	TNTC
HRC Tests							
Low HRC	10.00	1.61	1.61	0.67	3	3	32
					0	0	Spreader
					4	4	Spreader
High HRC	10.00	1.98	2.29	0.32	4	4	Spreader
					0	4	496
					0	0	184
Quality Control Tests							
Control	10.00	6.05	6.33	na	146	Spreader	TNTC
					496	Spreader	TNTC
					63	42	TNTC
Sterile	10.00	8.21	6.29	6.29	20	Spreader	TNTC
					75	Spreader	TNTC
					103	Spreader	TNTC

[A] Initial concentrations are assumed based on mass of PCP added to test tubes
[B] Aerobic total plate counts
[C] Anaerobic total plate counts
[D] Sulfate reducing bacteria
[E] Colony-forming units per milliliter
Spreader - Counts could not be done due to merging colonies
TNTC - Too numerous to count

The results of the HRC tests were as follows:

1. PCP in the sample with the high HRC dose was detected at 1.98 mg/L (80% reduction) at 30 days, 2.29 mg/L (77% reduction) at 60 days, and 0.321 mg/L (97% reduction) at 90 days. The increase between 30 days and 60 days may be due to laboratory variability, or variability in PCP dechlorination rates.
2. PCP in the sample with the low HRC dose was detected at 1.61 mg/L (84% reduction) at both 30 days and 60 days, and 0.672 mg/L (93% reduction) at 90 days.
3. The control and "sterile" samples were those of the ORC test above.
4. No 2,4-DCP and very little 2,4,6-TCP was detected in any of the samples. 2,4,6-TCP was detected in the 90-day samples at concentrations at or below 0.22 mg/L.

The microbial counts show that SRB-type microbes thrived in the site soil under both aerobic and anaerobic conditions. It is also interesting to note that anaerobic bacteria survived under aerobic conditions. This implies that the SRB and anaerobic microbes (and possibly the dechlorinators) are facultative. The results of the HRC plate counts also seem to indicate that the SRB are key to the reduction of PCP at the site.

Laboratory Test Conclusions. These results indicate that both HRC and ORC will degrade PCP under laboratory conditions. ORC was selected as the preferred method for the proposed field treatability test because it was shown to degrade PCP to a greater extent during the 90-day test, it is less likely that aerobic degradation will create toxic by-products, and aerobic conditions are more likely to provide the added benefit of breaking down the residual TPH concentrations in the soil at the site.

FIELD TEST DESIGN

Implementation. The field treatability test focused on the application of ORC within and around the footprint of the former dip tank. The effectiveness of the treatability study will be evaluated by collecting and analyzing groundwater samples from selected wells prior to and throughout the duration of the test. The groundwater-monitoring network initially consisted of one upgradient well, four shallow (15 ft [4.6 m]) downgradient wells and two deep (23 ft [7 m]) downgradient wells (Figure 1). A fifth shallow downgradient well (MW26S) was added after the baseline sampling round to better monitor PCP concentrations.

During well installation, soil samples were collected from the aquifer sediments and analyzed for nitrite, nitrate and phosphate to help determine available nutrient concentrations in the subsurface.

Baseline groundwater samples were collected using low-flow methods from each of the seven wells in the monitoring network to determine baseline aquifer characteristics and water quality. Analytical parameters for baseline groundwater sampling are tabulated in Table 2.

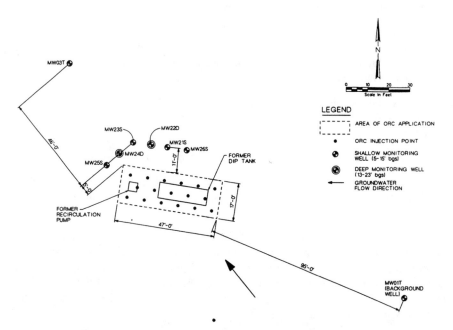

FIGURE 1. Plan view showing layout of ORC injection points and groundwater monitoring system at Site 13.

TABLE 2. Summary of analytical parameters for baseline and follow-up groundwater sampling rounds.

Parameters (Test Method)	Baseline	Rounds 1 - 4
TCL SVOCs and TICs* (OLC02)	All Wells (7)	All Wells (8)
Redox potential (A2580B)	All Wells (7)	All Wells (8)
TOC (SW846-9060)	All Wells (7)	All Wells (8)
Alkalinity (E310.1)	All Wells (7)	All Wells (8)
CO_2 (RSK-175)	All Wells (7)	All Wells (8)
Fe^{2+} (SM3500D)	All Wells (7)	All Wells (8)
Chloride (E300)	All Wells (7)	All Wells (8)
pH, Temp, Spec. Cond. (field probe)	All Wells (7)	All Wells (8)
DO (probe and Chemets field kit)	All Wells (7)	All Wells (8)
TAL Metals: total and diss. (ILM04)	4 wells (01T, 23S, 24D, 03T)	2 wells (23S, 24D)
TCL VOCs (OLC02)	4 wells (1T, 23S, 24D, 3T)	2 wells (23S, 24D)

TICs - tentatively identified compounds

Approximately 1,400 pounds of ORC powder was mixed with water and injected into the water table aquifer over a depth range of 6 to 23 ft (1.8 to 7 m) bgs using direct push (Geoprobe™) technology. The slurry was injected at 17 injection points over an area of 800 sq. ft (74 sq. m). This area centered around the former tank location (Figure 1). An estimate of a 5-to-8 ft (1.5-to-2.4 m) distribution radius from each injection point was anticipated.

The ORC application rates were conservatively designed based on the maximum concentrations of PCP and TPH found in both the groundwater and soil in the treatment area over the past 4 years. While ORC treats the contaminants in the dissolved phase, the sorbed phase concentrations were also taken into consideration to address the future leaching of these contaminants from the soil into the groundwater. The intent of using these conservative assumptions is to minimize the need of having to return for a repeat injection later on. The concentration of other oxygen-demanding factors, such as dissolved iron and TOC, are relatively low at this site and would not expect to significantly increase the amount of oxygen needed.

The oxygen requirements for degrading PCP and TPH are based on stoichiometry. Different application rates were applied to the upper 9 ft (2.7 m) and lower 8 ft (2.4 m) of the aquifer because site data indicated that concentrations of both TPH and PCP are significantly greater in the upper portion of the aquifer. Design concentrations are shown in Table 3.

TABLE 3. Site concentrations used for designing ORC injection rates.

Media and Depth	PCP	TPH	TOC	DO	NO$_3$	Fe^{+2}	SO$_4$	BOD	COD
Shallow Soil: 6-15 ft. (mg/kg)	36	39	3,540	na	na	na	na	249	11,600
Shallow Groundwater: 6-15 ft. (µg/L)	2,000	<1,500	1,400	1,000	190	940	18,400	na	na
Deep Soil: 15-23 ft. (mg/kg)	<0.4	<20	140	na	na	na	na	na	na
Deep Groundwater: 15-23 ft. (µg/L)	500	<1,000	2,700	1,300	550	500	27,000	<20,000	na

na: not analyzed

The theoretical (stoichiometry-based) oxygen demands for the aerobic degradation of PCP and TPH are as follows:

- 3 lbs (1.4 kg) of oxygen (30 lbs. or 14 kg of ORC) are required to completely degrade 1 lb (0.45 kg) of hydrocarbon.

- 0.54 lbs (.24 kg) of oxygen (5.4 lbs. or 2.45 kg of ORC) are required to completely degrade 1 lb. (0.45 kg) of PCP.

The total amount of ORC required for the upper portion of the aquifer was approximately 1,000 lbs. (453 kg), while approximately 400 lbs (181 kg) was required for the lower half of the aquifer. Most of this demand is related to the TPH concentrations.

The ORC was applied at a rate of 6.5 lbs of powder per linear foot of injection depth (9.68 kg/m) in the upper half of the aquifer and 3 lbs/ft (4.47 kg/m) in the lower half.

The injection of the 1,400 lbs (634 kg) of ORC required two days. The cost for injection was $18,800, including the cost of the product. These costs do not include design and consulting fees, the bench-scale tests, installation of monitoring wells and the baseline and post-injection monitoring.

Future Plan of Action. Four rounds of groundwater sampling are proposed over a 36-week period to evaluate performance: 6, 12, 24, and 36 weeks after the injection of the ORC. The parameters listed in Table 2 will be analyzed for in each round. Results shall be reported in a separate paper.

REFERENCES

Mahaffey, W.R. 1997. *Bioremediation of Pentachlorophenol, Literature Survey of Metabolic Pathways and Rate Constants.* Pelorus Environmental & Biotechnology Corporation, Evergreen, CO.

Palmer, T. 1999A. *Treatability Study, CH2M HILL PCP-HRC Study.* Applied Power Concepts, Anaheim, CA.

Palmer, T. 1999B. *Treatability Study, CH2M HILL PCP-ORC Study.* Applied Power Concepts, Anaheim, CA.

BIOMASS INFLUENCE ON BIOAVAILABILITY DURING PENTACHLOROPHENOL TRANSPORT THROUGH ARTIFICIAL AQUIFERS

Yves Dudal, Réjean Samson and Louise Deschênes
École Polytechnique de Montréal, QC Canada

ABSTRACT : The influence of micro organisms on the transport and bioavailability of pentachlorophenol (PCP), leaking from utility poles, is assessed in this project. A pre-chromatography column was packed with previously inoculated sand (with PCP degraders) and PCP was continuously injected through the column. The contaminant was monitored in the column effluent. The same experiment was carried out three consecutive times: first at a pore-water velocity of 2.61×10^{-3} cm.s^{-1}, then, after flushing the contaminant, at half that velocity, and finally, after another wash, at the initial velocity. Results show that through the course of each experiment, PCP concentration in the effluent decreases from 13 mg/L injected to a non detectable level (less than 0.5 mg/L), but at different rates: in 167h for the first, 4h for the second and 10h for the last. Micro organisms were able, in all cases, to gain full access to the contamination due to their growth. In the first experiment, PCP was initially barely bioavailable due to the transport, but the growth-associated biodegradation kinetic increased over time, allowing the bioavailability, expressed as the ratio of concentration to time, to increase over time.

INTRODUCTION

Pentachlorophenol (PCP) has been extensively used as a wood-preserving agent. It has been shown that PCP that leaks from a treated pole to the ground can contaminate the surrounding soil and can migrate to the groundwater (Valo et al., 1984). One of the possible ways to approach this problem is risk assessment, in which the hazard associated with PCP has to be quantified in terms of source, fate and effects (USEPA, 1992). Once the contaminant has leaked from the pole, its fate is governed by a number of interactions: hydrodynamic, physical-chemical and biotic. The aqueous phase concentration of contaminant, resulting from these interactions is often regarded as bioavailable, i.e. that biological organisms can access. Bioavailability is usually understood as the extent of exposure organisms get of the contaminant (Alexander and Alexander, 2000). Most studies on bioavailability take place in closed systems, where hydrodynamics cannot influence the different interactions responsible for bioavailability. In such systems, the bioavailable fraction of contaminant is often observed to decrease with time: this is the aging phenomenon (Hatzinger and Alexander, 1995; Kelsey and Alexander, 1997). However, the influence of water flow on bioavailability has not yet been the subject of much interest. Depending on the pore-water velocity, the micro organisms may not have enough exposure to the contaminant in order to uptake it from the solution and to metabolize it. The question arises to know if micro organisms are able to adapt to a limiting access of substrate due to hydrodynamics. The project presented here focuses on the impact of micro organisms on the fate of pentachlorophenol, during its transport through open systems, such as aquifers.

To fulfil this objective, a set of saturated column experiments was performed where PCP was continuously injected through columns packed with previously inoculated standard sand. PCP was monitored over time in the effluent. Two factors were tested: the pore-water velocity and the micro organisms adaptation capacity over time.

MATERIALS AND METHODS

The columns used for all the experiments were pre-chromatography columns (15cm long and 4.8cm diameter) made of borosilicate glass and Teflon endings (pore size of 20μm) (MARQUE) and packed with a model solid: standard Ottawa sand (Anachemia, Montreal). The sand was either abiotic or previously inoculated. The packed column was saturated with a mineral salt medium (MSM; Greer and Shelton, 1992). The same inoculated sand column was used for the following three experiments. NaPCP (Anachemia, Montreal) was injected dissolved in MSM with a pore-water velocity of 2.61×10^{-3} cm.s^{-1} using a peristaltic pump (Column 1). At the end of this first experiment, the column was washed with MSM only and then reinjected with PCP dissolved in MSM at half the velocity (Column 2). Finally, the column was washed once more and reinjected at the initial velocity (Column 3) to assess the time adaptation of the micro flora to the flow through situation. Another set of four columns was injected simultaneously with PCP dissolved in MSM and each column was sacrificed at different times. Material from the sacrificed columns was separated in three sections (top, middle and bottom, each 5 cm long) and the PCP-biodegradation activity was monitored on each section. This set of four columns was performed to assess the time and spatial evolution of microbial activity in the columns.

Production of the inoculated standard sand. A soil sample from a PCP-contaminated site was taken and placed in an aerated fed-batch bioreactor where MSM was added (3:1 w/w) and sequential additions of NaPCP were performed (Bécaert et al., 1999). The aqueous phase containing the microbial suspension was taken from this reactor and placed in a second identical reactor where standard sand was added (1:3, w/w). Sequential additions of NaPCP were also performed on this reactor from 1 to 100 mg/L. The inoculated sand obtained from this procedure consisted mainly of the *Sphingomonas sp.* (Beaulieu et al., *in press*).

PCP injection and monitoring. The inoculated sand was packed in the columns by sedimenting 50g of material at a time in MSM. Once packed, the columns were flushed for 3 hours with MSM, in order to remove non-attached bacteria. Then a solution of NaPCP dissolved in MSM (13 mg/L) was injected through the columns, until the effluent concentration was invariant with time. 4-mL samples were taken from the effluent, filtrated at 0.45µm, and their absorbances at 319 nm were read using a spectrophotometer (Varian, model DMR90). The results were confirmed by further HPLC analysis.

Biodegradation activity monitoring. Four control columns were packed with inoculated sand in the same manner as previously described. Each of these for columns was sacrificed at a given time (2.5h, 21.5h, 190h and 408h). Each sacrifice consisted of stopping the NaPCP injection, opening the column, emptying the mobile water and in carefully separating the packed inoculated sand into three sections, each representing 5 cm of the column. Each of these samples was then divided into two replicates (90g) and placed in a 500-mL amber glass bottle, amended with 250 mL of MSM and 13 mg/L of NaPCP. All the bottles were agitated at 200rpm. Disappearance of NaPCP from the aqueous phase was monitored daily by sampling 3 mL of liquid, centrifugating it at 13000 rpm for 10 minutes and filtering the supernatant at 0.45µm. Absorbance of the filtrate was read at 319 nm.

PCP fate modeling. A model, developed by Borden and Bedient (1986) was used here to fit the experimental results. The model equations, presented in equations 1 to 3, reflect that the contaminant fate and transport in the columns are governed by advection, dispersion and growth-associated biodegradation. However, the system is limited with the dissolved oxygen in the MSM that elutes with PCP through the columns.

$$\frac{\partial C}{\partial t} = D\frac{\partial^2 C}{\partial z^2} - v\frac{\partial C}{\partial z} - Y_{C/X}\mu_m X \frac{C}{K_C + C}\frac{L}{K_L + L} \quad (1)$$

$$\frac{\partial X}{\partial T} = \mu_m X \frac{C}{K_C + C}\frac{L}{K_L + L} \quad (2)$$

$$\frac{\partial L}{\partial T} = \frac{1}{Pe}\frac{\partial^2 L}{\partial Z^2} - \frac{\partial L}{\partial Z} - Y_{L/X}\mu_m X \frac{C}{K_C + C}\frac{L}{K_L + L} \quad (3)$$

The variables are: C, the contaminant concentration in the aqueous phase (mg/L), X, the biomass concentration (mg/L), L, the dissolved oxygen concentration in the aqueous phase, (mg/L), T, the time (s) and Z, the length along the column (cm). The hydrodynamic parameters are: v, the pore water velocity (cm/s) and D, the dispersion coefficient (cm^2/s). Finally, the growth-associated biodegradation parameters are: $Y_{C/X}$, the contaminant-biomass yield (mg/mg), μ_m, the maximal growth rate (s^{-1}), K_C, the half-saturation constant for the contaminant (mg/L), K_L, the half-saturation constant for the dissolved oxygen (mg/L) and $Y_{L/X}$, the dissolved oxygen-biomass yield (mg/mg).

This set of four coupled partial differential equations was first discretized using the finite difference method and then solved numerically using the Runge-

Kutta order 4 method with the set of initial and border conditions expressed in equations 4 to 6.

$$C(t=0, z>0) = 0 \quad C(t>0, z=0) = 1 \quad \frac{\partial C}{\partial T}(z \to \infty) = 0 \quad (4)$$

$$X(t = 0) = X_0 = 1 \quad (5)$$

$$L(t=0, z>0) = 0 \quad L(t>0, z=0) = 1 \quad \frac{\partial L}{\partial T}(z \to \infty) = 0 \quad (6)$$

All the parameters were estimated form fitting the model-generated curves to the experimental results using a non-linear regression method based on the Marcquardt-Levenberg algorithm. The hydrodynamic parameter D was estimated for each column after performing a non-reactive tracer test for that column, using a step elution of ^3H-H_2O based MSM. The value obtained was then used directly for the estimation of the other parameters.

RESULTS AND DISCUSSION

PCP fate and transport through sand.

Column weight analysis and tracer test analysis allowed the estimation of different hydrodynamic parameters. Their average values over all the columns used and standard errors are given as followed: porosity: 0.35±0.008, axial dispersion: $1.1\pm0.2\times10^{-4}$ cm^2/s and bulk density: 1.75±0.02 kg/L. Standard errors show that all the columns are comparable in terms of hydrodynamic characteristics.

Figure 1a presents the results of the PCP continuous injection through abiotic and inoculated sand. The sharp breakthrough obtained under abiotic conditions shows that the standard sand does not retain PCP. The model confirmed this fact, which was able to predict the breakthrough without using any retardation factor. Under biotic conditions, the initial breakthrough is similar as under abiotic conditions, but after a dozen pore volumes, the PCP concentration in the effluent starts to decrease. This decrease lasts over 100 pore volumes (167h) until PCP concentration in the effluent is non-detectable. At the early stage of the experiment, the micro organisms are not capable of degrading much of the contaminant, but acquire this capacity with time. Such increase results in an accelerating biodegradation kinetic able to compete with the water flow kinetic.

Similar observations are made for the same column further injected with PCP, at half the initial flow rate (Column 2, Figure 1b) and injected again at the initial flow rate (Column 3, Figure 1b). These times, the decreases are much faster and after only 3 and 6 pore volumes respectively, PCP concentration in the effluent reaches a non-detectable level. The increase in the biodegradation kinetic observed in Column 1 is responsible for this much faster disappearance of PCP in the effluent of Column 2 and 3. This performing kinetic combined with a lower flow rate leads to a short breakthrough of PCP for Column 2.

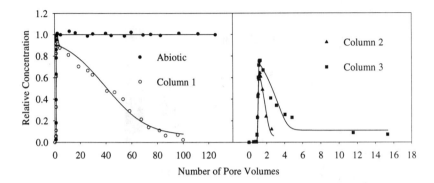

FIGURE 1. PCP experimental and model-fitted relative concentration in the columns effluent. (a) For Column 1 (standard sand) under abiotic and biotic conditions. (b) For Column 2 (half the pore-water velocity) and Column 3 (initial pore-water velocity).

Growth-associated biodegradation.

To explain this increase in the biodegradation kinetic over time within the columns, results from the sacrificed columns were analysed. Biodegradation from each section of the sacrificed columns was monitored and a biodegradation activity index (BAI) was computed from the results. This index is the ratio between the maximum biodegradation rate to the time it took to reach the point of inflexion. It integrates a possible lag time. The BAIs for the different columns are presented in Figure 2a. Over the 20 first hours of PCP injection (equivalent to 12 pore volumes), the biodegradation activity stays constant throughout the column, with time and space. After 190 hours, the activity has been dramatically increased, but is identical over the length of the column. Finally, after 408 hours, colonization of the Teflon ending was observed and its biodegradation activity was measured. It appeared to be the major source of activity within the whole column.

Such an increase in biodegradation activity over time can either be due to enzymatic adaptation of the micro flora or to microbial growth. The first hypothesis is less likely considering that the micro flora used was grown on PCP before inoculation on the sand. The model was also used to assess the space and time evolution of the biomass in the system. From the results of Column 1, simulations were run corresponding at the different sacrifice times. Results, presented in Figure 2b, clearly show biomass growth over time and a spatial shift of the population towards the source of substrate, the entrance of the column. The model parameters estimated from fitting the experimental results of Columns 1 to 3 (Figure 1a and b) are presented in Table 1.

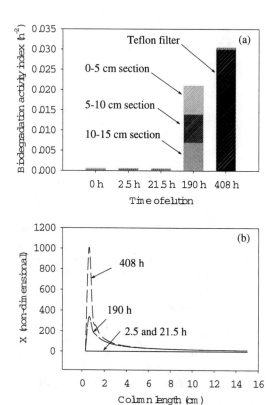

FIGURE 2. (a) Experimental biodegradation activity index as a function of time and column length. (b) Model-generated profiles of biomass within the column at the various sacrifice times.

TABLE 1. Model parameters estimated from experimental results.

Column	Experimental conditions	Y_x (mg/mg)	μ_m (h^{-1})	K_C (mg/L)	Y_L (mg/mg)	K_L (mg/L)
1	Inoculated sand	0.59	0.26	6.5×10^{-3}	0.56	6.07
2	id. ½ flow rate	0.60	0.47	6.5×10^{-3}	0.56	0.24
3	Id. initial flow rate	0.61	0.39	6.5×10^{-3}	0.54	0.2

The growth-describing parameters are similar for the three columns. Values of 0.6 for Y_X are similar to those obtained in a similar study of biodegradation of benzoate during transport (Brusseau et al., 1999). The half-saturation constant,

K_C, shows a good affinity from the micro flora to the contaminant, in accordance to the previous adaptation. Maximum growth rates, µm, between 0.26 and 0.47 per hour are also comparable with the literature. The only problem resides in the half-saturation constant for the limiting nutrient, K_L, that had to be changed from Column 1 to Column 2 and 3 in order to fit the experimental results. This change could reflect a physiological modification within the micro flora that adapted to the situation over the 167 hours of the first experiment.

Biomass influence on PCP bioavailability.

It is clear from the experimental and model results that growth is occurring within the columns due to continuous injection of PCP. The question arises to know the influence of this growth on the contaminant bioavailability. In Column 1, during the 12 first pore volumes, PCP is barely bioavailable to the microorganisms. The hydrodynamic kinetic is too strong for the microbes to compete and they can only partially access the contaminant that passes in front of them. Bioavailability can be quantified as 10% of the total contamination. But, with time, microorganisms gain access to the contamination, due to their growth on the bioavailable substrate. After 100 pore volumes, bioavailability is 100% of the total contamination. Biomass growth led to a dramatic (10 fold) increase in contaminant bioavailability. It allowed the microbes to compete with the hydrodynamic kinetic. Furthermore, this gain was still observable in the second and third injection, where it took only 3 and 6 pore volumes respectively for the microbes to gain full bioavailability. The difference between these two values is explained by the pore-water velocity, divided by two in the second injection, that allowed greater access for the micro organisms to the contaminant, in agreement with previous observations (Brusseau et al., 1999). Bioavailability, defined as the exposure between the contaminant and the living organism and known to be sorption-limited, is here shown to be dependent on the time of exposure imposed by the different kinetics (hydrodynamics, biodegradation), leading to the concept of bioavailability rate. Indeed, microbial limitations as well as strong hydrodynamics can limit the exposure, thus limiting bioavailability. Bioavailability is then not directly related to the aqueous-phase concentration of the contaminant, but to the time the microbial population is exposed to this concentration.

Such observation of biomass-related increase in bioavailability over time is included in a line of thought where living organisms are seen as capable of modifying a physical-chemical situation between the contaminant and its environment (Baveye and Bladon, 1999; Gevao et al., 2001; Wick et al., 2001). They are considered as possible sinks for the contamination, and the results presented here show a microbial sink for PCP. In the case where an efficient sink is present, and not limited by any external factor, the contamination aging process may not lead to a lesser extent of organisms' access to the contaminant.

ACKNOWLEDGMENTS

The authors acknowledge the financial support from the industrial Chair partners: Alcan, Bell Canada, Cambior, Canadian Pacific Railway, CEAEQ, Elf Aquitaine, Hydro-Québec, Natural Science and Engineering Research Council (NSERC), Gaz de France - Électricité de France, Ministère de la Métropole et des affaires municipals du Québec, Petro-Canada, Solvay, Ville de Montréal. Constructive discussions with Prs. Michel Perrier and Denis Dauchin as well as the technical help from Gabrielle Soucy were greatly appreciated.

REFERENCES

Alexander, R. R. and Alexander, M. 2000. "Bioavailability of genotoxic compounds in soils." *Environmental Science and Technology.* 34: 1589-1593.

Baveye, P. and Bladon, R. 1999. "Bioavailability of organic xenobiotics in the environment: a critical perspective."In P. Baveye, J.-C. Block and V. V. Goncharuk, Eds. *Bioavailability of organic xenobiotics in the environment - Practical consequences for the environment*, pp.227-248. Kluwer Academic Publishers, Dordrecht, The Netherlands.

Beaulieu, M., Bécaert, V., Deschênes, L. and Villemur, R. *in press*. "Evolution of bacterial diversity during enrichment of PCP-degrading activated soils." *Microbial Ecology.*

Bécaert, V., Beaulieu, M., Gagnon, J., Villemur, R., Deschênes, L. and Samson, R. 1999. "Activation of an indigenous microbial consortium for bioaugmentation of soil contaminated with wood-preservation compounds."In B. C. Alleman and A. Leeson, Eds. *Bioremediation of Nitroaromatic and Halogenated Compounds*, pp.89. Batelle Press, Columbus, OH.

Borden, R. C. and Bedient, P. B. 1986. "Transport of dissolved hydrocarbons influenced by oxygen-limited biodegradation. 1. Theoretical development." *Water Resources Research.* 22: 1973-1982.

Brusseau, M. L., Hu, M. Q., Wang, J.-M. and Maier, R. M. 1999. "Biodegradation during contaminant transport in porous media. 2. The influence of physicochemical factors." *Environmental Science and Technology.* 33: 96-103.

Gevao, B., Mordaunt, C., Semple, K. T., Piearce, T. G. and Jones, K. C. 2001. "Bioavailability of nonextractable (bound) residues to earthworms." *Environmental Science and Technology.* 35: 501-507.

Greer, L. E. and Shelton, D. R. 1992. "Effect of inoculant strain and organic matter content on kinetics of 2,4-dichlorophenoxyacetic acid degradation in soil." *Applied and Environmental Microbiology. 58*: 1459-1465.

Hatzinger, P. B. and Alexander, M. 1995. "Effect of aging of chemicals in soil on their biodegradability and extractability." *Environmental Science and Technology. 29*: 537-545.

Kelsey, J. W. and Alexander, M. 1997. "Declining bioavailability and inappropriate estimation of risk of persistent compounds." *Environmental Toxicology and Chemistry. 16*: 582-585.

USEPA (1992). Framework for ecological risk assessment. Washington, DC, Risk Assessment Forum, Office of Research and Development.

Valo, R., Kitunen, V., Salkinoja-Salonen, M. and Raisanen, S. 1984. "Chlorinated phenols as contaminants of soil and water in the vicinity of two finnish sawmills." *Chemosphere. 13*: 835-844.

Wick, L. Y., Colangelo, T. and Harms, H. 2001. "Kinetics of mass ransfer-limited bacterial growth on solid PAHs." *Environmental Science and Technology. 35*:.354-361.

ANAEROBIC IN SITU BIOREMEDIATION OF PENTACHLOROPHENOL

Andrew J. Frisbie and *Loring Nies*
Purdue University, West Lafayette, IN 47907-1284, USA

ABSTRACT: Anaerobic in situ bioremediation of pentachlorophenol (PCP) was successfully initiated at a former wood treatment site located in Tippecanoe County, Indiana, USA. The objective of this study was to determine factors limiting PCP biodegradation in the field. The soil mass to solution volume (m/V) ratio controlled PCP toxicity in laboratory experiments. Aqueous PCP concentrations greater than ~19 mg/L inhibited anaerobic dechlorination of PCP, while PCP concentrations greater than ~7 mg/L inhibited degradation of a primary metabolite, 3-monochlorophenol (3-MCP). Selective PCP toxicity to organisms with specific functions, such as monochlorophenol dechlorination, may explain why a lack of complete dechlorination of PCP is often observed. These results suggest that in situ where soil m/V ratio cannot be significantly altered, the dechlorination of less chlorinated phenols might need to be spatially separated from zones where high concentrations of PCP exist. Glucose significantly increased the extent of dechlorination when compared to not adding substrate. In situ anaerobic bioremediation was initiated by adding a glucose solution to the contaminated soil via an infiltration drainage system. Since initiating in situ bioremediation, new dechlorination products have appeared and the molar percent of PCP of the total chlorophenols present in recovery well water has been decreasing.

INTRODUCTION

Complete removal of all chlorines from PCP by microbial reductive dechlorination has been observed under anaerobic conditions (Mikesell and Boyd, 1986; Bryant et al., 1991; Juteau et al., 1995). The degradation of 3-monochlorophenol (3-MCP) is often the limiting step in the degradation of PCP (Madsen and Licht, 1992).

Several studies have identified PCP concentrations that are toxic even to PCP-degrading microorganisms. The aqueous solubility of PCP has been reported to be ~20 mg/L (Arcand et al., 1995). The PCP dissociation constant (pKa) is 4.75; therefore, the total actual aqueous solubility of the neutral and ionized PCP species is dependent on solution pH since the phenolate anion is more soluble than the neutral species (Lee et al., 1990). Since PCP is relatively hydrophobic (log K_{ow} = 5.24, Jafvert et al., 1990), the distribution between the aqueous and soil phase is significantly influenced by the soil mass to water volume (m/V) ratio. The aqueous concentration of PCP is controlled by both the total mass of PCP in the system and the m/V ratio. The aqueous concentration controls the PCP toxicity to microorganisms. Adding water to a unit mass of soil containing a fixed mass of PCP lowers the m/V ratio and results in redistribution of the PCP between solid and aqueous phases. By decreasing the m/V ratio the aqueous concentration can be reduced and the toxicity of PCP to the microbial population can be limited.

Under anaerobic conditions, the presence of electron donors can significantly affect reductive dechlorination. For example, glucose increased the amount of PCP degraded, the amount of dechlorination, and the biomass in one study (Hendriksen et al., 1992). Reductive dechlorination of 2-MCP has been shown to cease once the added electron donor has been utilized (Dietrich and Winter, 1990).

The PCP contaminated industrial site under investigation in this study had a wood dip-treatment operation from 1950 until the early 1960s. The PCP solution was allowed to run off of the treated wood over an uncontained area next to the pit. Previous work had demonstrated that both aerobic and anaerobic PCP degraders were present in the soil (Frisbie and Nies, 1997). The objective of this investigation was to determine whether PCP toxicity and electron donor deficiencies were limiting PCP biodegradation in the field.

MATERIALS AND METHODS

Laboratory experiments in which the PCP-contaminated soil m/V ratio was varied were used to determine the threshold toxicity of PCP to different anaerobic microorganisms, and whether additional electron donor would enhance the extent of PCP dechlorination. Based on the laboratory results, an existing pump-and-treat system at the site was modified to enhance in situ bioremediation of PCP.

Soil Sampling and Analysis. Core samples were taken throughout the PCP contaminated area using a manual core sampler. Two-foot sections were recovered in a Teflon tube, capped on each end with aluminum foil, and stored at 4°C. Soil PCP concentrations were measured by extracting approximately 2 grams of soil with 5 ml of acetonitrile and 5 ml of water on a rotator for 48 hours. The samples were then centrifuged and the supernatant was acidified with phosphoric acid prior to analysis.

Concentrations for all chlorophenols were determined by HPLC with a gradient flow of 30% acetonitrile/70% 25 mM phosphoric acid for 25 minutes, a linear ramp for 10 minutes to 50% acetonitrile/50% 25 mM phosphoric acid, and then held constant for the next 35 minutes on a C_{18} column. Retention times, areas, and spectra were compared against standards of all 19 chlorophenols. The compounds were quantified at a wavelength of 214 nm.

PCP Toxicity Experiments. The soil/water distribution of PCP at the site was first determined from a desorption isotherm using contaminated soil. Then soil was thoroughly mixed in an anaerobic chamber to obtain a homogeneous sample. The experiments were set up in 100-ml amber serum bottles capped with Teflon™ lined septa held in place with an aluminum crimp seal. A range of soil mass to liquid volumes, 0.05 kg/L, 0.1 kg/L, 0.2 kg/L, 0.3 kg/L, 0.4 kg/L, 0.5 kg/L, and 1 kg/L, were used to conduct the experiments. A competent PCP degrading consortia, derived previously from the site, was used to inoculate the experiments.

Electron Donor Experiments. PCP contaminated soil was placed in serum bottles as described above. The total mass of organic substrate added to the serum bottles was normalized according to their electron donating equivalents. The concentration of glucose added was 450 mg/L. Uncontaminated site soil, which had previously been shown to contain PCP-degrading microorganisms, was added to each sample on as an inoculum. Uncontaminated site soil was used as an inoculum because our competent PCP degrading organisms had been maintained on glucose and methanol for several years, which precluded their use in a substrate effects study.

RESULTS AND DISCUSSION

PCP Toxicity Experiments. Varying the soil-solution m/V ratio controlled the aqueous phase concentration of PCP, and therefore, the toxicity of PCP to the microorganisms responsible for PCP degradation. The effect that aqueous PCP concentration has on PCP biodegradation is illustrated in Figure 1. Concentrations above ~19 mg/L were toxic to anaerobic PCP-dechlorinating bacteria. The accumulation of 3-MCP occurred in samples with relatively high initial PCP concentrations. Aqueous PCP concentrations below approximately 7 mg/L were not toxic to any of the microorganisms responsible for complete reductive dechlorination of PCP. This leads us to believe that different groups of microorganisms with different tolerances to PCP are responsible for different steps in the PCP degradation pathway. Methane gas production occurred in the samples only after depletion of PCP occurred. Even if the methanogens are not directly involved in PCP dechlorination, methanogenesis may be necessary for a functioning anaerobic microbial ecosystem. These results suggest that in situ where soil m/V ratio cannot be significantly altered, the dechlorination of less chlorinated phenols might need to be spatially separated from zones where high concentrations of PCP exist.

Electron Donor Experiments. Anaerobic degradation of PCP may be enhanced by the addition of an easily degradable substrate to provide a biological oxygen demand (McAllister et al., 1996; Hendriksen et al., 1992). For use in anaerobic remediation at a contaminated site, a substrate must effectively stimulate dechlorination by the indigenous microorganisms while ideally, being inexpensive, safe to use and acceptable to regulatory agencies. Glucose will support a relatively complex microbial ecosystem, and is an almost universally biodegraded compound. When it is anaerobically degraded, glucose forms a variety of intermediates that may provide an effective electron donor for reductive dechlorination of PCP.

Following inoculation with uncontaminated soil, significant PCP degradation was observed. The effectiveness of different substrates for enhancing the anaerobic degradation of PCP was evaluated and compared by the extent of

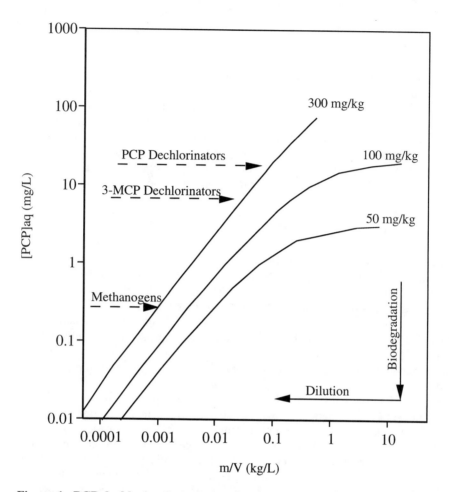

Figure 1. PCP dechlorinating microorganisms were tolerant to higher PCP concentrations than 3-MCP dechlorinators. The level of PCP toxicity is directly related to the aqueous phase concentration, which can be controlled in laboratory settings by adding water and reducing the soil mass/volume (m/V) ratio. The curves shown for different initial soil PCP concentrations are derived from desorption experiments from which the data fit the Freundlich model for soil-water partitioning.

PCP conversion into lesser-chlorinated phenols. Initially the contaminated soil contained primarily PCP. After substrate addition, 2,3,4,5-TeCP, 3,4,5-TCP, 2,3,5-TCP, 2,4,5-TCP, 3,5-DCP, 3,4-DCP, 2,5-DCP, and 3-MCP appeared over 323 days (Figure 2). A significant amount of the chlorophenols remaining was 3-MCP which accumulated in all samples and reached amounts as high as 60% of the total chlorophenols. In the samples with no substrate much less extensive dechlorination occurred.

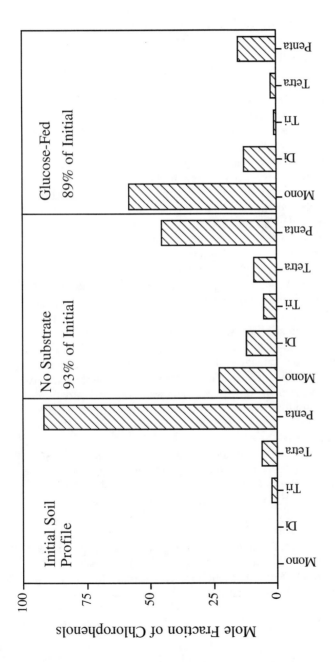

Figure 2. The addition of an electron donor enhanced the extent of PCP dechlorination relative to background soil levels. No 3-MCP dechlorination occurred because the initial aqueous phase PCP concentration was above the threshold tolerated by 3-MCP dechlorinators.

Organic substrates enhanced the extent of PCP dehalogenation in contaminated soil. Glucose is the best choice for an anaerobic bioremediation scheme. In addition to being the most effective substrate for dehalogenation of PCP, glucose is also a safe compound for use in the environment.

In Situ Bioremediation. A previously installed pump-and-treat system had recovery wells around the area with the highest contamination. The recovery wells pump the contaminated water that leaches through the contaminated soil into a remediation building containing aerated tanks.

Anaerobic PCP degradation occurred in the laboratory producing a potentially recalcitrant intermediate 3-MCP that only degraded when the microorganisms were not inhibited by high concentrations of PCP. The contaminated area is located in a heavily used section of an industrial site. Capping the area with concrete would allow for productive use of the area and limit exposure to the contaminated area. Concrete would also limit oxygen transfer and aid in maintaining anaerobic conditions. The cost of anaerobic remediation is less because an inexpensive source of glucose is located nearby. The aerobic PCP degrading microorganisms on-site are most likely facultative and should survive anaerobic conditions. Thus, if anaerobic remediation is not successful, the possibility still exists for aerobic remediation. Aerobic degradation was rapid at low PCP concentrations, but exhibited a significant lag period at higher concentrations. Aerobic degradation in the subsurface would require an input of oxygen; however, the soil at the site has high clay content that would limit oxygen transfer. In addition, injecting oxygen into the subsurface may decimate the existing anaerobic population, eliminating the possibility for future anaerobic remediation.

An anaerobic in situ treatment system was initiated by adding glucose to the recovered well water to stimulate anaerobic conditions and reductive dechlorination. An infiltration drainage system was installed over the area inside the four existing recovery wells. Initially, the recovered water is pumped into two 350 gallon aerated polypropylene holding tanks. The water is aerobically treated to biodegrade PCP concentrations to ~50 ug/L. Glucose is then added and the water is discharged to the infiltration system. Recovery well water samples were routinely collected and analyzed for all chlorophenols.

Anaerobic degradation was monitored by measuring concentrations of PCP and other chlorophenols in the well water. The chlorophenols concentration in the wells had wide fluctuations. In the two wells with the highest PCP levels the concentrations varied by a factor of six. The aqueous concentration of chlorophenols was influenced by precipitation; therefore, monitoring the molar percent of each chlorophenol was used as an indicator of anaerobic degradation. If anaerobic degradation occurs, the molar percent of PCP should drop and the molar percent of its degradation products should increase. In the best case, chlorophenol products from PCP dechlorination increased from ~11% to ~67% in less than one year (Figure 3). In the worst case, PCP dechlorination products increased only from ~6% to ~25%.

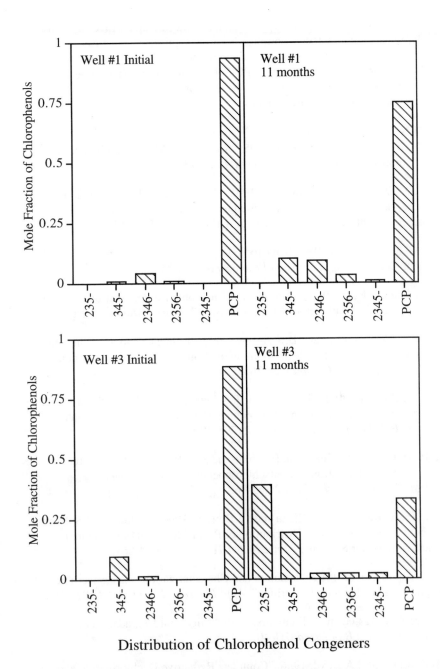

Figure 3. The mole fraction of all chlorophenol congeners extracted from two recovery wells prior to electron donor addition, and after 11 months of glucose addition, are shown. PCP concentrations approach concentrations toxic to PCP dechlorinators near well #1, while PCP concentrations are below the 3-MCP toxic threshold near well #3.

Since initiating anaerobic bioremediation at the site under investigation in this study, the aqueous PCP concentration has very significantly decreased in one recovery well. PCP dechlorination products have been detected in all recovery wells. In addition, the molar percent of PCP of the total chlorophenols detected has decreased, while molar percentage of the dechlorination products has increased during anaerobic bioremediation.

REFERENCES

Arcand Y., J.Hawari and S.R.Guiot. 1995. "Solubility of Pentachlorophenol in Aqueous Solutions: The pH Effect." *Wat.Res.* 29:131-139.

Bryant F.O., D.D.Hale and J.E.Rogers. 1991. "Regiospecific Dechlorination of Pentachlorophenol by Dichlorophenol-Adapted Microorganims in Freshwater, Anaerobic Sediment Slurries." *Appl.Environ.Microbiol.* 57:2293-2301.

Dietrich G. and J.Winter. 1990. "Anaerobic Degradation of Chlorophenol by an Enrichment Culture." *Appl.Microbiol.Biotechnol.* 253-258.

Frisbie A.J. and L.Nies. 1997. "Aerobic and Anaerobic Biodegradation of Aged Pentachlorophenol by Indigenous Microorganisms." *Bioremediation J.* 1:65-75.

Hendriksen H.V., S.Larsen and B.K.Ahring. 1992. "Influence and a Supplemental Carbon Source on Anaerobic Dechlorination of Pentachlorophenol in Granular Sludge." *Appl.Environ.Microbiol.* 58:365-370.

Jafvert C.T., J.C.Westall, E.Grieder and R.P.Schwarzenbach. 1990. "Distribution of Hydrophobic Ionogenic Organic Compounds Between Octanol and Water: Organic Acids." *Environ.Sci.Technol.* 24:1795-1803.

Juteau P., P.Beaudet, G.McSween, F.Lepine and J.-G.Bisaillon. 1995. "Study of the Reductive Dechlorination of Pentachlorophenol by a Methanogenic Consortium." *Can.J.Microbiol.* 862:862-868.

Lee L.S., P.S.C.Rao, P.Nkedi-Kizza and J.J.Delfino. 1990. "Influence of Solvent and Sorbent Characteristics on Distribution of Pentachlorophenol in Octanol-Water and Soil-Water Systems." *Environ.Sci.Technol.* 24:654-661.

Madsen T. and D.Licht. 1992. "Isolation and Characterization of an Anaerobic Chlorophenol-Transforming Bacterium." *Appl.Environ.Microbiol.* 58:2874-2878.

McAllister K.A., H.Lee and J.T.Trevors. 1996. "Microbial Degradation of Pentachlorophenol." *Biodegrad.* 7:1-40.

Mikesell M.D. and S.A.Boyd. 1986. "Complete Reductive Dechlorination and Mineralization of Pentachlorophenol by Anaerobic Microorganisms." *Appl.Environ.Microbiol.* 52:861-865.

ARTHROBACTER CHLOROPHENOLICUS A6 – THE COMPLETE STORY OF A 4-CHLOROPHENOL DEGRADING BACTERIUM

Karolina Westerberg (Stockholm University, Stockholm, Sweden and Södertörn University College, Huddinge, Sweden), Cecilia Jernberg (Södertörn University College and Karolinska Institutet, Huddinge, Sweden), Annelie M. Elväng (Stockholm University, Stockholm, Sweden) and Janet K. Jansson (Södertörn University College, Huddinge, Sweden).

ABSTRACT: An enrichment culture inoculated with soil was adapted to degrade high concentrations of 4-chlorophenol. From this enrichment culture, a bacterial isolate was obtained which was particularly efficient at degrading 4-chlorophenol. The isolate was previously shown to be a novel *Arthrobacter* species, and named *Arthrobacter chlorophenolicus* A6. The *A. chlorophenolicus* A6 strain was shown to mineralise 4-chlorophenol, and also to degrade a number of other *para*-subsituted phenols. Furthermore, the strain produced a melanin-like compound from several phenols. *A. chlorophenolicus* A6 was chromosomally tagged with either the *luc* gene, encoding firefly luciferase, or the *gfp* gene, encoding green fluorescent protein. The tagged strains were inoculated into 4-chlorophenol contaminated soil microcosms and were shown to rapidly degrade the 4-chlorophenol. During degradation, the introduced marker phenotypes provided useful information about the survival and activity of the inoculant.

INTRODUCTION
Employing the capabilities of microorganisms for removal of pollutants (bioremediation) is potentially more cost-effective than traditional methods such as excavation, capping or pump and treat (Atlas and Unterman, 1999). Bioremediation may be carried out by the indigenous microorganisms of a contaminated site and if necessary nutrients can be added to stimulate the process (Atlas and Unterman, 1999). However, there are cases when the intrinsic microflora is unable to degrade the pollutant(s) and inoculation of the contaminated site with cells capable of degradation is a promising alternative (bioaugmentation). For example, bioaugmentation may be required when high localized concentrations of pollutants inhibit natural degradation, when the indigenous microbial population is insufficient or unacclimated for pollutant degradation, when degradation pathway(s) are non-optimal or absent, or when there is a desire to speed up the rate of degradation (Atlas and Unterman, 1999; Pritchard 1992). However, the success of bioaugmentation is often uncertain and there is therefore a need to monitor bioaugmentation strategies to judge their efficacy.

Monitoring should include studies of fate and activity of the inoculant during degradation. Two very useful markers for this purpose are the firefly luciferase gene (*luc*) and the Green Fluorescent Protein gene (*gfp*). Firefly luciferase is ATP-dependent which means that it provides a measure of the activity of the microorganisms under the prevailing conditions (Prosser et al., 2000). In contrast, the GFP protein does not need any cofactors, therefore providing a means to monitor cells regardless of their energy status. With these marker genes, information may be gained about cell number, size, energy status, localization and persistance in soil.

We have chosen 4-chlorophenol (4-CP) as a model compound to study bioaugmentation using *A. chlorophenolicus* inoculum. 4-CP is formed from anaerobic degradation of more highly chlorinated phenols, from chlorination of waste water, in pulp mills and from partial hydrolysis of the herbicide 2,4-D (Prithchard et al., 1987; Woods et al., 1989).

The aim of this work was to develop a strategy for bioaugmentation of 4-CP contaminated soil and to monitor the inoculum during degradation.

MATERIAL AND METHODS

Culture conditions. *A. chlorophenolicus* A6 was grown on a mineral salts medium, GM, with 4-CP as sole carbon source, as described in Westerberg et al. (2000). When higher cell densities were desired, yeast extract was added to a final concentration of 0.1 – 0.3 % (w/v) (GMY medium). The *luc*-tagged strain *A. chlorophenolicus* A6L and *gfp*-tagged strain, A6G were grown on media supplemented with kanamycin (Elväng et al., 2001).

Enrichment on 4-CP and isolation of strain A6. The details of the process are described in Westerberg et al. (2000). Briefly, a soil suspension was enriched with gradually increasing concentrations of 4-CP (from 50 to >350 μg/ml).

^{14}C studies in soil and liquid medium. [ring-U-^{14}C] labelled 4-CP (11.61 mCi mM^{-1}, Pathfinder Laboratories Inc., St Louis, MO.) was used to monitor the conversion of ^{14}C - labelled 4-CP to ^{14}CO$_2$. Stock solutions of 4-CP were prepared with the labelled substrate as a tracer. For both pure culture and soil microcosm studies, a flow-through respiration system was used.

For studies of 4-CP degradation in liquid cultures, triplicate cultures were prepared with 100 ml medium contained in 500 ml Erlenmeyer flasks with sidearms equipped with a serum stopper. The 4-CP concentration was set to 150 μg/ml and the medium was either inoculated with *A. chlorophenolicus* A6 or left sterile for controls.

The soil used for respiration studies, was an Aridic haplustoll, sandy loam (pH 7.57, 68.6% sand, 16.1% silt and 15.3% clay). The soil was sieved (2 mm), air dried and stored. Each microcosm consisted of 50 g soil contained in 250 ml Erlenmeyer flasks and labelled 4-CP stock and sterile water were used to bring the moisture content to 1/3 bar with a final 4-CP concentration of 150 ppm (calculated on a soil dry weight basis). For inoculation into soil samples *A. chlorophenolicus* A6 was precultured in GMY (0.3%) medium. The cells were harvested, washed and resuspended in 1.5% NaCl to 1/10 of the original volume and 1 ml of this suspension was used to inoculate triplicate soil microcosms. Uninoculated soil microcosms were used as controls.

The flasks containing the soil or liquid medium were sealed with rubber stoppers that were fitted with glass tubes connected by tubing to the respiration apparatus and incubated at 23 °C. CO_2-free air was continuously passed through the flasks (respiration chambers) and the gases evaporating from the respiration chambers were trapped in 1M NaOH (primary CO_2 trap), followed by a second tube of 1 M $BaCl_2$ (secondary trap). The primary CO_2 trap was changed at specific intervals and emptied into a Mason jar along with distilled water used for rinsing the tubes. Scintillation vials, containing 10 ml of Scintsorb C (Isolab, Akron, OH), were also placed in the jars. The jars were then sealed and negative pressure was established with a vacuum source. 6 M HCl was injected into the jars to bring the pH to 1 and to release the CO_2 which was collected into the scintillation vials over night. Samples were then analysed for radioactivity in a Beckman liquid scintillation counter (Beckman Instruments, Fullerton, CA).

4-CP analyses. In pure cultures, 4-CP was analysed by UV spectroscopy by measuring the absorbance at 279 nm and comparing with a standard curve prepared with known amounts of 4-CP. From soil samples, 4-CP was analysed by gas chromatography (Elväng et al., 2001).

Studies of chemical characteristics of the pigment. *A. chlorophenolicus* A6 cells were inoculated into GM medium with 4-CP as sole carbon source, and incubated without shaking at 28 °C. Since the GM medium contains $FeCl_3$, cells were also grown using $FeSO_4$ as iron source, and these cultures were treated identically to the ones grown on GM medium. When pigment had developed, as shown by strong coloration of the medium and presence of a black precipitate, 0.6 ml samples were taken. Flocculation with $FeCl_3$ was tested by adding 200 µl of a 2g/l stock solution, decolorisation by addition of 200 µl H_2O_2 (25%, v/v), reaction with HCl by addition of 200 µl 5M HCl and resistance to extraction by organic solvents with chloroform, hexane and ethyl acetate.

Soil microcosms. Details of the experimental set-up was previously described (Elväng et al., 2001). Briefly, a sandy loam soil was contaminated with 4-CP, dispensed in 40g portions into bottles and inoculated with either *A. chlorophenolicus* A6L or A6G. Cells were extracted from the soil and plated on LB plates supplemented with kanamycin (75 µg/ml). Luminesence or fluoresence was analysed in a luminometer or flow cytometer, respectively.

RESULTS AND DISCUSSION

Enrichment culture and isolation of *A.chlorophenolicus* A6. Soil microorganisms were adapted to degrade high concentrations of 4-CP by gradually increasing the 4-CP concentration in an enrichment culture. The initial 4-CP concentration in the enrichment was 50 µg/ml, as concentrations higher than this were recalcitrant to degradation by the soil microflora. When 30-50% of the 4-CP had been degraded by the enrichment culture, new 4-CP was added to a higher final concentration (Figure 1). This process was repeated until at the end of the enrichment period, the 4-CP concentration was 350 µg/ml, all of which could be completely degraded by the enrichment culture. During the adaptation process, there was a four-fold increase in the rate of 4-CP degradation. Moreover, at the end of the adaptation process, 4-CP was removed without the lag phase seen in the earlier stages of the experiment (Figure 1). 4-CP concentrations below 60 µg/ml could not be measured in the second half of the experiment (Figure 1), due to accumulation of coloured products, which interfered with the spectrophotometric method used for 4-CP quantitation. Microorganisms that could grow on 4-CP as the sole carbon source were isolated from the enrichment culture and isolate A6 was particularly efficient at 4-CP degradation and was studied further. Strain A6 was recently described as a novel *Arthrobacter* species (Westerberg et al., 2000), and named *Arthrobacter chlorophenolicus* A6.

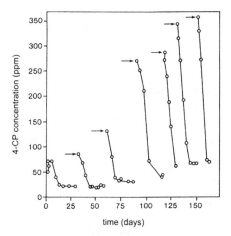

FIGURE 1. Removal of 4-CP by a soil community during enrichment with increasing concentrations of 4-CP. Arrows show the time points when the culture was spiked with 4-CP.

Degradation of 4-CP and other aromatics by *A.chlorophenolicus* A6. *A. chlorophenolicus* A6 mineralised 4-CP to CO_2 as demonstrated by studies using ^{14}C-labelled 4-CP as a substrate (Figure 2). Of added [ring-U-^{14}C] 4-CP to growth medium or soil, approximately 50% and 40%, respectively, of the added radioactivity showed up as $^{14}CO_2$ within 10 days of incubation (Figure 2). These results are consistent with complete or near-complete mineralisation of the 4-CP, the rest of the ^{14}C most likely being incorporated into biomass (Egli, 1992). The difference in $^{14}CO_2$ evolution between soil and pure culture studies may be the result of 4-CP absorbing to soil or production of recalcitrant or bound intermediates. In uninoculated soil, used as a control, $^{14}CO_2$ started to evolve late in the experiment (Figure 2), demonstrating that the *A. chlorophenolicus* A6 inoculum was responsible for initial 4-CP degradation in the inoculated samples.

A. *chlorophenolicus* A6 was able to grow on 4-CP as a sole carbon and energy source in pure culture, and no additional growth factors were required. It could grow on up to 350 μg/ml 4-CP, the lag time increasing with the 4-CP concentration, probably due to the toxicity of 4-CP (Westerberg et al., 2000). To our knowledge this is by far the highest 4-CP concentration to be degraded by a microorganism in a pure culture, without the requirement for any additional substrates or cofactors.

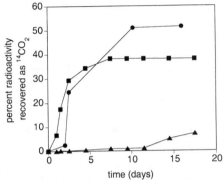

FIGURE 2. Percent $^{14}CO_2$ released from ^{14}C-labelled 4-CP during growth of A. *chlorophenolicus* A6 in liquid cultures and in soil. Symbols: liquid cultures (●), soil inoculated with A. *chlorophenolicus* A6 (■), uninoculated, nonsterile soil (▲). Data points are means of three replicates.

Presently, the 4-CP degradation pathway in A. *chlorophenolicus* A6 is being elucidated, and genes encoding key enzymes are being cloned. Preliminary results indicate that A. *chlorophenolicus* A6 employs an unusual 4-CP degradation pathway, which could partly explain the unusual tolerance to 4-CP exhibited by this strain (authors' unpublished observations).

In addition to 4-CP, A. *chlorophenolicus* A6 could degrade several other aromatic compounds, namely 4-bromophenol, 4-fluorophenol, 4-nitrophenol and phenol (Westerberg et al., 2000).

Pigment production by *A.chlorophenolicus* A6. A. *chlorophenolicus* A6 was found to produce a pigment from several phenolic compounds (Westerberg et al., 2000). The pigment was initially orange to red in colour, eventually darkening to black, forming an insoluble precipitate. Pigment was mainly seen in liquid cultures, rarely on plates. The pigment was produced from all the phenolics which could serve as growth substrates, and also from 2-chlorophenol, 3-chlorophenol and 2,4-dichlorophenol. To be able to study pigment formation from the latter compounds, which could not be used for growth, yeast extract was added to the medium. From those phenols that could be used as growth substrates, pigment was only formed when oxygen was limited (cultures incubated without shaking), whereas pigment was formed from non-growth substrates regardless of the oxygen level in the growth medium (Westerberg et al., 2000).

The pigment produced by *A. chlorophenolicus* A6 showed some similarities to melanin. Melanins are large polymers which are formed by enzymatic oxidation and subsequent polymerization of phenolic compounds (Ivins and Holmes, 1980). The pigment had some chemical characteristics in common with melanin, in that it was resistant to extraction with organic solvents, showed some precipitation upon addition of acid, decolorized when H_2O_2 was added and formed a flocculant precipitate upon addition of $FeCl_3$ (Ivins and Holmes, 1980).

Possibly, reactive quinones could be intermediates in 4-CP degradation in *A. chlorophenolicus* A6 and the pigment could be caused by polymerisation of these compounds. Such a process could conceivably be useful for decontamination of soil, by diminishing the toxicity of chlorophenols through cross-coupling and binding to humus, creating stable polymers (Bollag, 1992).

Soil microcosm studies. In order to track *A. chlorophenolicus* A6 cells in soil, cells were chromosomally tagged with either the *gfp* or the *luc* gene, (Elväng et al., 2001). The novel strains were denoted *A. chlorophenolicus* A6L and *A. chlorophenolicus* A6G, respectively, and showed the same growth behaviour on 4-CP as the wild type strain (Elväng et al., 2001).

Both the A6L and the A6G strains successfully degraded 4-CP in soil microcosms within 10 days (Figure 3A). In the uninoculated microcosms, the 4-CP concentration was stably maintained at around 100 ppm for the duration of the experiment, showing that it was indeed the inocula which were responsible for 4-CP degradation.

The CFU counts for both the tagged strains steadily declined in the soil microcosms (Figure 3B), although 4-CP was completely removed. In that respect, the two strains behaved similarly. Comparing the phenotypes of the two marker genes, however, told a different story. Consistent with the CFU results, there was a slow decline in luminesence of *A. chlorophenolicus* A6L cells (Figure 3C), indicating that energy was not plentiful in the population, although 4-CP was used as a substrate. In contrast, flow cytometric counting of *A. chlorophenolicus* A6G cells showed that the number of cells in the soil initially increased, and then were maintained at a constant level throughout the experiment (figure 3C). The increase at the beginning of the experiment may be due to residual division of the cells, or to fragility of the cells to extraction immediately after inoculation, or a combination of both factors. The decline in CFU of *A. chlorophenolicus* A6G over the course of the experiment was not reflected in the number of green fluorescent cells counted by flow cytometry. Obviously, the two phenotypes give different information. The discrepancy could be due to dead GFP-fluorescing cells that remain intact, and therefore are still counted in the flow cytometer. Another possibility is that the *A. chlorophenolicus* A6G cells are viable but not culturable (VBNC). In another study, Unge et al. (1999) saw a similar behavior for starved pseudomonads, and the research of Lowder et al. (2000) indicated that while

VBNC *Pseudomonas* cells retained green fluorescence, the phenotype was eventually lost from dead cells. These results would indicate that the discrepancy between CFU and green fluorescent cells in this study could be due to a VBNC state. Further research is needed to establish if this is really the case.

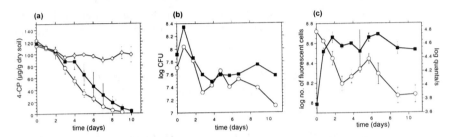

FIGURE 3. Soil microcosm studies. (a) 4-CP degradation by *A. chlorophenolicus* A6L(O), A6G (■) and uninoculated control (◊). (b) CFU counts of *A. chlorophenolicus* A6L (O) and A6G(■). (c) Number of green fluorescent *A. chlorophenolicus* A6G (■) and lumicesence output from *A. chlorophenolicus* A6L (O). Error bars show standard deviation (n=3). Adapted from Elväng et al (2001) (with permission).

CONCLUSIONS

We were able to isolate a novel *Arthobacter* strain which was particularly efficient at 4-CP degradation. The isolate, *A. chlorophenolicus* A6, is an interesting candidate for bioaugmentation, since it mineralizes 4-CP at high concentrations, which may otherwise be inhibitory to the indigenous microflora. Moreover, *A. chlorophenolicus* A6 degrades several other *para*-substituted phenols, and also produces a melanin-like pigment from many phenolic compounds. When *A. chlorophenolicus* A6 was tagged with either the *luc* or the *gfp* gene, much information could be gained about the behavior of the strain in 4-CP contaminated soil, and the strain was found to rapidly remove 4-CP from soil microcosms. In summary, 4-CP degradation by *A. chlorophenolicus* A6 is an interesting model to demonstrate the promise of bioaugmentation.

REFERENCES

Atlas, R. M. and R. Unterman. 1999. "Bioremediation." In A. L. Demain, J. E. Davies, R. M. Atlas, G. Cohen, C. L. Hershberger, W.S. Hou, D. H. Sherman, R. C. Willson and J.H. D. Wu (Eds.), *Manual of Industrial Microbiology and Biotechnology*, 2nd ed, pp. 666-681. ASM Press, Washington, D.C.

Bollag, J.-M. 1992. "Enzymes Catalyzing Oxidative Coupling Reactions of Pollutants." In H. Sigel and A. Sigel (Eds.), *Metal Ions in Biological Systems*, vol. 28, pp. 205-217. Marcel Dekker, New York, NY.

Egli, T. W. 1992. "General Strategies in the Biodegradation of Pollutants." In H. Sigel and A. Sigel (Eds.), *Metal Ions in Biological Systems*, vol 28, pp. 2-39. Marcel Dekker, New York, NY.

Elväng, A. M., K. Westerberg, C. Jernberg and J. K. Jansson. 2001. "Use of green fluorescent protein and luciferase biomarkers to monitor survival and activity of *Arthrobacter chlorophenolicus* A6 cells during degradation of 4-chlorophenol in soil." Environmental Microbiology 3:1-12

Ivins, B. E. and R. K. Holmes. 1980. "Isolation and Characterization of Melanin-Producing (mel) Mutants of *Vibrio chloerae*." Infection and Immunity 27(3):721-729.

Lowder, M., A. Unge, N. Maraha, J. K. Jansson, J. Swigget and J. D. Oliver. 2000. "Effect of starvation and the viable-but-nonculturable state on green fluorescent protein (GFP) fluorescence in GFP-tagged *Pseudomonas fluorescence* A506." Applied and Environmental Microbiology 66(8): 3160-3165.

Pritchard, P. H. 1992. "Use of inoculation in bioremediation." Current Opinion in Biotechnology 3:232-243.

Pritchard, P. H., E. J. O'Neill, C. M. Spain and D. J. Ahearn. 1987. "Physical and biological parameters that determine the fate of *p*-chlorophenol in laboratory test systems." Applied and Environmental Microbiology 53(8):1833-1838.

Prosser, J. I., A. J. Palomares, M. T. Karp and P. J. Hill. 2000. "Luminescence-Based Microbial Marker Systems and Their Application in Microbial Ecology." In J. K. Jansson, J. D. van Elsas and M. J. Bailey (Eds.), *Tracking Genetically-Engineered Microorganisms*, pp. 69-85. EUREKAH.COM/Landes Bioscience, Georgetown, TX.

Westerberg, K., A. M. Elväng, E. Stackebrandt and J. K. Jansson. 2000. "*Arthrobacter chlorophenolicus* sp. nov., a new species capable of degrading high concentrations of 4-chlorophenol." International Journal of Systematic and Evolutionary Microbiology 50(6):2083-2092.

Woods, S. L., J. F. Ferguson and M. M. Benjamin. 1989. "Characterization of chlorophenol and chloromethoxybenzene biodegradation during anaerobic treatment." Environmental Science and Technology 23(1):62-68.

Unge, A., R. Tombolini, L. Mølbak, and J. K. Jansson, 1999. "Simultaneous monitoring of bacterial number and activity in soil using a dual *gfp/lux* marker system." Applied and Environmental Microbiology 65(2): 813-821.

BACTERIAL CELL SURFACE HYDROPHOBICITY AFFECTING DEGRADATION OF PAHS IN NAPLS

Pia Arentsen Willumsen & Ulrich Karlson (National Environmental Research Institute, Roskilde, Denmark)

ABSTRACT: The scope of the present work was at laboratory scale in liquid batch systems to investigate the impact of cell surface hydrophobicity on degradation of HMW PAHs from hydrophobic liquid phases. Liquid paraffin, silicon oil and a dense NAPL extracted from tar lumps were used as non-aqueous-phase liquids. The beneficial effect of paraffin on fluoranthene mineralization increased with increasing cell surface hydrophobicity. The effect was associated with an improved cell-to-substrate contact and with growth on the NAPL phase. No extensive stimulatory effect of silicon oil was observed. Both the *Sphingomonas spp.* and the *Mycobacterium spp.* were able to degrade fluoranthene solubilized in tar-NAPL coated silica beads. With the *Sphingomonas spp.* the rates of mineralization of the tar-solubilized fluoranthene were only half of the rates obtained with crystalline fluoranthene, whereas for the *Mycobacterium spp.* the rates were almost doubled. Relative to the silicon- and paraffin-amended systems the overall extent of fluoranthene mineralization decreased for the *Sphingomonas spp.* and increased for the *Mycobacterium spp* in the tar-NAPL-solubilized fluoranthene system, indicating strain and NAPL specific differences in the bioavailability and in the efficiency of bacterial utilization of NAPL-solubilized PAHs, the cell surface characteristics playing an important role in the degradation performance of the cells.

INTRODUCTION

In coal tar contaminated soil and ground water, hydrophobic contaminants like the polycyclic aromatic hydrocarbons (PAHs) may be present in a non-aqueous phase liquid (NAPL), possible resulting in low bioavailability. Studies have been conducted on the bioavailability and degradation of low molecular weight PAHs solubilized in light model NAPLs, such as heptamethylnonane, silicon oil, diethylhexylpthalate or hexadecane (Deziel et al. 1999). However, relative few studies have been reported on the degradation of PAHs, and in particular of high molecular weight PAHs, in more realistic coal tar NAPL systems.

It has been suggested that bacteria with hydrophobic cell surfaces may overcome the low aqueous phase concentration of hydrophobic substrates by efficient substrate-to-cell contact mechanisms, thus increasing the flux of substrate into the aqueous phase and into the cells. Thus, inoculation with specialized bacterial cultures or consortia may be an option to increase biodegradation of hydrophobic contaminants in NAPL contaminated soil and ground water systems. However, an improved understanding of the interactions between the degrading bacterial communities and the multi-phase contamination is necessary to be able to select or stimulate the microorganisms active in the NAPL-PAH degradation.

The scope of the present work was at laboratory scale in liquid batch systems to investigate the impact of cell surface hydrophobicity on degradation of HMW PAHs from hydrophobic liquid phases. For this purpose the degradation of ^{14}C-labled fluoranthene solubilized in simple light NAPL systems (liquid paraffin and silicon oil, respectively) and in a realistic dense NAPL (extract from coal tar lumps) was studied using different fluoranthene-degrading bacterial strains.

MATERIALS AND METHODS

TABLE 1. Bacterial test strains

Bacterial strains	Cell surface hydrophobicity *
Sphingomonas sp. strain 10-1, DSM no 12247	28.8
Sphingomonas paucimobilis strain EPA505, NNRL no B-18512*	26.8
Mycobacterium frederikbergense strain Fan9, DSM no 44346	77.3
Mycobacterium gilvum strain GJ3P, DSM no 98-471	90.6
Mycobacterium gilvum strain LB307T (Bastianes et al. 2000)	94.0

* Measured on LB-grown cell cultures by Hauke Harms and Lukas Wick as described in Bastians et al. 2000. All bacterial strains were stored at –80 °C in glycerol (20% v/v). **NNRL Agricultural Research Culture Collection, International Depository, Peoria Ill.

Mineralization of NAPL-Solubilized Fluoranthene. The liquid light NAPL two-phase batch systems consisted of triplicate 250 ml Erlenmeyer flasks, containing 50 ml Bushnell-Haas (BH) mineral salts medium (Difco) amended with 20 mg/l crystalline fluoranthene alone as a mixture of cold and hot fluoranthene (final radioactivity 1130 Bq/mg fluoranthene) or supplemented with 5 µl/ml medium of paraffin, 0.88 kg/l (Merck) or silicon oil (RS1100 silicone, 350 cSt). Neither paraffin nor silicon oils are realistic tar NAPLs. However, the lighter-than-water NAPLs were selected as the model organic phase because of their simple chemical composition and for not being toxic to the bacterial test strains. The fluoranthene crystals were solubilized into the NAPLs by shaking the flasks at 150 rpm 24 hours before inoculation. Silicone stoppers sealed the flasks. After inoculation the flasks were incubated at 150 rpm at room temperature (23 °C). During the experiments both light NAPLs were floating on top of the aqueous phase.

To obtain a "realistic" complex dense tar NAPL, tar lumps collected from a former gas works site were extracted in DCM using a soxhlet apparatus. This procedure resulted in a very sticky, viscous, strong tar-smelling, dark brown liquid. The chemical composition of the tar NAPL was not further investigated. A reproducible system for testing the degradation of fluoranthene solubilized in the dense NAPL was prepared by mixing DCM-diluted tar NAPL and acetone-dissolved ^{14}C-fluoranthene and allow silica beads to absorb the mixture. The final concentration and radioactivity was 1 mg fluoranthene and 653 Bq, respectively, per 13 beads in 50 ml Bushnell-Haas (BH) medium. The solvents were allowed to evaporate before the beads were introduced to the mineral salts medium contained in triplicate 250 ml Erlenmeyer flasks.

Inocula for the degradations experiments were pregrown in LB-medium (Sambrook et al 1989) at 23 °C, harvested in late exponential growth phase, and further prepared as described in Willumsen et al. (1998). The number of culturable cells in the inoculum and in the aqueous phase during the mineralization experiments were determined by plating serial dilutions on LB-plates (LB medium supplied with 15 g of Bacto agar (Difco) per liter) using the drop plate counting technique. Initial biomass densities are mentioned in the result section.

Fluoranthene mineralization was quantified by determination of the amount of $^{14}CO_2$ produced during mineralization of the ^{14}C-labelled fluoranthene as described in Willumsen et al. (1998). The amount of $^{14}CO_2$ collected was quantified using a model LS 1801 scintillation counter (Beckman). Counts in controls were never above the background level.

Bacterial Growth on NAPLs. The ability of the bacterial strains to utilize the model NAPLs as sole sources of carbon and energy was assayed by inoculating triplicate 60 ml glass tubes to an initial OD_{420nm} of approximately 0.05. The tubes contained 15 ml BH-medium amended with single carbon sources at concentrations similar to the one used in the mineralization experiments. The tubes were incubated oblique on a shaker table at 200 rpm at 23 °C. Bacterial growth was assayed periodically by measuring the OD_{420nm}.

Bacterial toxicity assay. The effect of the model NAPLs on the bacterial activity were determined using the ^{14}C-glucose assay described in Willumsen et al. (1998).

RESULTS AND DISCUSSION

Mineralization of Fluoranthene Solubilized in Paraffin. The beneficial effect on the mineralization of paraffin-solubilized fluoranthene relative to crystalline fluoranthene by the five fluoranthene-degrading bacterial test strains increased with increasing cell surface hydrophobicity (Table 1 and Table 2).

TABLE 2. Mineralization of paraffin-solubilized fluoranthene

Strain	Mineralization rates (% of initial ^{14}C/h)			Overall mineralization (% of initial ^{14}C)		
	cry fla	paraffin-fla	paraffin/cry*	hours**	cry-fla	paraffin-fla
EPA505	1.19±0.00	1.22±0.00	1.03	189	77	79
10-1	0.72±0.07	0.84±0.07	1.17	189	63	78
FAn9	0.23±0.02	0.31±0.01	1.35	189	71	42
GJ3P	0.21±0.01	0.63±0.00	3.00	189	58	67
LB307T	0.28±0.00	0.88±0.00	3.14	189	46	50

Initial fluoranthene concentration 20 mg/l; * Ratio between rates obtained with paraffin–solubilized and crystalline fluoranthene; **Final experimental runtime; fla: fluoranthene; cry: crystalline fluoranthene. Initial cell densities: EPA505: $2.0*10^8$; 10-1: $4.0*10^7$; FAn9: $1.5*10^7$; GJ3P: $3.4*10^7$; LB307T: $4.1*10^7$ cfu/ml medium.

As the experiments progressed differences in culture broth appearance were observed. It was visually clear that the more hydrophobic *Mycobacterium* cells adhered to the hydrophobic paraffin surface, the transparent paraffin pearls on the aqueous surface turning yellow. No similar adherence was observed with the more hydrophilic *Sphingomonas* strains. During the mineralization experiment the number of cells of *Mycobacterium spp.* strains FAn9 and LB307T decreased significantly in the aqueous phase of the paraffin-amended cultures relative to the crystalline systems (data not shown). No significant differences in cell numbers in the aqueous phase in the two degradation systems were observed with *Mycobacterium* sp. strain GJ3P and *Sphingomonas* spp. strain 10-1 and EPA505. These results support the visual observations and indicate strain specific adherence to the NAPL-phase. In addition, significant growth of the *Mycobacterium* cells on the paraffin oil as sole source of carbon and energy was observed (data not shown).

Little beneficial effect of the NAPL phase was observed with the *Sphingomonas* strains (Table 2). It has previously been shown that *Sphingomonas spp* take up substrate through the aqueous phase (Willumsen et al. 1998). Thus the results indicates the flux of fluoranthene out of the paraffin phase into the aqueous phase being in the same range as the crystalline solubilization rate. This information suggests that the enhanced fluoranthene mineralization by the more hydrophobic *Mycobacterium* cells in the presence of paraffin oil was not related to enhanced fluoranthene partitioning into the aqueous phase. More likely the effect was associated with an improved cell-substrate contact accomplished by the cell attachment to or possible also growth on the NAPL phase resulting in an increased biomass. The experiment does not allow us to distinguish between the two processes.

Mineralization of Fluoranthene Solubilized in Silicon Oil. Different to the results obtained with paraffin-solubilized fluoranthene the beneficial effect of silicon oil on fluoranthene mineralization decreased with increasing cell surface hydrophobicity (Table 1 and Table 3). However, with exception of strain EPA505 no extensive stimulatory effect of the silicon oil was observed. It has previously been shown that *Sphingomonas spp* take up substrate through the aqueous phase (Willumsen et al. 1998). This may explain why little beneficial effect of a NAPL phase in general was observed with the *Sphingomonas* strains; the flux of fluoranthene out of the paraffin and silicon oil into the aqueous phase possibly being in the same range as the crystalline solubilization rate.

TABLE 3 Mineralization of silicon oil-solubilized fluoranthene

	Mineralization rates (% of initial ^{14}C /h)			Overall mineralization (% of initial ^{14}C)		
Strain	cry fla	silicon-fla	silicon/cry*	hours**	cry-fla	silicon-fla
EPA505	1.00±0.00	1.46±0.00	1.46	186	76	71
10-1	0.83±0.07	0.94±0.07	1.13	186/355	65/72	68/65
FAn9	0.23±0.02	0.22±0.01	0.96	186/355	37/44	35/49
GJ3P	0.21±0.01	0.20±0.00	0.95	186/355	35/48	33/37
LB307T	0.25±0.00	0.20±0.00	0.80	186/355	41/48	28/34

Initial fluoranthene concentration 20 mg/l ; * Ratio between rates obtained with silicon oil–solubilized and crystalline fluoranthene; **Experimental runtime; fla: fluoranthene; cry: crystalline fluoranthene. Initial cell densities: EPA505: $1.6*10^8$; 10-1: $5.0*10^5$; FAn9: $9.2*10^6$; GJ3P: $2.6*10^7$; LB307T: $1.3*10^7$ cfu/ml medium.

The silicon oil was non-toxic and did not support growth of any of the bacterial test strains (data not shown). Different from the paraffin oil it was not possible to detect visually adherence of the *Mycobacterium* cells to the silica oil drops or onto the tar-coated silica beads during the 15 days experiment. In addition, no significant decrease in cell number in the aqueous phase was observed in the silicon oil-amended relative to the crystal-amended degradation system. An exception was *Mycobacterium sp.* strain Fan9 where the no of cells in the aqueous phase increased one order of magnitude within the first 150 hours. Ascon-Cabrera & Lebeault (1993) observed adherence to silicon oil after agtation was stopped. It is possible that the applied speed of agitation (150 rpm) in our experiments was too high to support adherence of the test strains to the applied silicon oil.

In the paraffin- as well as in the silicon oil-amended degradation system the overall extents of mineralization were significantly lower (less of the compound converted to CO_2), for the *Mycobacterium spp.* than for the *Sphingomonas spp.* The data indicate differences in the efficiency of utilization of PAH and suggest that the light NAPL-solubilized fluoranthene is more available for the *Mycobacterium* spp. than for the *Sphingomonas* spp. enabling a more efficient incorporation of fluoranthene carbon into the biomass.

Mineralization of Fluoranthene Solubilized in DNAPL Tar Extract. Both the *Sphingomonas* spp. and the *Mycobacterium* spp. were able to degrade the fluoranthene solubilized in the tar coated silica beads (Table 4). With the *Sphingomonas* spp. the rates of mineralization of the tar-solubilized fluoranthene were only half of the rates obtained with crystalline fluoranthene, whereas for the *Mycobacterium* spp. the rates were almost doubled.

TABLE 4 Mineralization of tar-NAPL-solubilized fluoranthene

Strain	Mineralization rates (% of initial ^{14}C /h)			Overall mineralization (% of initial ^{14}C)		
	cry fla	tar-fla	tar/cry*	hours**	cry-fla	tar-fla
EPA505	1.00±0.00	0.50±0.00	0.50	186	76	65
10-1	0.83±0.07	0.43±0.02	0.52	186/355	65/72	51/74
FAn9	0.23±0.02	0.38±0.01	1.65	186/355	37/44	50/74
GJ3P	0.21±0.01	0.30±0.01	1.43	186/355	35/48	50/77
LB307T	0.25±0.00	0.40±0.01	1.60	186/355	41/48	61/90

Initial fluoranthene concentration 20 mg/l ; * Ratio between rates obtained with tar NAPL–solubilized and crystalline fluoranthene; **Experimental runtime; fla: fluoranthene; cry: crystalline fluoranthene. Initial cell densities: EPA505: $1.6*10^8$; 10-1: $5.0*10^5$; FAn9: $9.2*10^6$; GJ3P: $2.6*10^7$; LB307T: $1.3*10^7$ cfu(ml medium.

Different from the paraffin system no visually cell adherence to the tar-coated silica beads was observed during the 15 days experiment. The tar NAPL did not support growth by any of the test strains (data not shown), however, the effect of the tar NAPL on the activity of the cells, measured as effects on rate and extent of glucose mineralization, varied with the strains (Figure 1).

With the three less hydrophobic strains *Sphingomonas* spp strain EPA505 and 10-1 and *Mycobacterium* sp. strain FAn9 (Table 1), the maximum glucose degradation rates were unaffected, whereas the overall mineralization was significantly enhanced in the presence of the tar NAPL.

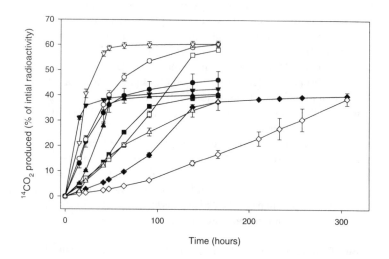

FIGURE 1. Effect of tar-NAPL on glucose mineralization (initial concentration 100 mg/l) by the five PAH-degrading bacterial strains. Solid symbols represent glucose-amended systems and open symbols glucose/tar-amended systems, mean±SC, n=3. ●: strain EPA505; ▼: strain 10-1; ■: strain Fan9; ♦: strain GJ3P; ▲: strain LB307T.

With the two more hydrophobic strains, *Mycobacterium* spp. strain GJ3P and strain LB307T the maximum mineralization rates as well as the overall extent of mineralization were significantly reduced. Some of the compounds in the NAPLs are themselves toxic or growth substrates and become available and potentially inhibitory as they partitioning into the aqueous phase of soils and aquifers. However, the fluoranthene mineralization rates were enhanced by the presence of the tar-NAPL excluding toxic or competitive substrate effects.

Relative to the silicon- and paraffin amended systems the overall extent of fluoranthene mineralization decreased for the *Sphingomonas* spp. and increased for the *Mycobacterium* spp in the tar-NAPL-solubilized fluoranthene system (Table 4). It is not clear why less of the substrate was mineralized by the latter two strains in the more viscous NAPL but similar observations have been reported by Birman & Alexander (1996). Due to the inhomogeneous distribution of the NAPLs in the test systems the amount of ^{14}C converted to biomass was not determined at the conclusion of the experiments. However, one may speculate that the fluoranthene in the tar-NAPL system was more bioavailable for the *Sphingomonas* spp. than the crystalline, silicon- and paraffin-solubilized fluoranthene resulting in more substrate being converted to biomass by these cells. The less ^{14}C-labelled fluoranthene converted to CO_2 by the *Mycobacterium* spp.

may be attributed to a greater persistence of tar-NAPL than of silicon oil and paraffin-solubilized fluoranthene.

In general it is misleading to extrapolate results obtained with light NAPLs, as silicon oil and paraffin, to activities taken place in "natural" NAPL-contaminated soil and ground water systems. However, the results obtained in this study with the dense coal tar NAPL absorbed to the silica bead, possible mimic the stimulatory and inhibitory multiple substrate interactions taken place in a tar contaminated soil system. The results of this work indicate major strain- and NAPL specific differences in degradation of NAPL-solubilized fluoranthene, the cell surface characteristics playing an important role in the degradation performance of the cells. Thus, in order to select bacterial cultures or consortia to increase biodegradation of hydrophobic contaminants in NAPL contaminated soil and ground water systems the bacterial cell surface characteristic should be taken into account.

ACKNOWLEDGEMENT

The work was financed by the Danish Strategic Environmental Research Program (BIOPRO) and from BIOVAB (EU grant no BIO4-CT97-2015) and BIOSTIMUL (EU grant no. QLK3-1999-00326). We thank Anne Grethe Holm-Jensen for technical assistance and Hauke Harms and Lukas Wick for performing the cell surface hydrophobicity analysis..

REFERENCES

Ascon-Cabrera, M., Lebeault, J-M. 1993. "Selection of xenobiotic-degrading microorganisms in a biphasic aqueous-organic system". *Appl. Environ. Microbiol.* vol59(6): 1717-1724.

Bastiaens L.D., Springael D., Wattiau P., Harms H., de Wachter R., Verachtert H., Diels L.. 2000. "Isolation of new polycylic aromatic hydrocarbon (PAH) degrading bacteria using PAH sorbing carriers". *Appl Environ Microbiol* 66: 1834-1843

Birman, I. Alexander, M. 1996. "Effect of viscosity of nonaqueous-phase liquids (NAPLs) on biodegradation of NAPL constituents". *Environ. Tox Chem*, Vol 15(10): 1683-1686.

Deziel E., Comeau Y., Villemur R. 1999. "Two-phase bioreactors for enhanced degradation of hydrophobic/toxic compounds". *Biodegradation* 10: 219-233.

Sambrook J., Fritsch E., Maniatis A 1989. "Molecular Cloning: A Laboratory Manual". Cold Spring Harbour Laboratory, Second edition, Cold Spring Harbour, N.Y.

Willumsen P.A., Karlson U., Pritchard H.P. 1998. "Response of fluoranthene degrading bacteria to surfactant systems". *Appl Microbiol Biotechnol* 50: 475-483

SEQUENTIAL REACTIVITY STUDIES OF B[a]P AND B[e]P BY CHEMICAL AND BIOLOGICAL PROCESSES.

Vito Librando, Francesco Castelli and Maria Grazia Sarpietro (Department of Chemistry, Viale Doria 6, 95127 Catania, Italy - e_mail: vlibrando@dipchi.unict.it)
Michele Aresta (Department of Chemistry, METEA Center, Via C. Ulpiani 27, 70126 Bari)

ABSTRACT: Benzo pyrene, a and e isomers (B[a]P, B[e]P), are polycyclic aromatic hydrocarbons (PAHs) originating from natural and anthropogenic sources and listed as priority pollutants being toxic, mutagenic, and recalcitrant compounds.

This paper focuses on the sequential treatment of these PAHs, involving chemical followed by biological oxidation to reach the following objectives: (i) to obtain mechanistic reactivity aspects of the ozone-initiated degradation of B[a]P and B[e]P in the aqueous phase, (ii) to test the biodegradability of resultant intermediates and (iii) to test the feasibility of the integrated chemical-biological treatment of the five-ring PAHs.

Batch reactors were used, and after different ozonation times, samples containing reaction intermediates and byproducts were collected. The samples were identified for organic contents, then biologically inoculated to determine their biodegradability. The O_3-pretreated samples were incubated from five to twentyfive days; afterward biochemical oxygen demand (BOD), chemical oxygen demand (COD), and toxicity tests were conducted. Further qualitative and quantitative determinations of B[a]P and B[e]P, intermediates, and reaction products by HPLC and GC/MS techniques were conducted. Major intermediates identified at different stages were ring-opened aldehydes, phthalic derivatives, and hydrocarbons. The degradation of B[a]P and B[e]P is primarily initiated via O_3-mediated ring-opening, followed by O_3 and hydrohyl radical fragmentation. The last step is the complete mineralization via $OH°$ radicals.

The proposed approach seems suitable for treating recalcitrant compounds, while pretreatment by chemical oxidation appears useful to generate soluble intermediates from otherwise highly insoluble, biologically unavailable B[a]P and B[e]P.

INTRODUCTION

Polycyclic aromatic hydrocarbons (PAHs), are hydrophobic organic compounds which are commonly found in the environment through the disposal of coal processing wastes, petroleum sludges, asphalt, creosote, and other wood preservative wastes. The decontamination of PAH-polluted sites is of major importance because many PAH compounds are either known or suspected carcinogens and mutagens. Most low molecular weight PAHs are biodegradable in the presence of suitable microbial populations, and a number of bioremediations programs have had some success in the decontamination of PAH contaminated sites, however, high molecular weight PAHs remain recalcitrant. Their hydrophobicity is cited as the primary reason for their persistence in soils and sediments. This is especially true for larger PAH such as benzo pyrene, a and b isomers, B[a]P and B[e]P. They have a very low water solubility (0.003 mg of B[a]P liter $^{-1}$).

Chemical oxidation using O_3 was seen as an alternative treatment for aqueous PAHs. The first studies of benzo[a]pyrene ozonation were carried out in different solvents, but reaction products were not identified. Kinetic studies showed that PAHs were removed within 30 min of ozonation. Pyrene and benz[a]anthracene, in separate studies, were ozonated in acetonitrile:water mixture producing 14 products (from pyrene) and 15 products (from benz[a]anthracene); these products appeared to be primarily from initial ring cleavage. Recently Zeng et al. (2000) identified 25 intermediates and products suggesting both ring cleavage and free-radical reactions after ozonation of excess pyrene in water environment. Integrated chemical and biological processes are potentially more effective than either one alone. Recently experiments were conducted to investigate the impact of ozonation on the degradability of wine distillery wastewater, usually called vinasses, with the goal of developing combined chemical-biological methods for their treatment; in such studies ozone was demonstrated to be an appropriate oxidizing agent to improve vinasses's biodegradability and organic matter removal. The feasibility of treating municipal wastewater by a combined ozone-activated sludge continuous flow system was also studied. Experiments of both single activated sludge and combined ozone activated sludge processes were carried out to determine the kinetic coefficients of the biological stage. The results indicated a clear improvement in the kinetic parameters of the aerobic oxidation when a pre-ozonation stage was applied.

This paper involves the chemical oxidation as a pretreatment and biological treatment in the subsequent step to evaluate the feasibilty of a combined chemical-biological treatment scheme intended for highly insoluble, recalcitrant benzo[a]pyrene, benzo[e]pyrene and other high molecular weight PAHs..

MATERIALS AND METHODS
Chemicals and media - B[a]P and B[e]P were purchased from Aldrich (Milwaukee, Wis.). Pyrene was purchased from Supelco (U.S.A.). Solvents were obtained in analytical grade from BDH Laboratory Supplies (Poole England). Bacteriological media were purchased from Oxoid (Unipath Ltd., Hampshire, England). [^{14}C]Pyrene was purchased from Sigma Chemical Company (St. Louis MO).

The basal salt medium (BSM) was prepared and then autoclavated. Stock solutions of PAH compounds were prepared. When used singularly, PAHs were added to BSM to give 50 mg L^{-1} final concentrations. To test the degradation of the two PAHs supplied in mixtures, BSM was supplemented with a mixed-PAH stock solution to obtain a final concentration of 20 mg L for each PAH.

Chemical treatment – Analytical methods, equipment and procedures were similar to those of Zeng et al. (2000).

Bacteria enrichment and isolation - PAH-contaminated soil samples were collected from an abandoned factory. Enrichment and isolation of PAH-degrading bacteria was carried out using pyrene as a sole carbon and energy source. The bacterial consortium was enriched by shaking 30 g (wet weight) of contaminated soil overnight in a 100 ml volume of Ringer's solution. Part of the supernatant was used to inoculate BSM containing PAH. After several subcultures, the cultures were isolated following the spray plate method. The cultures were tested for their ability to grow on various PAH concentrations. Compounds tested were added to BSM as a sole carbon and energy source. Inoculated BSM without a carbon source as well as uninoculated BSM containing a carbon source served as controls.

Metabolism studies – Cultures of the bacterial isolate were used to study degradation patterns of $[^{14}C]$Pyrene by serum bottle radiorespirometry.

Analysis of metabolites – After the growth period, the cultures were centrifuged to remove cells and then extracted twice with ethyl acetate. The ethyl acetate layers were evaporated with a rotary evaporator and then dryed under nitrogen. Dried samples were redissolved in chloroform methanol (1:1, V/V) for HPLC analysis.

HPLC analysis – Reverse-phase chromatography was performed on a Waters liquid chromatographic system. A linear methanol-water gradient (50 to 100% methanol in 30 min) and a flow rate of 1ml/min were used.

RESULTS
COD and BOD measurements were made for the samples before and after ozonation. COD values were higher in the ozonated samples (both from B[a]P and B[e]P) which indicates dissolution of daughter compounds of B[a]P and B[e]P as a result of ozonation. The intermediates after ozonation were identified and quantified by GC/MS techniques. Many compounds were identified as intermediates and products including ring opened aldheydes, phthalic derivatives, alkanes and alkenes. The composition of compounds varied with ozonation times, compounds with longer column retention time which were present in the early stage of ozonation disappeared in the latest stage indicating a further oxidation of early intermediates in to other products. Long chain aliphatic alkanes compounds disappeared after the first period of ozonation indicating they were fragmented by secondary free-radical oxidants.

In the degradation pathways of B[a]P and B[e]P subject to ozonation, the earlier reaction stage was rich in aromatic and oxygenated intermediates, the latter phase in alkanes and alkenes. The resolved compounds suggest that the degradation is initiated by electrophilic attach of O_3 on of the conjugate rings of the BPs molecules giving the formation of ring opening products and aldehydes. Subsequent reactions of intermediates with O_3 or its concomitant oxygenated radicals resulted in additional oxygenated intermediates. The production of alkenes and long chain aliphatic alkanes is attributed to oxidation reactions prompted by O_3 and $OH°$ or other free radicals.

The biodegradability of intermediates of the ozonated B[a]P and B[e]P were tested using the ozonated column effluent for a 25 day period; the intermediates were qualitatively and quantitatively determined. The effluents were collected from a B[a]P packed column and a B[e]P packed column fed with ozonated water. The effluent did not contain O_3, indicating its complete consumption in the column. The early products of ozonation either decreased or disappeared over the incubation period; the later products increased in the first days of incubation to decrease or disappear by the last days. The results suggest that ozonation makes available water-soluble compounds that are biodegradable from otherwise recalcitrant BP.

Both COD and BOD measurements changed during the incubation period, with BOD values increasing and COD values decreasing (Fig. 1). These results suggest that biodegradable organic compounds were biodegraded over the incubation period.

From the toxicity test conducted during the 25 days incubation period arises that the biodegradable intermediates and by products, including biodegradation products, possess no acute toxic effects to the bacteria (Fig. 2).

In order to determine the efficacy of ozone treatment 160 mg B[a]P and 160 mg B[e]P, each, on a suitable packed column were used. Ozonated water was eluted through the columns. The amount of B[a]P and B[e]P after the ozonation was respectively 126.72 mg (20.8% reduction) and 112 mg (30% reduction). The combined chemical-biological treatment promotes efficient use of chemical oxidant for pretreatment and viable biodegradation of the resulting biodegradable, water soluble intermediates.

Evaluation of effectiveness of B[a]P degradation by bacteria, before and after ozonation, was performed in 30 ml reaction vials containing 10 ml of broth. An appropriate amount of B[a]P or intermediate products of the first time of B[a]P ozonation was added to the vials. The experiments were started by inoculating each vial with washed pyrene-grown cells. The vials were incubated on a rotary shaker at 30° C and 175 rpm. Samples were sacrificed for analysis. Extracts of samples taken from cultures containing B[a]P or intermediate products of the first time of B[a]P ozonation and from killed cell control were analysed by HPLC. The HPLC profiles demonstrated that B[a]P was not degradated whereas intermediate products of the first time of ozonation were degradated by bacteria.

The results agree with those obtained by the COD and BOD measuraments.

The reactions of O_3 and its concomitant $OH°$ radical with recalcitrant B[a]P and B[e]P produce ring cleavage and hydroxylated intermediates that are more soluble and thus more available to further chemical or biological degradation in the aqueous solution.

Microbes can degrade or transform high molecular weight PAHs after their ozonation. These results can have direct implications for PAH containing wastewater degradation.

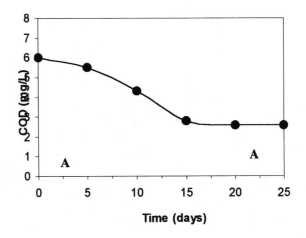

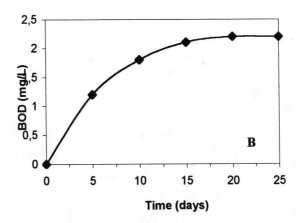

FIGURE 1. (A) COD and (B) BOD changes during the 25-day biological incubation of the ozone-pretreated column effluent.

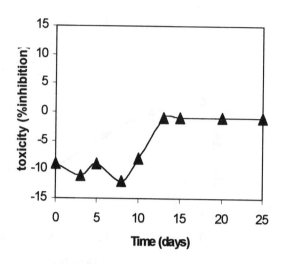

FIGURE 2. Toxicity assessed during the 25-day biological incubation of the ozone-pretreated column effluent (non-toxic range: -10% to +10%).

Acknowledgements

V. Librando is indebted for financial support with MURST, Cluster Project C11 no. 25, Toxicological Impact of power production plants.

REFERENCES

Beltran F.J., Garcia-Araya J.F., Alvarez P.M., 1999. "Wine distillery wastewater degradation. 1. Oxidative treatment using ozone and its effect on the wastewater biodegradability." *J. Agric. Food Chem.* 47: 3911-3918.

Beltran F.J., Garcia-Araya J.F., Alvarez P.M., 2000. "Estimation of biological kinetic parameters from a continuous integrated ozonation-activated sludge system treating domestic wastewater." *Biotechnol. Prog.* 16: 1018-1024.

Cerniglia C.E., 1992. "Biodegradation of polycyclic aromatic hydrocarbons." *Biodegradation* 3: 351-368.

Juhasz A.J., Britz M.L., Stanley G.A., 1996. "Degradation of high molecular weight polycyclic aromatic hydrocarbons by Pseudomonas cepacia." *Biotechnol. Lett.* 18: 577-582.

Kiyohara H., Nagoa K., Yana K., 1982. "Rapid screen for bacterial degrading water soluble, solid hydrocarbons on agar plates." *Appl. Environ. Microbiol.*, 43: 454-457.

Knaebel D.R., Vestal J.R., 1988. "A comparison of double vial to serum bottle respirometry to measure microbial mineralization in soils." *J. Microbiol. Methods* 7:309-317.

Librando V., Liberatori A., and Fazzino D.A., 1992. "Determination of Nitro-Polycyclic Aromatic Hydrocarbons in Fly Ash of Sanitary Wastes Inceneritor." *J. Polycyclic Arom. Compounds* 3: 587-594.

Librando V., and Fazzino D.S., 1993. "The quantification of polycyclic aromatic hydrocarbons and their nitro-derivatives in the atmospheric particulate matter of Augusta city." *Chemosphere* 27(9): 1649-1654.

Librando V., D'Arrigo G., and Spampinato D., 1994. "Reverse phase extraction of arenes and nitroarenes in environmental samples." *Analusis* 22: 340-342.

McElroy A.E., Farrington J.W., Teal J.M., 1989. "Bioavailability of polycyclic aromatic hydrocarbons in the aquatic environment." In Varanasi U. Ed. *Metabolism of polycyclic aromatic hydrocarbons in the aquatic environment*, pp. 1-39. CRC Press Inc., Boca Raton FL..

Trapido M., Veressinina Y., Munter R., 1995. "Ozonation and advanced oxidation processes of polycyclic aromatic hydrocarbons in aqueous solutions: a kinetic study." *Environmental Technology* 16: 729-740.

U.S. Department of Health and Human Services. 1990. "Toxicological profile for polycyclic aromatic hydrocarbons" U.S. Department of Health and Human Services, Washington, D.C.

Wright A.S., 1980. "The role metabolism in chemical mutagenesis and chemical carcinogenesis." *Mutat. Res.* 75: 215-241.

Yao J.J., Huang Z.H., Masten S. J., 1998. "The ozonation of pyrene: pathway and product identification" *Water Research* 32: 3001-3012.

Yao J.J., Huang Z.H., Masten S. J., 1998. "The ozonation of benz[a]anthracene: pathway and product identification" *Water Research* 32: 3235-3244.

Zeng Y., Hong P.K.A., Wavrek D.A., 2000. "Chemical-biological treatment of pyrene." *Water Res.* 34: 1157-1172.

ASSESSMENT OF MASS-TRANSFER LIMITATIONS DURING SLURRY BIOREMEDIATION OF PAHS AND ALKANES IN AGED SOILS

Michael H. Huesemann (Battelle, Sequim, Washington)
Tom S. Hausmann (Battelle, Sequim, Washington)
Tim J. Fortman (Battelle, Sequim, Washington)

ABSTRACT: It is commonly assumed that mass-transfer limitations are the cause of slow and incomplete biodegradation of petroleum hydrocarbons in aged soils. In order to test this hypothesis, the biodegradation rate and the abiotic release rate were measured and compared for selected PAHs and n-alkanes in three different soils. It was found that hexadecane biodegraded rapidly and extensively although very little, if any, was released abiotically from the soil, indicating that n-alkanes do not have to be present in the aqueous phase in order to be taken up and metabolized by alkane degrading microorganisms. By contrast, the abiotic release rate of pyrene was at all times much greater than the respective biodegradation rate in two different soils, indicating that mass-transfer limitations do not control pyrene biodegradation kinetics. In fact, only ca. 35% of pyrene biodegraded in one of the soils despite the fact that it was almost completely bioavailable as implied by the large fraction which was released abiotically. It can therefore be concluded that the slow and incomplete biodegradation of PAHs is not caused by mass-transfer limitations but rather by microbial factors. Consequently, the residual PAHs that remain after extensive bioremediation treatment are still bioavailable and, for that reason, could pose a greater risk to environmental receptors than previously thought.

INTRODUCTION

It is currently assumed that the slow and incomplete biodegradation of polynuclear aromatic hydrocarbons (PAHs) in aged soils is caused by the extremely slow desorption or dissolution rates of these hydrophobic organic contaminants (Alexander, 1995; Huesemann, 1997; Pignatello and Xing, 1996). As a result of these mass-transfer limitations, it has been claimed that the residual PAHs which remain in the soil or sediments after intensive bioremediation treatment are generally not bioavailable to hydrocarbon degrading microbes or other environmental receptors and therefore pose no significant risk to human health or the environment.

It is the objective of this research to determine whether mass-transfer mechanisms limit the rate of PAH and n-alkane biodegradation in different aged petroleum hydrocarbon contaminated soils. More specifically, both the biodegradation rates and the abiotic release rates are measured and compared for selected PAHs and n-alkanes in three different soils undergoing long-term slurry bioremediation treatment.

MATERIALS AND METHODS

Aging Procedure for Soils. A high organic matter soil (Belhaven, TOC=20.2% wt.) and a clay (Cullera, TOC=3.1% wt.) were air dried, sieved (<2mm), sterilized using γ radiation, moisture adjusted, and spiked with a mixture of 16 model hydrocarbons (i.e., n-alkanes and PAHs at a final soil concentration of ca. 1000 mg/kg each) dissolved in ethyl-benzene. After evaporation of the solvent, the moisture of both soils was adjusted to 70% of field capacity. The spiked soil and clay were then transferred to glass jars that were tightly capped and subsequently kept in the dark at room temperature for a 51 month aging period. In addition to these laboratory spiked and aged soils, an aged (> 5 years) Bunker C contaminated sandy field soil from a Navy site at Port Hueneme, CA, was also used in this study.

Slurry Bioreactor Operation and Sampling. All three aged soils were bioremediated in 2L slurry reactors for 91 days. The slurry consisted of ca. 20% solids by weight, and the aqueous medium contained ammonium nitrate and potassium phosphate to maintain a C/N/P ratio of 100/1/0.2 (wt) (Huesemann, 1994). Calcium chloride (1.11 g/L) was also added as a dispersion agent to avoid the clotting of clayey materials, and the pH was maintained between 6 – 7 throughout the treatment via addition of NaOH. The slurry was mixed at around 500 rpm and aerated with humidified air to maintain a dissolved oxygen concentration above 2 mg/L. At specified time intervals, representative slurry samples (ca. 25 mL) were taken from the reactors, transferred to 30 mL glass centrifuge tubes, and centrifuged at 2000 rpm for 30 minutes. The supernatant was carefully removed and stored for future use in the abiotic release rate assay (see below). The wet soil pellet in the bottom of the centrifuge tube was mixed vigorously with a spatula and a 1.5 g subsample was taken for the gravimetric determination of the moisture content (Method 2540 B, 1989). Another representative 1.5 g of wet soil was removed from the centrifuge tube, transferred to a 250 mL Quarpak glass jar and extracted for subsequent hydrocarbon (HC) analyses as described below. The wet soil that remained in the centrifuge tube was used for the determination of abiotic HC release rates as outlined next.

Abiotic Release Rate Assay. The rate of release of PAHs and n-alkanes from soil slurries under abiotic conditions was measured using procedures that were adapted from those reported by Williamson et al. (1998). Approximately 8 g of pre-wetted (= 3 g dry wt) Amberlite XAD-2 resin (Supelco, Bellefonte, PA) and 0.3 g of sodium azide (microbial poison) were added to the centrifuge tube which contained the remaining soil pellet. A specified volume of the reserved supernatant was returned to the centrifuge tube to recreate a slurry with the same solids density as the original bioreactor slurry (i.e., ca. 20% solids by weight). A teflon-lined cap was tightly screwed onto the centrifuge tube which was then rolled on a modified rock roller at ca. 100 rpm. During this intensive mixing process, the XAD resin absorbs "instantaneously" any hydrophobic contaminants that are released into the aqueous phase from the soil particles by either desorption or dissolution. The centrifuge tube is removed from the rollers at the same time as the next slurry sample is taken from the bioreactor. This enables the direct comparison of both biodegradation and abiotic release reates during exactly the same time interval (see also Figures 1-3). In order to facilitate the separation of XAD resin from the soil particles, 3 g of NaCl was added to the centrifuge tube which was then centrifuged three times at 2000 rpm for 30 minutes. Between each centrifugation, the XAD resin floating on top of the supernatant was gently mixed to remove any potential soil particles still adhering to the resin without disturbing the soil pellet. After the third centrifugation, the floating XAD resin and most of the supernatant were poured into a 250 mL Quarpak glass jar. Most of the supernatant was transferred back to the centrifuge tube via a pipette and the clean XAD resin was extracted and analyzed for hydrocarbons as described below. The soil pellet and the supernatant remaining in the centrifuge tube were spun at 2000 rpm for 30 minutes and after carefully decanting the supernatant, the soil pellet was transferred to a 250 mL Quarpak glass jar for extraction and analysis of hydrocarbons as described next.

Hydrocarbon Analyses. For each slurry bioreactor sampling event, a set of three different samples were generated for subsequent HC analyses: a) soil from the bioreactor; b) soil remaining after exposure to XAD resin; and c) XAD resin. Both soils and XAD resin were extracted by adding 100 g of sodium sulfate (drying agent) and 100 mL of methylene chloride to the respective 250 mL Quarpak glass jars and by rolling them for 12 hours at ca. 100 rpm. For the Belhaven and Cullera soils and the respective XAD resins, the resulting solvent extracts were analyzed for PAHs and n-alkanes via GC-MS using standard EPA methods 3540, 3630C, and 8270 as a guideline (EPA, 1986). For the Bunker C contaminated Navy field soil and the respective XAD resin, the extracts were separated into a PAH and alkane fraction using silica column chromatography as described in detail by Wang et al. (1994). Using the gravimetric moisture content of the original wet soil pellet, all HC concentrations were calculated as ng per g of dry soil and then normalized to the initial bioreactor slurry concentration. A comparison of the HC masses in the initial soil slurry sample with the respective HC masses in both the XAD resin and the soil slurry after XAD exposure indicated that with few exceptions, a mass balance of better than 100±10% was achieved for most HCs.

RESULTS AND DISCUSSION

As shown in Figure 1, the normal alkane hexadecane (C_{16}) biodegraded rapidly and more than 80% in Belhaven soil during the first 60 days of treatment. However, when bioreactor slurry samples were subjected to abiotic release rate tests, hardly any hexadecane was desorbed or dissolved from the soil particles and taken up by the XAD resin. In fact, hexadecane biodegradation rates are at all times significantly greater than the respective abiotic release rates, indicating that hexadecane (and other n-alkanes, data not shown) does not have to be transferred into the aqueous phase in order to be biodegraded by soil microorganisms. Apparently, bacteria are able to overcome these serious mass-transfer limitations by either solubilizing n-alkanes with the help of excreted biosurfactants or possibly by facilitating the direct uptake of these highly insoluble compounds across the lipophilic cell membrane (Alexander, 2000, Miller and Bartha, 1989).

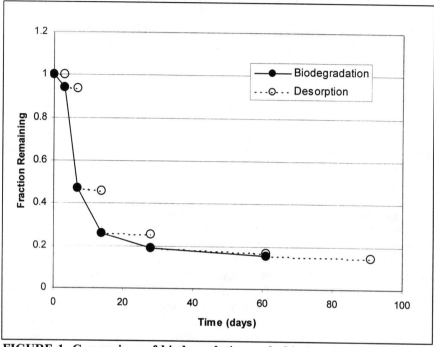

FIGURE 1. Comparison of biodegradation and abiotic desorption rates for C_{16} n-alkanes in Belhaven soil.

A rather different situation was observed in the case of pyrene biodegradation in the aged Bunker C contaminated Navy field soil. As shown in Figure 2, abiotic desorption and/or dissolution rates of pyrene are always greater than the respective biodegradation rates indicating that pyrene biodegradation was not mass-transfer limited at any time during the treatment. Despite the ready bioavailability of pyrene at the beginning of the slurry treatment (i.e., more than 80% is released from the soil abiotically), pyrene biodegradation proceeds rather slowly – most likely because certain microbial factors such as low cell counts or enzymatic inhibitions are limiting biodegradation. It is currently not clear whether pyrene biodegradation has truly leveled off after 60 days of treatment. Additional slurry samples will be taken after 180 and 360 days to determine whether pyrene biodegradation has ceased. If it has, it would be another indication that the incomplete pyrene biodegradation is not caused by lack of bioavailability but rather microbial factors.

Polycyclic Aromatic Hydrocarbons

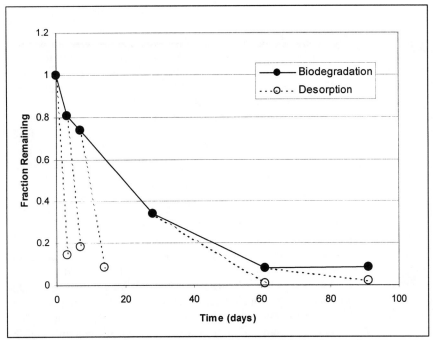

FIGURE 2. Comparison of biodegradation and abiotic desorption rates for pyrene in Bunker C contaminated Navy field soil.

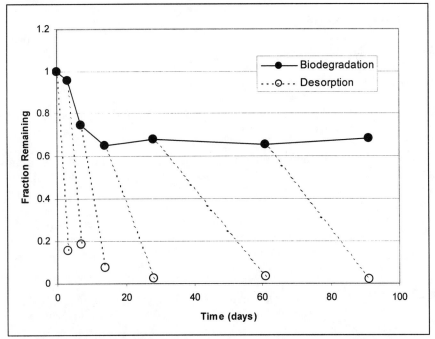

FIGURE 3. Comparison of biodegradation and abiotic desorption rates for pyrene in Cullera soil.

Despite the rapid and extreme abiotic release of pyrene at all times, it is only at most 35% biodegraded in the Cullera clay (see Figure 3). Again, it is clear that bioavailability limitations are not responsible for the incomplete biodegradation of pyrene in this soil. Most likely, some unknown microbial limitations are responsible instead. (Note: The concentration of total heterotrophs and hydrocarbon degraders in this bioreactor was above 10^8/mL, indicating the presence of large microbial populations. However, since we did not specifically assay for the number of pyrene degraders in this bioreactor, it is possible that a lack of pyrene degrading microorganisms is responsible for the slow and incomplete biodegradation of this PAH).

In summary, it can be concluded that mass-transfer limitations have no significant influence on hydrocarbon biodegradation in the aged soils used in this study. The extremely slow release of insoluble alkanes from aged soils into the aqueous phase apparently does not restrain their relatively fast biodegradation by alkane degraders who apparently have developed mechanisms to enhance the uptake of these lipophilic saturated hydrocarbons (Miller and Bartha, 1989). Pyrene is readily bioavailable at all times in the two different soils used in this study. Thus, incomplete pyrene biodegradation cannot be due to mass-transfer limitations but rather to yet unknown microbial factors. As a result, residual PAHs that remain after extensive bioremediation treatment are still bioavailable and for that reason could pose a greater risk to environmental receptors and human health than previously thought.

ACKNOWLEDGEMENTS

This research was supported by the U.S. Department of Energy - National Petroleum Technology Office.

REFERENCES

Alexander, M. 1995. "How Toxic Are Toxic Chemicals in Soil?" *Environmental Science and Technology*, 29(11):2713-2717.

Alexander, M. 2000. "Aging, Bioavailability, and Overestimation of Risk from Environmental Pollutants", *Environmental Science and Technology*, 34(20):4259-4265.

Cornelissen, G., H. Rjgterink, M.M.A. Ferdinandy, and P.C.M. Van Noort. 1998. "Rapidly Desorbing Fractions of PAHs in Contaminated Sediments as a Predictor of the Extent of Bioremediation", *Environmental Science and Technology*, 32(7):966-970.

EPA SW-846, *Test Methods for Evaluating Solid Waste, Physical/Chemical Methods*, U.S. Environmental Protection Agency, Office of Solid Waste, 3rd edition, 1986.

Huesemann, M.H. 1994. "Guidelines for Land-Treating Petroleum Hydrocarbon-Contaminated Soils", *J. of Soil Contamination*, 3(3):299-318.

Huesemann, M.H. 1997. "Incomplete Hydrocarbon Biodegradation in Contaminated Soils: Limitations in Bioavailability or Inherent Recalcitrance?", *Bioremediation Journal*, 1(1):27-39.

Method 2540 B. 1989. "Total Solids Dried at 103 to 105 °C", In *Standard Methods for the Examination of Water and Wastewater*, 17th edition, pp. 2-72-73.

Miller, R.M., and R. Bartha, "Evidence from Liposome Encapsulation for Transport-Limited Microbial Metabolism of Solid Alkanes", *Applied and Environmental Microbiology*, 55(2):269-274, 1989.

Pignatello, J.J., and B. Xing. 1996. "Mechanisms of Slow Sorption of Organic Chemicals to Natural Particles", *Environmental Science and Technology*, 30(1):1-11.

Wang, Z., M. Fingas, and K. Lee. 1994. "Fractionation of a Light Crude Oil and Identification and Quantification of Aliphatic, Aromatic, and Biomarker Compounds by GC-FID and GC-MS, Part I and Part II", *J. of Chrom. Science*, 32:361-382.

Williamson , D.G., R.C. Loehr, and Y. Kimura. 1998. "Release of Chemicals from Contaminated Soils", *J. of Soil Contamination*, 7(5):543-558.

BIODEGRADATION OF ANTHRACENE AND PYRENE IN A LABORATORY-SCALE BIOBARRIER

C. Sartoros (Biotechnology Research Institute-NRC/Université du Québec à Montréal, Montreal, QC, Canada)
L. Yerushalmi (Biotechnology Research Institute-NRC, Montreal, QC, Canada)
P. Béron (Université du Québec à Montréal, Montreal, QC, Canada)
S. R. Guiot (Biotechnology Research Institute-NRC, Montreal, QC, Canada

ABSTRACT: The ability of a developed enrichment culture to degrade two PAHs, anthracene and pyrene, was investigated at $10^{\circ}C$. The overall biodegradation of 2 mg/L PAHs with free cell suspension was over 97%, where 19% anthracene and 62% pyrene were mineralized to carbon dioxide and water. At the initial PAH concentration of 20 mg/L, over 96% PAHs were degraded, where only 9% anthracene and 5% pyrene were mineralized. A laboratory-scale biobarrier, inoculated with the enrichment culture, was evaluated for the removal of PAHs from contaminated water. High removal efficiency of PAHs was maintained in the biobarrier as well as in a control system which was not inoculated with the PAH-degrading consortium. After stabilization of the experimental system, between 94% and 100% removal of PAHs was obtained in both systems. PAH elimination rates ranged from 0.15 to 3.21 mg/L·d for influent PAH concentrations of 0.93 to 6.21 mg/L. The high removal efficiencies obtained in the biobarrier and in the control system were primarily due to the superior sorption capacity of the packing material. Independent mineralization experiments, using the microbial culture withdrawn from the biobarrier, confirmed the biological degradation of contaminants in the biobarrier.

INTRODUCTION

The contamination of soil and groundwater by polycyclic aromatic hydrocarbons (PAHs) is of increasing environmental concern. PAHs persist in the environment due to their low aqueous solubility, low volatility, and their resistance to biological degradation. The treatment of PAH-contaminated sites is imperative because of the potential toxicity, carcinogenicity, and mutagenicity of these hydrocarbons.

In Canada, approximately 67 tons of PAHs are dumped onto soils every year (Environment Canada, 1994). Sources of PAHs in the environment include petroleum industries and refineries, wood preserving industries using creosote, and the coal industry. PAHs may be introduced into the environment through the combustion of fossil fuels, disposal of wastes, or by accidental spills.

The most popular method for the treatment of contaminated soil and groundwater is pump-and-treat. However, field studies have shown that this technology has little success for the treatment of PAH-contaminated sites (Brubaker and Stroo, 1992). Other methods of site decontamination include air sparging, which leads to the volatilization of contaminants, and biosparging, a technology which uses indigenous microorganisms to degrade contaminants.

An alternative technology for the *in situ* treatment of contaminated groundwater is the biobarrier, a permeable bioreactive wall, which is placed across the path of a contaminated plume. It consists of a removable cassette filled with highly permeable packing material, on which a high density of microbial cells is immobilized. As contaminated groundwater moves by natural hydraulic gradient through the biobarrier, the contaminants are removed by a combination of biological degradation and physical/chemical processes. In previous studies for the removal of gasoline from contaminated water, the biobarrier demonstrated removal efficiencies between 86.6% and 99.9% (Yerushalmi et al., 1999).

The objective of the present study was to evaluate the performance of a laboratory-scale biobarrier in removal of PAHs from contaminated water at 10°C. A mixed microbial culture, isolated by enrichment techniques and capable of degrading two model PAHs, anthracene and pyrene, was used in this study.

MATERIALS AND METHODS

Chemicals. Pyrene and anthracene were obtained from Aldrich Chemical Co. (Oakville, ON, Canada). The purity of these chemicals was greater than 99%. Radiolabelled chemicals [4,5,9,10-^{14}C]pyrene (58.7 mCi/mmol) and [1,2,3,4,4A,9A-^{14}C]anthracene (1.0 mCi/mmol) in N,N-dimethylformamide were obtained from Sigma Chemical Co. (Oakville, ON, Canada).

Enrichment of Microbial Culture. The microbial culture used in this study was isolated by enrichment techniques from a sample of PAH-contaminated soil. The enrichment procedure is described elsewhere (Yerushalmi and Guiot, 1998). A stock solution of 5 g/L PAHs in N,N-dimethylformamide (DMF) was used as the carbon and energy source during the enrichment process.

Biodegradation Experiments. Biodegradation experiments were performed in sterile 1-L Erlenmeyer flasks, with a working liquid volume of 250 mL and a 10% biomass inoculum, closed with Teflon^{TM}-lined caps. Initial biomass concentrations were 41-53 mg/L dry weight. PAHs were added from a stock solution of 5 g/L PAHs in DMF for experiments with an initial PAH concentration of 2 mg/L, and 10 g/L PAHs in DMF for experiments starting with 20 mg/L, where pyrene and anthracene were present in equal concentrations. All experiments were carried out in duplicate.

Analysis of PAHs. PAHs were extracted by dichloromethane and were quantified by high-performance liquid chromatography (HPLC). The system included a model W600 pump and a model W717 autosampler from Waters (Milford, MA, USA), a column heater model TCM from Waters, a fluorescence detector model Spectroflow 980 from ABI Analytical (Ramsey, NJ, USA), and an ultraviolet detector model W490 from Waters. A volume of either 20 or 100 μL was separated on a Supelcosil LC-PAH C18 column (15cm X 4.6 mm, with a particle size of 5 µm, Supelco, Bellafonte, PA, USA) with a 40:60 (v/v) acetonitrile:water mobile phase for 5 minutes. This was followed by a 25-minute period in which the acetonitrile fraction was increased to 100%, at a flow rate of 1.5 mL/min. Quantification was performed by UV detection at 254 nm and by fluorescence detection at 280 nm. The detection limit was 0.001 mg/L.

Mineralization Experiments. The ability of the enrichment culture to mineralize PAHs was investigated, where mineralization refers to the complete transformation of PAHs into carbon dioxide and water. Using radiolabelled anthracene and pyrene, the generated $^{14}CO_2$ was measured by liquid scintillation counting. Mineralization experiments were performed at 10°C in 120-mL serum bottles, as described before (Yerushalmi and Guiot, 1998).

Sorption Experiments. The sorption capacity of granulated peat moss for anthracene and pyrene was investigated, where sorption is reported as the amount of PAH sorbed per unit mass of peat moss. Sorption experiments were carried out in 120-mL serum bottles, with Teflon™-lined septa, containing 80 mL MSM and 1 g granulated peat moss. Anthracene and pyrene were added from an 8 g/L PAH stock solution in DMF, to give 10 different concentrations ranging from 1 to 25 mg/L. The experimental procedure and the estimation of sorption capacity of peat moss are described elsewhere (Yerushalmi et al., 1999).

Sorption data were analyzed according to the Langmuir model, as described by the following equation:

$$(x/m) = abC_e /(1+bC_e) \qquad (1)$$

where (x/m) is the amount of PAH sorbed per unit weight of granulated peat moss, a and b are empirical constants, and C_e is the equilibrium concentration of PAHs in solution after sorption. The time to breakthrough for PAHs in the biobarrier was calculated from the following equation:

$$t_b = (x/m)_b M_c /[Q(C_i - (C_b/2))] \qquad (2)$$

where t_b is the time to breakthrough (d), $(x/m)_b$ is the breakthrough capacity of the granulated peat moss (mg/g), M_c is the mass of peat moss in the biobarrier (g), Q is the flow rate (L/d), C_i is the influent PAH concentration (mg/L), and C_b is the breakthrough PAH concentration (mg/L) (Tchobanoglous and Burton, 1991).

Biobarrier Operation. A laboratory-scale biobarrier, inoculated with the PAH-degrading enrichment culture was operated at 10°C in parallel with a control system. The biobarrier was a stainless steel structure (25x20x10 cm) with a rectangular cross-sectional area and a total volume of 5 litres, equipped with two glass windows and several ports (Figure 1). The packing material used in the biobarrier was granulated peat moss (Produits Recyclables Bio-forêt, Quebec, Canada), with particle sizes of 2-4 mm, which was sterilized twice before use. The free space in the biobarrier was 45%, where free space is defined as the volume of void space to the total volume occupied by the packing material.

The control system had the same structure as the biobarrier, but was not inoculated with the enrichment culture. Sources of nitrates and phosphates were removed from the control feed and there was no extra aeration so as to discourage microbial growth.

The performance of the two systems for the treatment of PAH-contaminated water was evaluated by their removal efficiency and elimination rate as described elsewhere (Yerushalmi et al., 1999).

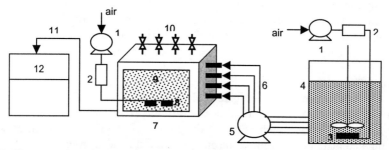

FIGURE 1. Schematic diagram of the biobarrier set-up during continuous operation. 1-pump, 2-air filter, 3-air diffuser, 4-feed tank with PAHs, 5-feed pump, 6-feed lines, 7-biobarrier, 8-air diffusers, 9-packing, 10-top ports, 11-effluent line, 12-effluent tank.

RESULTS AND DISCUSSION

PAH Biodegradation. At a total initial PAH concentration of 2 mg/L, the enrichment culture degraded 99.3% of anthracene and 97.2% of pyrene, after 32 days (Figure 2a). The maximum specific rates of biodegradation were 2.04 and 1.81 mg/g initial biomass·d for anthracene and pyrene, respectively. At a PAH concentration of 20 mg/L, the efficiency of anthracene and pyrene biodegradation was 96.4% and 97.1%, respectively (Figure 2b). The maximum specific

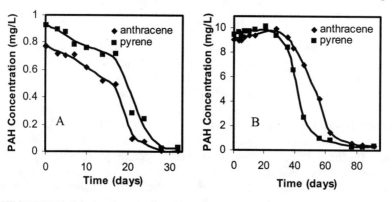

FIGURE 2. Biodegradation of 2 mg/L PAHs (a) and 20 mg/L PAHs (b) by free cell suspension at 10°C.

biodegradation rate increased to 8.99 and 9.49 mg/g initial biomass·d for the two respective PAHs.

With the increase of PAH concentration, there was an extended initial lag phase of 28 days resulting in complete biodegradation within 84 days. The extended lag phase may be due to the toxicity of PAHs to the microorganisms at higher concentrations, a phenomenon also observed by Yuan et al. (2000). However, the enrichment culture still maintained high biodegradation efficiencies with the increase in PAH concentration. Such high removal efficiencies had previously been reported in studies on PAH biodegradation, only performed at higher temperatures between 20°C and 30°C (Bouchez et al., 1996; Rehmann et al., 1998). There are not many reports of PAH biodegradation at low temperatures.

PAH Mineralization. The ability of the enrichment culture to mineralize PAHs at 10°C was investigated. At an initial PAH concentration of 2 mg/L, the culture mineralized 18.5% anthracene and 61.5% pyrene. The maximum specific rates of mineralization were 0.34 and 0.78 mg/g initial biomass·d for anthracene and pyrene, respectively. The higher mineralization efficiency of pyrene may be due to the higher aqueous solubility of pyrene and its metabolites.

The increase of PAH concentration to 20 mg/L had an adverse effect on PAH mineralization. Mineralization efficiency was only 8.7% for anthracene and 5.1% for pyrene. The maximum specific rate of anthracene mineralization increased to 2.49 mg/g initial biomass·d, while that of pyrene decreased to 0.53 mg/g initial biomass·d. PAH mineralization efficiency decreased with the increase in PAH concentration, even though biodegradation efficiencies remained very high. Therefore, the culture retained the ability to degrade PAHs at a higher concentration, but mineralization of PAHs was inhibited. This may be due to the suppression of activity of certain aerobic enzymes as a result of the presence of toxic metabolites, causing incomplete biodegradation of PAHs to carbon dioxide.

Sorption of PAHs on Granulated Peat Moss. Sorption experiments demonstrated that with the increasing equilibrium concentration of anthracene in the solution (C_e), the mass of anthracene sorbed per unit weight of peat moss (x/m) increased. For anthracene concentrations between 0 and 6.3 mg/L, the sorption isotherm followed the Langmuir relationship, as expressed by:

$$(x/m) = (0.517*1.417*C)/[1+(1.417*Ce)] \qquad (3)$$

At anthracene concentrations above 6.3 mg/L, the sorption of anthracene on peat moss followed a linear relationship, expressed by:

$$(x/m) = 0.0765*C_e \qquad (4)$$

The results for pyrene followed the Langmuir model at all concentrations studied, as expressed by the following equation:

$$(x/m) = (1.264*2.035*C_e)/[1+(2.035*C_e)] \qquad (5)$$

The time required for the breakthrough of anthracene in the biobarrier was estimated, using Equation 2. The biobarrier operating conditions included an inlet flow rate of 1.2 L/d and an overall mass of peat moss of 965 g. Assuming an influent PAH concentration of 1 mg/L and a breakthrough equilibrium concentration of 0.2 mg/L anthracene, the predicted time for breakthrough is 102 days. The increase of influent concentration to 2.5 mg/L shortened the breakthrough time to approximately 38 days. The breakthrough time calculated for pyrene in the biobarrier was 327 days for an influent pyrene concentration of 1 mg/L, and 123 days for an influent concentration of 2.5 mg/L.

Biobarrier Performance. During the continuous operation of biobarrier, the hydraulic retention time was maintained at approximately 2 days, which corresponds to a linear liquid velocity of 12.5 cm/d. With influent PAH concentrations of 0.93-2.68 mg/L, corresponding to PAH loading rates of 0.53-6.23 mg/L·d, the biobarrier maintained removal efficiencies of 83.4% to 100% (Figure 3). Lower removal efficiencies were obtained during the first 3 weeks of operation, after which a stabilization of the system yielded removal efficiencies over 97.9%. The elimination rates for this period ranged between 0.15 and 2.07 mg/L·d. The control system also showed high removal efficiency of PAHs. With PAH influent concentrations of 1.09-3.01 mg/L, corresponding to PAH loading rates of 2.57-6.81 mg/L·d, the control system maintained removal efficiencies of 84.1%-100%, with efficiencies of over 98.5% after the initial stabilization. The elimination rate ranged from 0.40 to 1.28 mg/L·d in the control system.

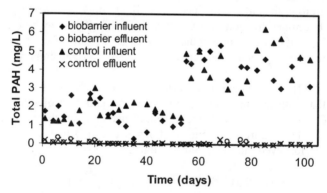

FIGURE 3. Performance of the biobarrier in removal of PAHs from contaminated water.

After 53 days of operation, due to the high removal of PAHs by the biobarrier, the influent PAH concentration was increased to 3.04-5.34 mg/L, corresponding to PAH loading rates of 5.93-10.26 mg/L·d (Figure 3). Removal efficiencies were maintained between 94.0% and 100%. The elimination rates for this concentration range were increased to 1.52-2.78 mg/L·d. For the control system, with the increase of PAH influent concentrations to 2.83-6.21 mg/L, corresponding to loading rates of 5.13-12.02 mg/L·d, removal efficiencies were maintained between 94.6% and 100%. Under these conditions, the elimination rates ranged from 1.53 to 3.21 mg/L·d.

In general, the biobarrier showed a high capacity for the removal of PAHs under the examined PAH concentrations. Similar high removal efficiencies have been demonstrated previously for the biobarrier in the removal of gasoline from contaminated water (Yerushalmi et al., 1999). The biobarrier also demonstrated an ability to adapt to the changes in PAH concentration. High removal efficiency was maintained with the increase in PAH concentration. Similarly, the control system demonstrated a high PAH removal efficiency due to the high sorption capacity of the packing material. This is supported by the long breakthrough times estimated for both systems.

Biodegradation of PAHs in the Biobarrier. The concentration profile of PAHs in the biobarrier is depicted in Figure 4a. PAHs accumulated at the first port of biobarrier with time. After 85 days of operation and with an influent PAH concentration of 4.1 mg/L, the total concentration of PAHs was 96.7 mg/L in liquid samples withdrawn from the first port of the biobarrier. This concentration decreased to 2.8 mg/L at the second port of the biobarrier and continued to decrease with increasing distance from the inlet. In the control system, the accumulation of PAHs was also observed at the first port (Figure 4b), reaching a value of 216.9 mg/L after 85 days, at an influent PAH concentration of 5.9 mg/L. The rate of accumulation of PAHs in the control system was much higher than that in the biobarrier, suggesting the enhanced removal of PAHs by biological degradation in the biobarrier, leading to less accumulation of PAHs.

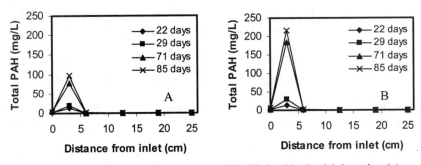

FIGURE 4. Concentration profile of PAHs inside the biobarrier (a) and the control system (b).

In order to confirm the activity of the microbial culture in biodegradation of PAHs, independent mineralization experiments were performed with suspended biomass withdrawn from the biobarrier, at 25°C. At an initial PAH concentration of 2.3 mg/L, the overall mineralization efficiency of anthracene was 71.8% and that of pyrene was 64.7%, after 42 days. At an initial PAH concentration of 21.9 mg/L, the overall efficiency of anthracene mineralization was 54.1%, and that of pyrene was 3.5%, after 19 days. The mineralization efficiency of anthracene by the microbial biomass withdrawn from the biobarrier was greatly enhanced compared to the results obtained with the inoculum under similar conditions (data not shown), demonstrating the adaptation of microbial culture with time. Results for pyrene mineralization were similar to those obtained with the inoculating culture (data not shown). Negligible mineralization was observed with samples withdrawn from the control system.

CONCLUSIONS

The developed enrichment culture demonstrated high biodegradation efficiencies of over 97% for anthracene and pyrene at concentrations of up to 20 mg/L PAHs and at 10°C. Such a superior performance has important implications for the use of enrichment culture for *in situ* bioremediation.

A laboratory-scale biobarrier inoculated with the PAH-degrading consortium demonstrated PAH removal efficiencies ranging from 94% to 100%. Removal of PAHs was primarily due to the high sorption capacity of the packing material. However, biodegradation of PAHs by the microbial biomass in the biobarrier was confirmed by mineralization experiments. This shows that the biobarrier may be a viable technology for *in situ* treatment of PAH-contaminated sites.

REFERENCES

Bouchez, M., D. Blanchet, and J.-P. Vandecasteele. 1996. "The microbiological fate of polycyclic aromatic hydrocarbons: carbon and oxygen balances for bacterial degradation of model compounds." *Appl. Microbiol. Biotechnol.* 45:556-561.

Brubacker, G. R., and H.F. Stroo. 1992. "*In situ* bioremediation of aquifers containing polyaromatic hydrocarbons." *J. Hazard. Mater.* 32:163-177.

Environment Canada. 1994. *Rejets de HAP à l'environnement au Canada: 1990. Document de support n°1 pour le rapport d'évaluation national des HAP.*

Rehmann, K., H.P. Noll, C.E.W. Steinberg, and A.A. Kettrup. 1998. "Pyrene degradation by a Mycobacterium sp. strain KR2." *Chemosphere 36*: 2977-2992.

Tchobanoglous, G., and F.L. Burton (Eds). 1991. *Wastewater Engineering– Treatment, Disposal, and Reuse/Metcalf and Eddy, Inc.* 3rd Edition, McGraw-Hill, Inc., New York, NY.

Yerushalmi, L., and S.R. Guiot. 1998. "Kinetics of biodegradation of gasoline and its hydrocarbon constituents." *Appl. Microbiol. Biotechnol. 49*(4):475-481.

Yerushalmi, L., M.F. Manuel, and S.R. Guiot. 1999. "Biodegradation of gasoline and BTEX in a microaerophilic biobarrier." *Biodegradation 10*(5):341-352.

Yuan, S.Y., S.H. Wei, and B.V. Chang. 2000. "Biodegradation of polycyclic aromatic hydrocarbons by a mixed culture." *Chemosphere 41*(9):1463-1468.

THE STUDY OF POLYCYCLIC AROMATIC HYDROCARBON BIOAVAILABILITY IN SOILS

Lorton, D. M, S. Smith, A.D.G. Jones and J. R. Mason
(King's College London, UK)

ABSTRACT: In this study the efficacy of biological methods for measuring the bioavailability of PAHs was examined by means of luminescent biosensors and ^{14}C-PAH mineralisation. The *lux*-tagged biosensor *Pseudomonas fluorescens* HK44 was able to grow on naphthalene as a sole carbon source. Bioluminescence was induced only by naphthalene. The naphthalene degradation kinetics and substrate specificity of *P. fluorescens* HK44 were investigated further by means of ^{14}C-naphthalene mineralisation assays. A Michaelis-Menten naphthalene K_s value of 45µM was derived for *P. fluorescens* HK44. Naphthalene bioavailability was assessed in the presence of a well-characterised un-contaminated sandy loam soil. Over a period of 1 hour incubation with added culture the mineralisation of added naphthalene was reduced by 30%. Further sorption of naphthalene in the presence of un-inoculated soil for 1 week resulted in an 80% reduction of the initial rate of CO_2 generation by comparison with a non-soil control. In addition the total stoichiometric amount of CO_2 evolved over a longer assay period was also substantially reduced upon sorption. These data may be interpreted in terms of a considerable reduction in naphthalene bioavailability.

INTRODUCTION

Polycyclic aromatic hydrocarbons (PAHs) are ubiquitous environmental contaminants commonly found at former manufactured coal gas production plants (MGP). Many organisms are known to degrade PAHs (Stringfellow, and Aitken, 1994, Bidaud, and Tran-Minh, 1998) and there is strong interest in applying bioremediation to MGP sites. The rate and extent of PAH biodegradation in soils can be limited by the toxicity of contaminants, unfavorable microbial growth conditions and the non-availability of pollutants to microorganisms. PAHs have very low aqueous solubility and exhibit strong partitioning into non-aqueous phases and sorption to organic particles. The partitioning of a compound into solid organic phases, remote micropores or adsorbing onto soil particles is commonly termed sequestration (Hatzinger and Alexander, 1995). There is often an initial rapid and reversible process followed by a period of slow partitioning which reduces bioavailability (Hatzinger and Alexander, 1995).

Interest in bioavailability has increased over recent years and the various factors affecting availability have been studied using a number of approaches (Chung and Alexander, 1998; Carmichael *et.al.*, 1997; Guerin and Boyd, 1997). Extraction methodologies are extensively used to investigate bioavailability, however, a number of studies have shown that they can over-estimate the available fraction (White *et.al.*, 1997; Kelsely *et.al.*, 1997). The measurement of the mineralisation of ^{14}C-labeled PAH compounds to $^{14}CO_2$ allows an indication of the amount of compound that was accessible to the microorganism, i.e. the bioavailable fraction. The induction of degradation pathways can also be used as an indicator of a pollutant's availability (Heitzer *et.al.*, 1992). The luminescent reporter bacterium *P. fluorescens* HK44 has a *nah-lux* reporter plasmid, containing a transcriptional gene fusion between a *lux*CDABE gene cassette from *Vibrio fischeri* and the *nah*G gene (salicylate hyroxylase) of the salicylate operon of the naphthalene degradation pathway (King *et.al.*, 1990). This study aimed to examine the efficacy of microbiological methods and to identify if they provide a more relevant approach to studying bioavailability.

MATERIALS AND METHODS

Preparation of media. Growth media used were Luria Bertani (LB) and Hutner's minimal medium (HMM) (Cohen-Bazire, 1957). Tetracycline (14 mgl^{-1}) was added to all media and agar plates for positive selection of the plasmid pUTK21 which contains the gene for tetracycline resistance, in *P. fluorescens* HK44. The concentration of sodium salicylate in all luminescent assays was 1mM, naphthalene (>99% purity, BDH, UK) was added in crystal form.

Luminescent biosensor assays. Resuscitated *P. fluorescens* HK44 culture was incubated in 25ml LB medium at 30°C and 180 rpm for 24 h. The culture was centrifuged (Sigma Laborzentrifugen 4K15) for 1 minute at 5 500 rpm, and re-suspended in HMM and fumarate (final concentration 10mM) and grown to an OD$_{520nm}$ of 0.35. The cells were centrifuged for 2 min at 5500 rpm, washed and re-suspended in 25ml HMM. To each flask 25 ml of test solution was added and the flasks placed in a water bath shaker (New Brunswick innova 3100) at 30°C, 180 rpm. At 30 minute intervals 1 ml samples were taken and the luminescence measured in a BioOrbit 1251 Luminometer for 30 s. The light readouts were obtained as millivolts. The sample was transferred to a spectrophotometer cuvette and optical density measured at 520nm (Beckman Spectrophotometer DU650).

^{14}C-labeled naphthalene mineralisation assays. *P. fluorescens* HK44 grown overnight from frozen stock was inoculated into LB or HMM with pyruvate (10mM) where specified pre-induction was achieved by the addition of 1mM salicylate, cultures were grown to an OD$_{520nm}$ of 1. The cells were washed as before and resuspended in HMM to an OD$_{520nm}$ 1.2. Naphthalene solution had been prepared by adding 400µl of naphthalene in methanol (6.5gl^{-1}) and 160µl naphthalene-UL-^{14}C (Sigma Aldrich Co, Milwaukee)/methanol solution (780µM naphthalene, 4.97 x 10^{7} dpm ml^{-1}) to 40ml HMM in a Wheaton vial (Sigma Aldrich Co., Milwaukee) shaken and left to mix on ice. Mineralisation assays were conducted by adding 2ml of HMM or soil slurry to a Katz flask. Soil slurry was prepared by mixing 1g soil with 2ml HMM and equilibrated for 2 h. Each Katz flask had a filter paper wick (Whatman No.1 ⌀70mm) in the central glass well soaked with 250µl of 2M NaOH. 2ml of the naphthalene solution was added, the lids tightly sealed and allowed to equilibrate for 1h. Prepared culture (1ml of OD$_{520nm}$ 1.2) was injected through the Teflon septa seal of the Katz flask. Naphthalene concentration in the Katz flask was 200µM and the total dpm per

Katz flask 330 000. Samples were incubated at 30°C and 180 rpm. At intervals, triplicate flasks were destructively sampled by injecting 5ml of 2M HCL through the Teflon septa into the culture/soil slurry phase and the flasks incubated for 30 min before the lids were removed and the flasks vented for 2h. The hydroxide wick was removed to a scintillation vial and the well washed with 0.5ml 2M NaOH and the hydroxide transferred to the same scintillation vial. 15ml of scintillant (Cocktail T 'ScinTran'. BDH, UK) was added and assayed in a scintillation counter.

Kinetic Analysis. 135µl of naphthalene-UL-^{14}C solution (800µM, 5.55 x10^7 dpm ml^{-1}) and 400µl of naphthalene solution (50mM in methanol) were added to 40ml of HMM on ice. The final naphthalene concentration was 250µM. Substrate concentrations of 5, 10, 25, 50, 100 and 250µM were prepared by adding 0.16, 0.4, 0.8, 1.6 and 4ml of labeled HMM respectively into Katz flasks, and made up to 4ml with HMM. 1ml of culture was added to each Katz flask (final OD$_{520nm}$ of 0.26). Samples were destructively sampled in triplicate at 15, 25, 35 and 60 min.

Assessment of naphthalene sorption to soil. An un-contaminated sandy loam soil, (total organic content of 1.5%) was used as a model soil. Soil was sterilized by double cycle autoclaving at 121°C for 25 min, with a 5 day, 25°C incubation period. Plate counts on LB agar showed a 99% kill. Naphthalene solution was prepared by adding 400µl of naphthalene in methanol (6.5gl^{-1}) and 160µl naphthalene-UL-^{14}C/methanol solution (780µM naphthalene, 4.97 x 10^7 dpm ml^{-1}) to 40ml HMM in a Wheaton vial. 40ml naphthalene solution was mixed with 40ml soil slurry and then 4ml of naphthalene/soil mixture transferred aseptically to Katz flasks, the lids sealed and incubated at 30°C, 180rpm. After seven days of sorption, a mineralisation assay was performed by inoculating 1ml of *P. fluorescens* HK44 (OD$_{520nm}$ 1.2) through the Teflon septa. Due to the limited period of the assay and the pre-sterilised nature of the soil tetracycline was not added as a selective agent. Naphthalene distribution between the sediment and aqueous phases was assessed by centrifuging 5ml of soil slurry/naphthalene solution for 20 min at 3000 rpm and measuring the ^{14}C-naphthalene in the supernatant.

RESULTS
Growth and Bioluminescence in *P. fluorescens* HK44. PAH-induced bioluminescence occurred with naphthalene (Figure 1) with peak bioluminescence at 3 hours after a 1 hour lag period. Induction of bioluminescence was not affected by the presence of 10mM pyruvate. No luminescence induction was exhibited with phenanthrene, fluoranthene or pyrene supplemented with 10mM pyruvate.

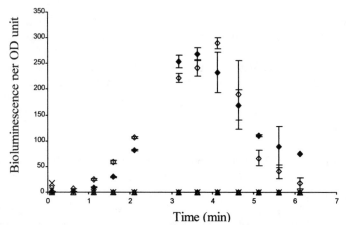

FIGURE 1 Induction of Bioluminescence in *P. fluorescens* HK44 grown at 30°C in HMM saturated with naphthalene (♦) or naphthalene with 10mM pyruvate (◊), phenanthrene with 10mM pyruvate (Δ), fluoranthene with 10mM pyruvate (O), pyrene with 10mM pyruvate (×), and 10mM pyruvate alone (□). All measurements were taken in duplicate; the data points are mean values with ± 1 standard deviation.

^{14}C-PAH Mineralisation Assays. Growth and induction conditions were optimized, and were pre-growth in HMM supplemented with pyruvate (10mM) and pre-induction with 1mM salicylate. Naphthalene mineralisation by *P. fluorescens* HK44 incubated for 1 hour in HMM with 200μM naphthalene and in the presence of soil slurry was observed (Figure 2). The rate of ^{14}C-naphthalene mineralisation in HMM was 24μgh and in the presence of soil slurry the mineralisation rate reduced to 15 μg/h.

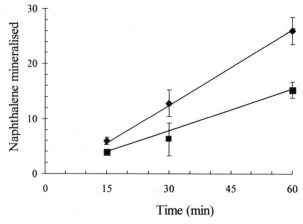

FIGURE 2 Mineralisation of naphthalene by *P. fluorescens* HK44 incubated at 30°C, 180 rpm with 200μM naphthalene in HMM (υ) and 200μM naphthalene in soil slurry (v). Error bars are ±1 standard deviation.

Kinetic Analysis. Naphthalene mineralisation by *P. fluorescens* HK44 followed Michaelis-Menten kinetics (Figure 3). Eadie-Hoftsee (insert figure 3) linear regression analysis derived a K_s value of 45μM. Results where adjusted for losses due to volatilisation.

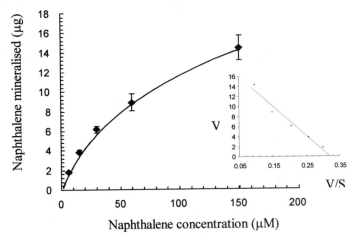

FIGURE 3 Effect of substrate concentration on ^{14}C-naphthalene mineralisation rates by *P. fluorescens* HK44 was incubated at 30°C, 180 rpm for 1 hour with naphthalene from 5 to 250μM. K_s value of 45μM was derived from the Eadie-Hofstee plot (Inset). Error bars are ±1 standard deviation.

Effect of Sorption of Naphthalene to Soil. Measurements of ^{14}C-naphthalene incubated in soil slurry show that there was an initial rapid loss from the amount measured in the aqueous phase (Figure 4) that could not accounted for by volatilisation losses.

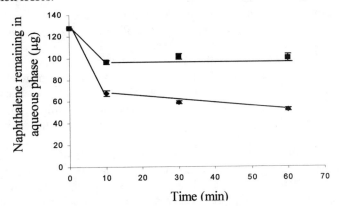

FIGURE 4 Distribution of ^{14}C-naphthalene remaining in aqueous phase of supernatant following incubation in Hutner's minimal medium (v) and soil slurry (♦). Error bars are ±1 standard deviation.

Estimation of Available Naphthalene. Mineralisation of naphthalene by *P. fluorescens* HK44 was measured after inoculation into HMM with 200μM naphthalene, soil slurry where 200μM naphthalene had been added 1hour prior to the assay and where the 200μM naphthalene-soil slurry had been incubated for one week. Considerable differences where observed in the rate of mineralisation and total amount of ^{14}C-naphthalene mineralised (Table 1).

Assay medium	Evolved CO_2 (μmol)	Initial Naphthalene mineralisation rate (μg/h)
Hutner's minimal medium	393	24
Soil slurry and naphthalene (sorbed for 1 hour)	262	15
Soil slurry and naphthalene (sorbed for 1 week)	37	3

TABLE 1 Mineralisation of Naphthalene by *P. fluorescens* HK44

DISCUSSION

Bioluminescent assays. Naphthalene induced bioluminescence was demonstrated in the presence or absence of pyruvate, which indicates that pyruvate does not catabolite repress the induction of bioluminescence. Pyruvate was used as a non-repressing carbon source to investigate the induction of luminescence by other PAHs that are not able to support growth. However, the high PAH-substrate specificity of *P. fluorescens* HK44 luminescent induction precluded its use as an indicator for PAHs other than naphthalene.

^{14}C-PAH Mineralisation Assays. As an alternative assessment of PAH bioavailability the kinetics of naphthalene mineralisation by *P. fluorescens* HK44 was examined using ^{14}C-naphthalene mineralisation assays. Michaelis-Menten kinetics for individual strains have been reported previously (Stringfellow and Aitken, 1995; Goshal and Luthy, 1998). Growth and induction conditions were standardised and the effect of naphthalene concentration on mineralisation by *P. fluorescens* HK44 was shown to follow Michaelis-Menten kinetics with a K_s value of 45μM. The low K_s value infers a high substrate affinity, indicative of an organism adapted to efficient scavenging of naphthalene from the environment.

Effect of sorption on naphthalene mineralisation in soil slurries. A number of studies have reported a reduction in the availability of PAHs aged in soil (Alexander 1995; Kelsey *et al.*, 1997). In this study the reduction in total CO_2 evolved and mineralisation rates infer a reduced bioavailability of naphthalene. The initial rate of naphthalene mineralisation may be influenced by the substrate concentration and Michaelis-Menten kinetics. However, the overall amounts of CO_2 evolved would not be affected by a Michaelis-Menten kinetic effect but by reduced bioavailability. The results of the present study would indicate that the process of naphthalene sorption in soil slurry is considerably rapid with up to an 80% decrease in the bioavailable fraction of naphthalene occurring after only 1 week sorption.

ACKNOWLEDGEMENTS

EPSRC – for financial support (Grant – GR/L74606)
Prof. C Thurston and Dr. C Robinson for constructive advice and encouragement.

REFERENCES

Alexander, M. 1995. How toxic are toxic chemicals in soil? Environmental Science and Technology. 29(11): 2713-2717

Bidaud, C., and Tran-Minh, C. 1998. Polycyclic aromatic hydrocarbons (PAHs) biodegradation in the soil of a former gasworks site: selection and study of PAHs-degrading microorganisms. Journal of Molecular Catalysis B: Enzymatic. 5: 417-421

Carmichael, L.M., Christman, R.F., and Pfaender, F.K. 1997. Desorption and mineralisation kinetics of phenanthrene and chrysene in contaminated soils. Environmental Science and Technology. 31(1): 126-132

Chung, N., and Alexander, M. 1998. Differences in sequestration and bioavailability of organic compounds aged in dissimilar soils. Environmental Science and Technology. 32(7): 855-860

Cohen-Bazire, G., W. R. Sistrom, and R. Y. Stanier. 1957. Kinetic studies of pigment synthesis by non-sulphur purple bacteria. J. Cell. Comp. Phys. 49: 25-68.

Goshal, S., and Luthy, R.G. 1998. Biodegradation kinetics of naphthalene in noaqueous phase liquid-water mixed batch systems: comparison of model predictions and experimental results. Biotechnology Bioengineering. 57: 356-366

Guerin, W.F., and Boyd, S.A. 1997. Bioavailability of naphthalene associated with natural and synthetic sorbants. Water Research. 31(6): 1504-1512

Hatzinger, P.B., and Alexander, M. 1995. Effect of sorption of chemicals in soil on their bioavailability and extractability. Environmental Science and Technology. 29: 537-545

Heizter, A., Webb, O.R., Thonnard, J.E., and Sayler, G.S. 1992. Specific and quantitative assessment of naphthalene and salicylate bioavailability by using a bioluminescent catabolic reporter bacterium. Applied and Environmental Microbiology. 58(6): 1839-1846

Kelsey, J.W., Kottler, B.D., and Alexander, M. 1997. Selective chemical extractants to predict bioavailability of soil-aged organic chemicals. Environmental Science and Technology. 31(1): 214-217

King, J.M.H., Digrazia, P.M, Applegate, B., Burlage, R., Sanseverino, J., Dunbar, P., Larimer, F., and Sayler, G.S. 1990. Rapid sensitive bioluminescent reporter technology for naphthalene exposure and biodegradation. Science. 249: 778-781

Luthy, R.G., Aiken, G.R., Brusseau, M.L., Cunningham, S.D., Gschwend, P.M., Pignatello, J.J., Reinhard, M., Traina, S.J., Weber, W.J., and Westall, J.C. 1997. Sequestration of hydrophobic organic contaminants by geosorbants. Environmental Science and Technology. 31: 3341-3347

Stringfellow, W.T., and Aitken, M.D. 1994. Comparative physiology of phenanthrene degradation by two dissimilar pseudomonads isolated from a creosote-contaminated soil. Canadian Journal of Microbiology. 40: 432-438

Stringfellow, W.T., and Aitken, M.D. 1995. Competitive metabolism of naphthalene, methylnaphthalenes and fluorene by phenanthrene-degrading Pseudomonads. Applied and Environmental Microbiology. 61(1): 357-362

White, J.C., Kelsey, J.W., Hatzinger, P.B., and Alexander, M. 1997. Factors affecting sequestration and bioavailability of phenanthrene in soils. Environmental Toxicology & Chemistry. 16(10): 2040-2045

BIOLOGICAL TREATMENT OF PAH CONTAMINATED SOIL EXTRACTS

Benoît Guieysse and Bo Mattiasson (Lunds University, Lund, Sweden)
Staffan Lundstedt and Bert van Bavel (Umeå University, Umeå, Sweden)

ABSTRACT: The biological treatment of a soil extract sample from a former Manufacturing Gas Plant was tested in several two-phase liquid systems. Silicone oil, dodecane, tetradecane and hexane were first tested as organic phase for the removal of phenanthrene, acenaphthene and pyrene by a well-adapted phenanthrene degrading community. The contaminants were removed in the same order of magnitude in all the biphasic systems showing that the four solvents were suitable. The same microflora was then tested for the treatment of a real soil extract sample in silicone oil. Phenanthrene was totally removed after 20 days of incubation showing no inhibition effect of other extract components on the microbial community. A new microbial consortium, isolated from the same contaminated soil, was then tested with the four organic solvents for the treatment of soil extract. Only silicone oil allowed PAH removal and strong evidence was found that organic compounds other than the PAHs were present in the extract and competed with the pollutants as carbon source. This should be taken into consideration in future studies aiming at combining soil extraction with biological treatment.

INTRODUCTION

Polycyclic Aromatic Hydrocarbons (PAHs) constitute or two- or more fused aromatic rings. They originate from the incomplete combustion of hydrocarbons and are therefore found in association with many industrial activities such as coal gasification and wood preservation (Bouchez et al., 1996a). Several PAHs are toxic, mutagenic and/or carcinogenic and can persist in the environment over very long periods of time. This explains why the US-EPA named 16 PAHs on their priority pollutant list and why the remediation of contaminated sites must be undertaken (Keith and Telliard, 1979).

Although many PAH-utilizing organisms have been isolated in the last decades (Cerniglia, 1992), the natural attenuation of PAHs is often limited by their low aqueous solubility, high boiling temperatures and binding capacity to soil particles (Bouchez et al., 1996b). In some cases, their extraction might become necessary to remediate contaminated sites but these techniques generate a new type of pollution in the form of a concentrated mixture of contaminants that must also be treated. Two-phase partitioning reactor systems, also denominated biphasic reactors, offer a logical and particularly adapted response to this problem: logical since soil extracts are often non soluble in water and particularly adapted since biphasic reactors allow to limit toxicity effects and to improve substrate dispersion.

Biphasic systems are based on the use of a non-miscible and safe organic solvent that is introduced together with the aqueous phase in the biological reactor (Daugulis, 1997). Minerals and microorganisms are present in the aqueous phase whereas the organic solvent serves to deliver a toxic pollutant at sub-inhibitory levels in the aqueous phase. By properly selecting the organic phase, a self-regulated *in-situ* solvent delivering system is created, allowing the degradation of toxic pollutants at unprecedented loads. Thus, the feasibility of the combined physical-biological treatment (soil extraction and biological two-liquid-phase reactor) was recently proven for BTX compounds (Collins and Daugulis, 1999). Biphasic reactors were then tested for the degradation or recalcitrant and scarcely water-soluble pollutants (Marcoux et al., 2000). Since the biodegradation of such compounds is often limited by their poor bioavailability, it was believed that the organic phase could increase their mass transfer to the aqueous phase (Déziel et al., 1999). Thus, several cases of improved degradations rates have been reported (Marcoux et al., 2000). However, certain authors have also shown that the presence of non-miscible organic phase could disturb the process by sequestering the pollutants or competing with them as carbon source.

In this context, a new attempt was done for the degradation of PAHs in two-phase biphasic reactors. An extract of the pollutants was provided by Accelerated Solvent Extraction (ASE) of a soil sample from a former MPG (Husarviken site, Stockholm, Sweden). Four organic phases were tested: silicone oil, dodecane, tetradecane, and hexadecane.

MATERIALS AND METHODS

All experiments were performed using the same mineral salt medium (MSM) of the following composition (mg/l): K_2HPO_4 4×10^3, Na_2HPO_4 4×10^3, KNO_3 3×10^3 $CaCl_2 \cdot 7H_2O$ 0.075, $MgSO_4 \cdot 7H_2O$ 200, $FeSO_4 \cdot 7H_2O$ 1, $MnCl_2 \cdot 4H_2O$ 5.5, $ZnCl_2$ 0.68, $CoCl_2 \cdot 6H_2O$ 1.2, $NiCl_2 \cdot 6H_2O$ 1.2, $CuCl_2 \cdot 2H_2O$ 0.85, H_3BO_3 3.1×10^{-3}, $Na_2MoO_4 \cdot 2H_2O$ 1.2×10^{-2}, $NaSeO_3 \cdot 5H_2O$ 1.3×10^{-2}, $NaWO_4 \cdot 2H_2O$ 1.65×10^{-2}. The pH was adjusted to 7.5 with 0.5 M HCl.

Solvent screening. A first series of tests was performed using a well-adapted phenanthrene degrading bacterial consortium (Guieysse et al., 2000). Acenaphthene, phenanthrene and pyrene were provided as model contaminants as crystals, or dissolved in silicone oil (Poly(dimethylsiloxane) 200 Fluid, 20 centistokes), dodecane, tetradecane or hexadecane. All tests were performed in 10 ml test tubes sealed with Teflon coated caps. Four tubes were prepared for each test, two being inoculated and two being provided with NaN_3 as growth inhibitor to estimate the potential pollutant abiotic loses.

The tubes were prepared as follows: 25 µl of a 5g/l stock solution of the three PAHs in acetone was added to each tube and the acetone was let for evaporation. Then 250 µl of organic phase (when needed) and 4750 µl of MSM were then added to each tube. The tubes were allowed to equilibrate for approx. 10 min before they were inoculated with 250µl of inoculum or provided with 250 µl of a 10 g/l solution of NaN_3. The tubes were incubated at room temperature for 5 days before being extracted with 5 ml of pentane containing 50 mg/l fluoranthene as an internal standard. The tubes were shaken manually for one min, ultrasonicated at 40 °C for 15 min and finally cooled on ice and centrifuged. The pentane extracts were then analyzed by HPLC-UV.

TABLE 1: PAH concentration and composition in the Husarviken soil.

PAHs	Concentration (mg/kg TS)	Percentage of the total PAH%
Naphthalene	3.9	0.5
2-Methylnaphthalene	3.7	0.5
1-Methylnaphthalene	3.1	0.4
2,6-Dimethylnaphthalene	2.7	0.3
Acenaphthylene	8.4	1.0
Acenaphthene	0.9	0.1
2,3,5-Trimethylnaphthalene	0.5	0.1
Fluorene	14.1	1.7
Phenanthrene	96.5	11.9
Anthracene	23.0	2.8
1-Methylphenanthrene	5.5	0.7
Fluoranthene	154.3	19.0
Pyrene	108.1	13.3
Benzo(a)anthracene	58.2	7.2
Chrysene	54.5	6.7
Benzo(b)fluoranthene	60.7	7.5
Benzo(k)fluoranthene	41.4	5.1
Benzo(e)pyrene	37.7	4.7
Benzo(a)pyrene	45.4	5.6
Perylene	14.1	1.7
Dibenz(a,c)anthracene	10.6	1.3
Indeno(c,d)pyrene	32.0	3.9
Benzo(g,h,i)perylene	29.7	3.7
Total	810.5	100

Soil extract treatment feasibilty. The phenanthrene-utilizing microbial community was also tested for the treatment of a soil extract from the former MGP facility of Husarviken (Stockholm, Sweden). Forty grams of dried and finely ground soil were extracted according the procedure described by Lundstedt et al. (2000) using a Dionex ASE 200 Accelerated Solvent Extractor with a mixture of hexane and acetone (1:1 vol.). Extraction was performed at 150°C and 14 MPa. The soil extract was then reduced to a volume of approximately five ml and mixed with 100 ml of silicone oil. The composition of the extract was determined by GC-MS and is presented in Table 1. Degradation tests were performed in two 50 ml flasks sealed with Viton caps and filled with 10 ml of MSM, 5 ml of silicone oil-extract. One flask was then provided with 0.5 ml of inoculum and one flask with 0.5 ml of a 10g/l NaN_3 solution to serve as control experiment. The flasks were inoculated at room temperature during two weeks. After the phases were let for separation, one ml of organic phase was withdrawn from each flask and transferred to two 10-ml test tubes. The contaminants were then extracted with five ml of an acetone:ethanol:chloroform mixture (10:10:2

v/v). The samples were shaken manually for two mins and cooled on solid carbon dioxide to accelerate the phase separation.

Soil extract treatment. A second inoculum was isolated from a contaminated soil sample from the former MGP site of Husarviken. The inoculum was first tested in a silicone oil biphasic system on naphthalene, phenanthrene and pyrene (300 mg/l each). After one month of cultivation, 70% of pyrene remained in the system, the naphthalene and phenanthrene were completely removed.

45 grams of contaminated soil were extracted using the ASE procedure described above with the difference that extraction was performed with hexane only. The extract was then transferred into 100 ml of acetone. The composition of this extract was found similar to the one listed in Table 1. Portions of two ml of acetone extract were then transferred to 10 ml vials, mixed with four ml of silicone oil, dodecane, hexadecane or tetradecane. The acetone was evaporated before the extracts were transferred to 50 ml test tubes sealed with Teflon coated caps. They were then mixed with 10 ml MSM and inoculated with one ml of inoculum. One control test was prepared with dodecane as organic phase and provided with one ml of a 10 g/l NaN_3 solution as growth inhibitor. The tubes were incubated during 21 days and were aerated approximately two hours every day. Samples of 0.1 ml were removed from the organic phase once every third day and mixed with 0.9 ml pentane for HPLC analysis.

RESULTS AND DISCUSSION
Solvent screening. Figure 1 shows the remaining levels of acenaphthene, phenanthrene, and pyrene when provided as crystal or dissolved in an organic phase. Phenanthrene was totally removed from all the biphasic systems but remained at approximately 5 mg/l in the test in which it was provided in crystalline form. Additional experiments later showed that, when added as crystals, part of the phenanthrene tended to adsorb on the biomass, and could thus become less available for biodegradation. In this sense, the biphasic systems prevent PAH bioaccumulation onto cells. The use of an organic phase did not enhance the degradation of acenaphthene and pyrene, both of which that were removed in the same order of magnitude in all systems (Figure 1). This also shows that the contaminants were not sequestered into the organic phase since they were degraded as much as when they were furnished as crystals.

Additional inoculated tests were performed in PAH-free biphasic systems for the each solvent. Signs of microbial growth were detected when the alkanes were used as organic phase, but not with the silicone oil. However, this did not seem to affect the removal of the PAHs showing that the alkanes were still suitable as organic phases. An explanation could be that the PAHs are normally preferred as substrates by this well-adapted microflora. It was therefore decided to test the four solvents for the treatment of the real soil extract sample.

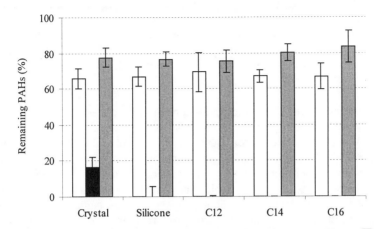

FIGURE 1. Remaining levels (%) of acenaphthene (□), phenanthrene (■) and pyrene (▨) provided as crystal or dissolved in silione oil, dodecane (C12), tetradecane (C14) or hexadecane(C16).

Soil extract treatment feasibility. The biological treatment of a mixture of contaminants is often much more difficult to achieve than the degradation of each individual compound (Stringfellow and Aitken, 1995). In addition, soil extract can contain many compounds other than PAHs, contaminant or natural components that might interfere with the degrading process. It was therefore decided to test the feasibility of the process with only one solvent. Silicone oil was chosen since it had been proven to be suitable for the degradation of PAHs (Marcoux et al., 1995). The high substrate-specificity of the phenanthrene degrading community was confirmed as only phenanthrene was almost completely degraded after two weeks of incubation. By comparison, anthracene was removed to 50% and the other PAHs were not affected (Figure 2).

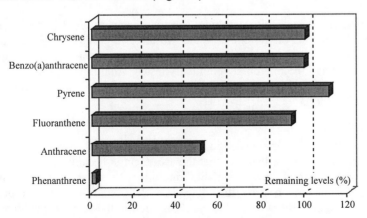

FIGURE 2. Remaining level of PAHs (%) in the soil extract after biological treament in the presence of silicone oil.

The microbial community did not seem affected by the presence of other PAHs or any other organic compounds that might have been extracted from the contaminated soil. It is also important to notice that silicone oil would not be suitable for direct soil extraction and was only used here as model. It was also difficult to transfer the acetone/hexane extract into silicone oil, and precipitation occurred.

Soil extract treatment. The biological treatment of the soil extract appeared to be limited by the low ability of the microflora used to degrade the high molecular weight PAHs (4 rings and more). Hence, a new attempt was done with a newly isolated microbial consortium. Microbial growth occurred in all the inoculated tests and was clearly observed after 6 days of incubation when the organic phases started to emulsify. None of high molecular weight PAHs was removed from any of the systems (data not shown). Silicone oil was the most suitable solvent since it supported the complete removal of phenanthrene in 9 days (Figure 3). By comparison, phenanthrene was removed to 25 %, 0 % and 45 % in the tests performed with dodecane, tetradecane and hexadecane respectively. HPLC analysis revealed the presence of two peaks of high intensity at the beginning of the chromatograms. These two unknown peaks eluted in the same time range as the 2-rings PAHs but it was latter shown from GC-MS analysis that they did not represent any PAHs. It was unfortunately impossible to characterize these substances with the analytical instrument used during this experiment. They were named Substance A and B and were likely low molecular weight aromatics that originated from the soil organic matter. However, the interesting fact is that the first substance was clearly biologically degraded in the tests performed with alkanes as organic phase. It is therefore believed that the competitive metabolization of this substance could explain why the phenanthrene was only partially removed in the tests with alkanes (Morrisson and Alexander, 1997). This could also be the reason why the phenanthrene removal rate in the hexadecane system started to decrease after day 6, precisely when degradation of the substance A began.

There is no clear explanation why substrate competition effects were not observed in the silicone oil system were the two unkown products were found in similar amounts. This might be attributed to the properties of silicone oil in regard to the pollutants and the microbial community tested. Additionally, the potential biodegradation of the alkanes should also be taken into consideration (Morrisson and Alexander, 1997).

CONCLUSIONS

This study represents in our knowledge the first attempt of biological remediation of real soil extract samples. It shows that this treatment is feasible in an artificial model system, provided that a suitable microflora is used. However, two important problems remain. First, the competitive degradation of natural organic substances by the microflora limits the metabolization of the target contaminants. Secondly, silicone oil still appears to be the most suitable solvent in such application since it does not tend to sequester the contaminants or to be competitively used as organic source. Unfortunately, silicone oil is also much more expensive and is therefore non-realistic for large-scale environmental applications.

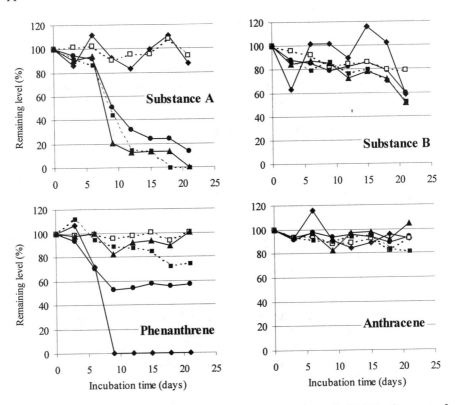

FIGURE 3. Remainings levels (%) of substances A and B, phenanthrene and anthracene as function of the incubation time (days) in the control (□), silicone oil (♦), dodecane (■), tetradecane (▲), hexadecane (●) biphasic systems.

ACKNOWLEDGMENTS

This work was supported by MISTRA (Foundation for Strategic Environmental Research) and was part of the COLDREM project (Soil Remediation in a Cold Climate.

REFERENCES

Bouchez, M., F. Blanchet, and J. P. Vandecasteele. 1996a. "Les hydrocarbures aromatiques polycycliques dans l'environnement: Propiété, origines, devenir." Revue de l'Institut Français du Pétrole. 51(3):407-419.

Bouchez, M., F. Blanchet, and J. P. Vandecasteele. 199b. "Les hydrocarbures aromatiques polycycliques dans l'environnement: La degradation par voie microbienne." Revue de l'Institut Français du Pétrole. 51(6):797-828.

Cerniglia, C. E. 1992. "Biodegradation of Polycyclic Aromatic Hydrocarbons." Biodegradation 3:351-368

Collins, L. D., and A. J. Daugulis. 1999. "Simultaneous Biodegradation of Benzene, Toluene, and p-Xylene in a Two-Phase Partitioning Bioreactor: Concept Demonstration and Practical Application." Biotechnol. Prog. 15:74-80.

Daugulis, A. J. 1997. "Partitioning Bioreactors." Curr. Opin. Biotechnol. 8:169-174.

Déziel, E., Y. Comeau and R. Villemur. 1999. "Two-Liquid-Phase Bioreactors for Enhanced Degradation of Hydrophobic/Toxic compounds." Biodegradation 10:219-233.

Efroymson, R. A., and M. Alexander. 1995. "Reduced Mineralization of Low Concentrations of Phenanthrene because of Sequestering in Nonaqueous-Phase Liquids." Environ. Sci. Technol. 29(29):515-521.

Guieysse B., I. Bernhoft, B. E. Andersson, T. Henrysson, and B. Mattiasson. 2000. "Degradation of Acenaphthene, Phenanthrene and Pyrene in a Packed-Bed Biofilm Reactor.". Appl. Microbiol. Biotechnol. 24:826-831.

Keith, L. H., and W. A. Telliard. 1979. "Priority Pollutants I-a Perspective View." Environ. Sci. Technol. 13:416-423.

Lundstedt, S., B. van Bavel, P. Haglund, M. Tysklind, and L. Öberg. 2000. "Pressuured Liquid Extraction of Polycylic Aromatic Hydrocarbons from Contaminated Soils." J. Chromatogr. A 883(2000)151-162.

Marcoux, J., E. Déziel, R. Villemur, F. Lépine, J. G. Bisaillon, and R. Beaudet. 2000. "Optimization of High-Molecular-Weight Polycyclic Aromatic Hydrocarbons Degradation in a Two-Liquid-Phase Bioreactor." J. Appl. Microbiol. 88:655-662.

Morrison, D. E., and M. Alexander. 1997. "Biodegradability of Non-Aqueous-Phase Liquids affects the Mineralization of Phenanthrene in Soil because of Microbial Competition." Environ. Toxicol. Chem. 16(8):1561-1567.

Stringfellow, W., and M. D. Aitken. 1995. "Competitive Metabolism of Naphthalene, Methylnaphthalenes and Fluorene by Phenanthrene-degrading Pseudomonads." Appl. Environ. Microbiol. 61:357:362.

AVAILABILITY AND BIOSLURRY TREATMENT OF PAHS IN CONTAMINATED DREDGED MATERIALS

Jeffrey W. Talley (University of Notre Dame, South Bend, Indiana)
Upal Ghosh and Richard G. Luthy (Stanford University, Stanford, California, USA)

ABSTRACT: This work applied new investigative techniques to assess the locations, distributions, and associations of polycyclic aromatic hydrocarbons (PAHs) in dredged harbor sediment. Dredged materials from the Milwaukee Confined Disposal Facility were collected and homogenized to provide sufficient sample for four month bioslurry treatment testing and for PAH analyses on various size and density fractions before and after biotreatment. Sediment PAH analyses included both whole-sample measurements and, most importantly, the determination of PAH distribution by sediment particle size and type. Physicochemical analyses included room temperature Tenax bead aqueous desorption experiments and thermal program desorption-MS studies to assess PAH binding energies on sediment particle types. Thermal programmed desorption-MS experimental protocols and data reduction techniques were developed to evaluate apparent PAH binding activation energies on sediment particles. Microbial ecology testing used polar lipid fatty acid (PLFA) and DNA procedures and radiolabel microcosm studies. Earthworm bioassays studied the acute toxicity effects and PAH bioaccumulation from untreated and biotreated PAH-impacted dredged materials. Overall, the results were used to synthesize and correlate data to assess the availability and treatability of PAHs in dredged sediments.

INTRODUCTION

Over three hundred million cubic yards of sediment are dredged from United States ports, harbors, and waterways each year. It is estimated that approximately 10% of these dredged materials are impacted with organic and inorganic contaminants (Winfield et al. 1999). Contaminated dredged materials are generally discharged into confined disposal facilities (CDFs). Many CDFs are filled or reaching their design capacity, and the diminishing capacity for contaminated dredged material disposal is a significant concern. Various treatments in engineered systems are being considered to extend the life of CDFs by producing clean material for reuse as fill or soil. However current treatment technologies have many problems, and conservative estimates of the technical and economic feasibility of these technologies suggest that realistic solutions for contaminated materials currently placed in CDFs will not be developed for many years. Bioremediation has emerged as a serious candidate for consideration to fulfill these needs.

Objective. The purpose of this study was to apply various investigative technologies to assess the locations and associations of PAHs in dredged harbor sediment, and to use this information to infer the availability of PAHs in the sediment and interpret the results of bioslurry treatment and earthworm bioassay testing. This investigation explores geochemical processes controlling PAH sequestration using physical, physicochemical,

chemical, and biological techniques to examine the distributions, associations, and sequestration energies affecting the bioavailability and bioaccumulation of PAHs in dredged Milwaukee Harbor sediments. Aqueous desorption and thermal program desorption mass spectrometry (TPD-MS) experiments were performed to evaluate PAH release and apparent sequestration energies. Bioslurry treatment tests provided samples for physicochemical, bioassay, and microbial testing. The combined results were used to develop a conceptual framework for the assessment and understanding of the bioavailability of PAHs in contaminated dredged materials.

MATERIALS, METHODS, AND RESULTS

One 55-gallon drum of dredged material from the Milwaukee confined disposal facility (MCDF) was collected and transported to the Environmental Laboratory, U.S. Army Engineer Research and Development Center (ERDC) (Bowman et al. 1996). The dredged material was homogenized and samples collected for bioslurry treatment testing and for PAH analyses on various size and density fractions.

Direct Observation of PAHs at the Sub-Particle Scale. Microprobe two-step laser desorption ionization mass spectroscopy (μL^2 MS) was used to identify and characterize PAHs found on laboratory and aged field samples. The data showed lateral variation in the extent of sorption at the sub-particle scale, which indicates that sorption phenomena are heterogeneous at the sub-particle dimensions (Gillette et al., 1999).

Microscale Location, Characterization, and Association of PAHs. Microprobe two-step laser desorption laser ionization mass spectroscopy was used for PAH measurements, infrared microspectroscopy was used for organic carbon measurements, and scanning electron microscopy with wavelength dispersive X-ray spectroscopy was used for elemental microanalysis. PAH concentrations on coal- and wood-derived particles were found to be several orders of magnitude higher than on silica particles in the Milwaukee Harbor sediment samples. A cryomicrotome sectioning procedure was employed for particle cross-sectional investigations and it was found that most PAHs are concentrated on external surface regions indicating near surface sorption mechanisms (Ghosh et al. 2000).

Physical Analyses. Physical analyses involved separating the whole sediment into four size fractions (>1000 μm, 1000-250 μm, 250-63 μm, and <63 μm) for purposes of determining if PAH presence and availability differ within the sediment. These size fractions (except >1000 μm) were further separated out by density resulting in heavy (clay/silt) and light (coal/wood) density fractions. The heavy sediment fraction in the sediment, which includes clays, silt, and sand, constituted 95% of the sediment by weight. The light sediment fraction comprising coal-derived material and debris constituted the remaining 5% of the sediment weight.

Chemical Analyses. Chemical analyses included sediment PAH analysis by soxhlet extraction with GC/MS and PAH analyses using GC-FID for determination of PAH distribution by particle size and type. Total PAHs in the whole sediment before bioslurry treatment were 115 mg/kg; however, concentrations for the lighter, coal-derived material in the sediment ranged from slightly less than 1000 mg/kg (1000-250 μm) to almost 4000 mg/kg (<63 μm). PAHs in the heavy fractions were primarily limited to the <63μm size fraction with concentrations at 35 mg/kg. More than 60% of the PAHs were in the light fractions, principally associated with coal-derived material.

Bioslurry Treatment. The bioslurry reactor studies consisted of six reactors and were operated for four months. Two active bioslurry reactors were controls (anaerobic and poisoned) and the remaining were active, (aerobic) bioslurry reactors. Each reactor had 1.5 liters of sediment (30% by volume) and 3.5 liters (70% by volume) of a modified Stanier's Basal Media with no added carbon source. Chemical analyses after bioslurry treatment indicated that total PAH concentrations declined by 50%, with some individual PAHs showing reductions greater than 65% (i.e., anthracene, fluorene, and acenaphthene). A marked distinction was observed between the biodegradation of PAHs on the heavy and light sediment fractions. PAH degradation was mainly achieved in the <0.063 mm clay/silt fractions, which exhibited slightly over 75% reduction over the 4-month bioslurry treatment. The coal/wood material in the <0.063 mm to 1 mm size fractions showed no reduction of PAH concentrations. This marked difference in PAH biodegradation between the clay/silt and coal-derived sediment fractions was shown to result from substantially different availabilities of the PAHs on these materials.

Physicochemical Analyses. Physicochemical analyses included room temperature Tenax bead aqueous desorption experiments and thermal program desorption (TPD) studies. Room temperature desorption kinetic studies using Tenax beads as PAH extractants were conducted to measure the physical availability of PAHs from the different sediment fractions. The intermediate size fraction of the light coal-derived particles (0.063-25 mm) and the heavy silt and clay fractions (<0.063 mm) were used for the desorption kinetic tests. PAH desorption kinetic studies on these separated fractions revealed a relatively low availability of PAHs from the coal-derived (light) fraction and a high availability from the clay/silt (heavy) fraction. As concluded from the TPD studies described below, PAHs associated with clays and silts desorbed faster at room temperatures and are characterized by low binding activation energies, whereas PAHs associated with coal-derived derived material desorbed at a much slower rate and are characterized by high binding activation energies.

Thermal Programmed Desorption. The initial TPD effort investigated the release of PAHs from solid surfaces and sediment particles using thermal desorption mass spectrometry. An experimental protocol was developed to obtain real-time PAH desorption data through use of a TPD probe that places the sample directly in the ion volume of a mass spectrometer then gradually heats the sample at a predetermined rate (Talley et al. 2001). Thermal desorption profiles of milligram-size samples were analyzed in order to explore the release of PAHs from mineral and organic surfaces and to compare the release of PAHs with increasing molecular weight. This showed that the release of PAHs is dependent both on PAH molecular weight and the character of the sorbent material.

TPD tests were then used to examine semi-quantitatively the apparent PAH binding activation energies, as interpreted from TPD rate responses, for Milwaukee CDF sediment before and after bioslurry treatment. TPD rate responses for PAH MW 202, 252 and 276 from the <0.063 mm light, <0.063 mm heavy, and the 0.063 – 0.25 mm light size fractions were evaluated. PAH thermograms from both density fractions shifted to the right with increasing molecular weight indicating that higher molecular weight PAHs have higher binding energy. Peak temperatures for pre- and post-bioslurry treatment also remain similar, with the exception that the biotreated heavy fraction was low in ion count indicating a loss of PAHs. The fact that peak temperatures did not change for the lighter, coal-derived particles from the bioslurry treatment shows that biotreatment did not change the character of the sorbent.

TPD-MS measurements showed that PAHs on the heavy fraction sediment particles were released at higher temperatures than PAHs on the light fraction sediment particles PAHs. This phenomenon is attributed to lack of significant activated diffusion processes for this size fraction. At room temperatures, the PAHs associated with the heavy fraction (clays/silts) desorb faster than the light (coal-derived) fraction. The PAHs sorbed on the heavy fractions are characterized by much lower binding activation energies, and thus the release is much less temperature dependent. In contrast, as the light (coal-derived) fraction sample is heated in the TPD probe the release of PAHs is very temperature dependent owing to the large desorption activation energies (Ghosh et al. 2001). It is envisioned that this behavior is due to the coal-like structure becoming less rigid and polymeric with increasing temperature. The heavy fraction (clay/silts) organic matter structure does not significantly change with elevated temperatures as that for the coal-like material. The overall conclusion from the results is that the bioslurry treatment did not change significantly the character of the sorbent with respect to binding the PAHs, and that the bioavailability of the PAHs is described by two markedly different responses owing to differences in binding activation energies.

Biological Analyses. The biological analyses consisted of microbial analyses and earthworm bioassays. The microbial analyses included radiolabeled microcosm studies, polar lipid fatty acid analyses (PLFA) characterization, and DNA analysis. The microcosm studies used radiolabeled PAHs to confirm that PAH-degrading microorganisms were present in the sediment. Mineralization rates for each of the PAHs examined were greatest at times T_1 month and T_2 month, corresponding to the greatest biomass levels and the introduction of nutrients.

The PLFA characterization identified the level of biomass present in the sediment and provided general information about the microbial community structure. The DNA analysis was utilized to target the presence of select genes known to be present during active PAH degradation. Ester-linked phospholipid fatty acid analysis revealed a significant increase in microbial biomass and a shifting microbial community structure with the observed decrease in total PAH concentration. Nucleic acid analyses revealed that copies of genes encoding PAH-degrading enzymes (extradiol dioxygenases, hydroxylases and meta-cleavage enzymes) increased by as much as four orders of magnitude during the bioslurry treatment with the shift in gene copy numbers correlating with shifts in microbial community structure and PAH reduction (Ringelberg et al. 2001).

Earthworm Bioassays. Earthworm bioassays evaluated the adverse effects of the PAHs associated with the sediment to earthworms (*Eisenia fetida*). Both acute toxicity and bioaccumulation tests were performed. The acute toxicity was a 14-day exposure test that measured the survivability of the earthworms. The bioaccumulation test was a 28-day test that measured PAH uptake within earthworm tissue. Earthworm acute toxicity tests with Milwaukee Harbor dredged sediment showed a 100% survival rate indicating that the PAHs present initially in the sediment were not toxic to induce death. It is unclear if this is due to PAH concentration below earthworm toxicity thresholds or if the PAHs were limited in their availability due to strong binding.

Earthworm bioaccumulation tests were performed on sediment before and after bioslurry treatment. These data comprised an average of the results from 192 earthworms exposed to sediment before treatment and 144 earthworms exposed to triplicate sediment samples after four months of bioslurry treatment. The uptake of total PAHs in earthworm tissue from the untreated sediment was 8.1 mg/kg. Earthworm uptake after bioslurry treatment was reduced to 2.0 mg/kg. This represents a 75% reduction in total PAH uptake. There were seven PAHs (naphthalene, 2-methylnaphthalene, acenaphthene, fluorene, fluoranthene, pyrene, and benzo(b)fluoranthene) with concentration reductions over 80%. Both the PAH content in the clay/silt fraction and the bioaccumulation of PAHs by earthworms decreased 75% as a result of bioslurry treatment. Since there was no reduction of PAHs on the coal-derived particle fractions, there is a strong correlation between the individual PAHs biodegraded on the clay/silt fraction and those reduced in earthworm tissue uptake. This observation suggests that only the PAHs in the clay/silt fraction comprise the readily available fraction to both microorganisms and earthworms (Talley et al. 2001).

SIGNIFICANT FINDINGS

The significant findings of this work are: the release of PAHs is dependent both on PAH molecular weight and the character of the sediment sorbent material; most PAHs are concentrated on external surface regions indicating near surface sorption mechanisms; two principal sediment particle classes dominated the distribution and release of PAHs; clay/silt and coal-derived; PAHs were found preferentially on coal-derived particles; clay/silt particles released PAHs more readily than coal-derived particles; bioslurry treatment reduced PAHs on the clay/silt fraction but not the coal-derived fraction; PAH reduction in clay/silt fractions by biotreatment resulted in significant reduction in earthworm PAH bioaccumulation; PAHs on coal-derived particles were associated with high binding activation energies; and changes in the phenotype and genetic potentials of the extant microbiota can be used to assess intrinsic biodegradative potential.

Benefits. The benefits of this work include: improved assessment of toxicity and risk for PAH contaminants in sediments by use of particle-scale techniques to assess PAH distribution and behavior; improved assessment for the potential success of biotreatment through understanding of factors contributing to the available and unavailable PAH fractions; improved decision making regarding sediment quality criteria for PAHs and the biotreatment of PAH-impacted sediments; and reduced treatment costs and greater likelihood for reuse of dredged sediments through knowledge of the underlying processes affecting PAH locations, availability, treatability, and toxicity.

This work shows that the PAHs associated with the fine clay/silt fraction are potentially the PAH contaminants of greatest concern. PAHs associated with coal-like materials are much less available and potentially much less of a concern for protection of the environment, even though the PAHs in this fraction comprise the majority of the PAHs in the sediment. Thus, decisions about dredged sediment quality criteria, and sediment treatment and material reuse, should focus on the PAHs associated with the clay/silt fraction, not on total PAHs in bulk sediment.

As the requirements for dredging continue and disposal options become more restrictive, improved assessment of toxicity and risk for contaminants in dredged materials in CDFs is needed. Existing CDFs must find ways to improve decision-making regarding the treatment of PAH-impacted sediments. The investigative approach presented in this work provides better understanding of contaminant availability and thereby improves the assessment for the potential success of technologies such as biotreatment. By demonstrating that both available and unavailable fractions of PAH contamination exist in sediment, and showing the connection to sediment treatment and toxicity, it may be possible to allow the use of less costly sediment treatments while increasing the likelihood for the reuse of already dredged and deposited sediment.

Limitations and Future Work. Direct inserted probe TPD-MS is not ready as a stand-alone tool for the assessment of the availability and treatability of PAHs in sediments. Numerous limitations still exist which prevent application beyond the research and development arena. Small sample size heterogeneity, fluctuating TPD responses for varying materials and low-molecular weight compounds, and difficulties in data reduction all contribute to the need for more work. More fundamental TPD research needs to be done on well-defined substrates. Using more homogenous substrates, varying PAH concentrations, and varying temperature ramps could increase our understanding of release mechanisms. This knowledge could improve our ability to properly model and assess PAH sequestration mechanisms from TPD responses. TPD-MS/MS capabilities could be utilized to further examine the structural effects and differences of binding for PAHs with the same molecular weight. These results could then be compared to various sub-sample physical and geochemical characteristics, PAH location and distribution, and PAH treatment and toxicological effects, which together could further define the issues of risk and exposure for residual and less mobile PAHs.

The overall approach presented in this work should be extended to other sediments and soils to determine if similar observations and conclusions can be made. Other contaminants, such as TNT and PCBs, should be examined. These compounds are tightly bound in soils and sediments and may have availability and treatability characteristics similar to PAHs. The knowledge that PAHs in the Milwaukee CDF sediments were preferentially bound to coal-derived particles, suggests that such materials may be used beneficially to in-situ stabilize the labile and mobile fraction of PAHs. Sediment dredging in conjunction with particle separation should explore large-scale, selective separation techniques to process the PAH-laden light sediment fraction. Future remediation strategies should target the heavy, fine-grained, sediment fraction, which represents the available and treatable PAHs.

ACKNOWLEDGEMENTS

We acknowledge funding and technical support for this research from the U.S. Army Engineer Research Development Center (ERDC), and the Department of Defense through the Strategic Environmental Research and Development Program (SERDP). We would like to thank Dr. Richard Zare and Dr. Seb Gillette of the Department of Chemistry, Stanford University, for their μL^2 MS work; Samuel Tucker and John S. Furey of the Environmental Laboratory (ERDC) for their assistance with thermal program desorption and earthworm measurements; Mr. David Ringelberg, Dr. Ed Perkins, and Dr. Herb Fredrickson of the Environmental Laboratory (ERDC) for providing PFLA and DNA analyses; and Ms. Margaret Richmond and Ms. Deborah Felt of the Environmental Laboratory (ERDC) for providing support and technical collaboration, which made this research possible. Additionally, we acknowledge Dr. Steven L. Larson and the Environmental Chemistry Branch (ECB), also at the Environmental Laboratory (ERDC) for their analytical and technical support.

REFERENCES:

Bowman, D.W., Brannon, J.M., and Batterman, S.A. **1996**. "Evaluation of Polychlorinated Biphenyl and Polycyclic Aromatic Hydrocarbon Concentrations in Two Great Lakes Dredged Material Disposal Facilities," Water Quality 96 Proceedings of the 11th Seminar, 26 February – 01 March 1996, Seattle, Washington, Miscellaneous Paper W-96-1.

Ghosh, U.; Luthy, R. G.; Gillette, J. S.; Zare, R. N. **2000**. *Environ Sci Technol*, 34: 1729-1736.

Ghosh, U., Talley, J.W., Luthy, R.G. **2001**. "Particle-Scale Investigation of PAH Desorption Kinetics and Thermodynamics from Sediment," *Environmental Science & Technology* (In Press).

Gillette, S., Luthy, R., Clemett, S., and Zare, R. **1999**. "Direct Observation of Polycyclic Aromatic Hydrocarbons on Geosorbents at the Subparticle Scale," *Environ Sci Technol.* 33: 1185-1192

Ringelberg, D., Talley, J.W., Perkins, E., Tuckers, S., Luthy, R.G., Fredrickson, H. and Bouwer, E. **2001**. "Defining In-Situ PAH Biodegradation Activity in Terms of Microbial Phenotype and Genetic Potential," *Applied and Environmental Microbiology,* 67 (4)

Talley, J.W., Ghosh, U., Tucker, S., Furey, J., Ghosh, U., and Luthy R.G. **2001**. "Particle-Scale Understanding of Bioavailability of PAHs in Sediments," *Environmental Science and Technology* (Accepted).

Talley, J.W., Ghosh, U., Tucker, S., Furey, J., and Luthy, R.G. **2001**. "Thermal Programmed Desorption (TPD) of PAHs From Mineral and Organic Surfaces," *Environmental Engineering Science* (Submitted).

Winfield, L.E., and Lee, C.R. **1999**. "Dredged material characterization tests for beneficial use suitability," DOER Technical Notes Collection (TN DOER-C2), U.S. Army Engineer Research Development Center, Vicksburg, MS.

THE EFFECT OF NUTRIENT AMENDMENTS ON PAH DEGRADATION IN A CREOSOTE CONTAMINATED SOIL

Kimberley A. Pennie (Stella-Jones Inc., Truro, N.S., Canada)
Glenn W. Stratton (Nova Scotia Agricultural College, Truro, N.S., Canada)
Gordon B. Murray (Stella-Jones Inc., Truro, N.S., Canada)

ABSTRACT: A laboratory biotreatability study was conducted on aged creosote contaminated soil obtained from a wood preservation site in Atlantic Canada. The purpose was to compare the removal of polycyclic aromatic hydrocarbons (PAH) from soil during a four-week incubation with unamended soil and soil containing two organic amendments (dairy manure and municipal solid waste compost) and one inorganic amendment (nitrate fertilizer). Both chemical and biological parameters were monitored to determine treatment efficacy. Although increasing trends were noted in total PAH levels for both the unamended and nitrogen amended soils, the most significant increases (up to 776%) occurred with nitrogen treatments. Mean reductions of up to 94% were achieved with soils receiving solid organic amendments. The results of this study indicate that significant decreases in total PAH levels can be achieved by adding organic amendments to this particular soil, while nitrogen addition may facilitate the microbial mobilization of bound PAH residues.

INTRODUCTION

Industrial processes, such as wood preservation, have resulted in considerable point source PAH contamination of soil (Mueller et al., 1991). As a result, it has been suggested that there may be at least 256,000 m^3 of moderately to highly PAH-contaminated soil in Canada (Kieley et al., 1986). Biodegradative processes have been studied as a means of addressing the contamination located at these sites. In order to increase the success of biological remediation of contaminated soil, the diversity of microorganisms present, as well as the environmental conditions conducive to degradation, should be enhanced. Some research has been conducted using organic matter additions as a method of achieving this (Crawford and Crawford, 1996). In some instances, these additions may act as a source of microorganisms as well as micro and macronutrients.

To address the PAH contamination of an industrially contaminated soil, a biopile was constructed in Atlantic Canada by a former wood preserving company during the summer of 1997. The goal of this remedial program is to decrease the PAH content to an acceptable level which will allow the soil to be used for industrial activities. As the pile was constructed soil was collected for use in laboratory biotreatability studies.

Objective. There were two objectives in this study. The first was to determine whether or not PAH degradation would occur in this soil without the addition of nutrient amendments. The second was to investigate the effect that different nutrient additions to this soil would have on PAH degradation. This included solid amendments which could be applied during the construction of future biopiles, as well as a water soluble amendment suitable for application to existing biopiles via irrigation systems. The 15 PAHs monitored during the course of this study are listed in Table 1.

TABLE 1. The 15 United States Environmental Protection Agency (USEPA) priority pollutants monitored during the biotreatability study.

PAH		
Acenaphthylene	Fluoranthene	Benzo(k)fluoranthene
Acenaphthene	Pyrene	Benzo(a)pyrene
Fluorene	Benz(a)anthracene	Dibenz(ah)anthracene
Phenanthrene	Chrysene	Benzo(ghi)perylene
Anthracene	Benzo(b)fluoranthene	Indeno(1,2,3-cd)pyrene

MATERIALS AND METHODS

The soil used for the laboratory study was obtained from a biopile during its construction on the site of a wood preservation plant in Atlantic Canada that had been in operation for approximately 67 years. This soil was river sediment that had been excavated from the perimeter of the treatment facility. It was mixed and screened to remove solids greater than 5 cm prior to construction of the biopile. During this time, samples of soil were removed and stored in 150 L drums for later use. The statistical design was a randomized complete block with repeated measures, with each treatment combination replicated five times.

Physical, Chemical and Biological Soil Characterization. Physical characterization consisted of percent moisture, pore space, pH, bulk density, particle density and water holding capacity (WHC) as well as soil texture (Table 2). These were determined using the standard methods described by the Canadian Society of Soil Science (Carter, 1993) as well as the American Society of Agronomy and the Soil Science Society of America (Page et al., 1982). Chemical characterization consisted of carbon, nitrogen and phosphorous determination (Table 2). Total organic carbon and nitrogen were determined using a Leco CNS-1000 elemental analyzer, while a Mehlich III extraction and ICAP quantification was used to determine the phosphorous level.

The PAH content of the soil was determined using EPA soxhlet extraction Method 3540-B followed by HPLC analysis with UV and fluorescence detection. The biological status of the soil was determined by microbial enumerations and soil respiration. Microbial enumerations consisted of population counts of bacteria, fungi, actinomycete, and pseudomonad populations. Soil respiration was determined by monitoring CO_2 evolution using gas chromatography.

TABLE 2. Physical characterization and C:N:P status of the biopile soil.

Physical Characteristic	
% moisture	14
% pore space	44.4
Bulk density	1.48 g·cm^{-3}
Particle density	2.66 g·cm^{-3}
WHC	0.356 mL·g^{-1}
pH[1]	7.4
% organic matter	2.4
Texture[2]	37% sand
	51% silt
	12% clay
C:N	14:1
C:P	157:1

[1] pH was measured in water (1:2); [2] Textural classification is that of a silt loam

Amendments. The three amendments used were municipal solid waste compost (abbreviated MSWc), obtained from the Lunenburg (Nova Scotia) Regional Recycling and Composting Facility in 1997, dairy manure (fresh, no wood fibers; abbreviated RAC) obtained from the Nova Scotia Agricultural College's Ruminant Animal Centre and ammonium nitrate fertilizer (34:0:0; abbreviated N). The masses of each amendment applied to the soil and the resulting nutrient ratios are reported in Table 3. Amendments were added to the soil and incubated at 20 °C in 5 L plastic pails for 4 wk. Each treatment was sampled at time-zero (immediately after amendment addition), wk 1, wk 2 and wk 4. Soil moisture was monitored and adjusted accordingly by weighing each bucket twice per week and comparing this mass to the theoretical mass at 60% WHC.

TABLE 3. Masses of each amendment and resulting soil nutrient ratios.

	Treatment		
	MSWc	RAC	N
Amendment mass / g_{ODW}[1]	270	304	2.06
Soil mass / g_{ODW}	2371	2371	2371
% C (w/w)[2]	23.7	40	-
C:N	11:1	17:1	10:1
C:P	97:1	300:1	157:1

[1] ODW = oven dry weight; [2] w/w = weight/weight

RESULTS AND DISCUSSION

Microbial Enumerations. The microbial enumerations were conducted both before and after the 4 wk incubation (Table 4). The organic amendments resulted in statistically significant increases in both the actinomycete and fungal populations when compared to the control (no amendment) and N treatment. The application of nitrogen fertilizer resulted in the largest stimulation of bacterial populations, an increase from $1.52\pm0.3 \times 10^6$ to $1.91\pm0.12 \times 10^9$ CFU·g^{-1}soil$_{ODWE}$.

TABLE 4. Initial and post incubation microbial enumerations for each treatment.

Microorganism	Unamended T-0[1]	Post Incubation[1]
Bacteria	$1.52 \pm 0.3 \times 10^6$	
Control		$1.83 \pm 0.31 \times 10^7$
N		$1.91 \pm 0.12 \times 10^9$
MSWc		$2.44 \pm 0.85 \times 10^7$
RAC		$1.55 \pm 0.25 \times 10^8$
Pseudomonad	ND[2]	
Control		ND
N		$4.00 \pm 0.24 \times 10^2$
MSWc		$7.00 \pm 0.10 \times 10^2$
RAC		$4.80 \pm 0.24 \times 10^2$
Actinomycete	870 ± 10	
Control		386 ± 2
N		574 ± 5
MSWc		$4.82 \pm 1.2 \times 10^5$
RAC		$7.30 \pm 0.52 \times 10^4$
Fungi	$3.48 \pm 0.13 \times 10^3$	
Control		338 ± 10
N		$3.40 \pm 0.95 \times 10^3$
MSWc		$6.5 \pm 0.9 \times 10^4$
RAC		$4.56 \pm 1.0 \times 10^5$

[1] Units are $CFU \cdot g^{-1} soil_{ODWE}$ where ODWE=oven dry weight equivalent
[2] ND = Not detected

The population increases in both the N and control systems represent a stimulation of indigenous populations which would have acclimatized to the presence of PAHs and, therefore, may be capable of PAH degradation. The stimulation of fungal populations observed with the MSWc and RAC treatments could be attributed to the presence of the substrate lignin within the organic matrix of these two amendments. The presence of lignin has been shown to stimulate the production of enzymes required for degradation of PAHs (Datta et al., 1991). Therefore, although not the direct target of metabolic activities, the PAH degradation potential of the soil could be enhanced due to fungal metabolism of aromatic substrates such as lignin.

Soil Respiration. Overall, the same general trend in CO_2 evolution was observed for each treatment group which consisted of a decrease in respiration from wk 1 to wk 3 followed by an increase by wk 4 (Figure 1). This could be attributed to an over stimulation of microbial activity as a result of the initial flush of available carbon to the system. Increased microbial activity can cause carbon starvation once the initial carbon supplies are depleted (Dosoretz et al., 1993). The dairy manure supplied the largest initial carbon reserves and also had CO_2 evolutions at least one order of magnitude greater than the other three treatments, yet the same trend in respiration was observed (Figure 2).

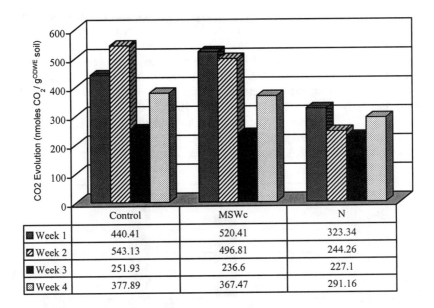

FIGURE 1. Mean CO_2 evolutions for the control, N and MSWc treatments.

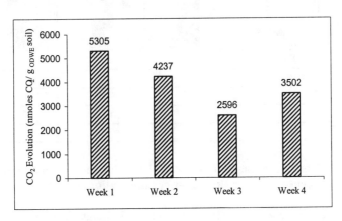

FIGURE 2. CO_2 evolution for the dairy manure amended soil (RAC).

Although there were no significant increases in respiration for the N amended soil during the four wk period, there was also no apparent suppression of microbial metabolic activities. Therefore, it can be concluded that unlike other studies (Wang et al., 1990) nitrogen application will not inhibit the metabolic activities of the indigenous microorganisms within this particular soil system.

PAH Characterizations. Although the concentrations of 15 PAHs were monitored, the total PAH (tPAH) concentration was used to provide a more accurate determination of the overall effect of each treatment. Correction factors were used to compensate for the weight/weight dilution effect that the addition of solid amendments (MSWc and RAC) would have on PAH concentrations in the soil. To reduce the statistical effect that inherent variations in PAH levels (due to a lack of soil homogeneity) would have, data for each week were reported as a percentage of the time zero tPAH (T-0 tPAH) level for each replicate. PAH concentrations that were below detection limits were replaced by the detection limit for that individual PAH divided by two, which is the method recommended by the US EPA for reporting non-detected values for the purpose of conducting statistical analysis (McBean and Rovers, 1999). A graphical representation of the changes in mean tPAH levels during the four week study is shown in Figure 3.

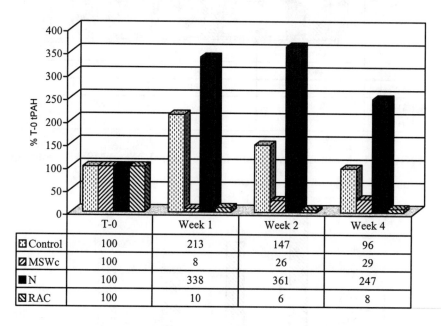

	T-0	Week 1	Week 2	Week 4
Control	100	213	147	96
MSWc	100	8	26	29
N	100	338	361	247
RAC	100	10	6	8

FIGURE 3. Mean percentages of T-0 tPAH levels for each amendment and the control during the four-week incubation period.

Both the MSWc and RAC treatments resulted in significant reductions in tPAH levels. After only 1 wk of incubation the initial PAH levels had decreased by 92% and 90% for the MSWc and RAC soil, respectively. There were no further significant changes in tPAH levels for the RAC treatment from wk 1 to wk 4, however, there was an increase recorded for the MSWc amended soil by wk 2, which was maintained to wk 4. These substantial reductions may be due to microbial degradation or, more likely, the sorption of solvent extractable PAHs. A similar decreasing trend has been reported to take place in the presence of organic matter (Guerin, 2000). The presence of lignin in manure could facilitate PAH sorption thereby decreasing their solvent extractability from amended soil. Garbarini and Lion (1986) found that the sorption of toluene and trichloroethylene to lignin and cellulose was positively related to carbon content.

In contrast, there were significant increases in tPAH levels in the N amended soil, while the control exhibited an overall increasing trend. The PAHs which decreased were the less recalcitrant PAHs, while the more structurally complex ones increased. The fact that there were increases in the tPAH levels suggests that there were bound residues in this soil that were not initially susceptible to solvent extraction. Bound residue formation with PAHs has been shown to occur after only 1 wk of exposure (Herbes and Schwall, 1978). Since the age of contamination in this soil is 50 to 60 years, it is likely that extraction resistant residues would be present. Solubilization of hydrophobic compounds such as PAHs has been shown to increase their cellular uptake by microorganisms (Reid et al., 2000) and a number of PAH degrading microorganisms have been reported to produce extracellular biosurfactants which aid in the transfer of PAHs from the sorbed to NAPL fractions in soil (Stelmack et al., 1999). If the observed increases are due to an increase in biosurfactant production, then it can be concluded that incubation at 20 °C alone (control) will not be sufficient to significantly increase the mobilization of the bound residues within this soil system. In order to achieve this, additional nitrogen sources may be required. The increase in tPAH levels from wk 1 to wk 2 that occurred in the MSWc soils could indicate the onset of residue mobilization with this amendment.

Although stimulation of microbial populations occurred within the control soil, no significant change in PAH levels occurred during this study. The treatments that showed the most promise were the MSW compost and ammonium nitrate fertilizer. The MSWc amendment not only has the ability to improve soil physical characteristics, such as WHC and porosity, but was shown to increase the microbial populations paramount to PAH degradation. However, there are several advantages associated with the use of the nitrogen fertilizer. First, since this nutrient amendment is highly water-soluble it can be applied to existing biopiles on an on-going basis while solid amendments can only be incorporated during the construction phase. Secondly, this amendment does not supply exogenous organisms to the soil which may compete with acclimatized indigenous microorganisms capable of PAH degradation. Finally, the nitrogen fertilizer was shown to initiate substantial mobilization of bound residues thereby facilitating a more complete remediation of the contaminated soil.

REFERENCES

Carter, M. R. 1993. *Soil Sampling and Methods of Analysis*. Canadian Society of Soil Science. Lewis Publishers, Ann Arbour, MI.

Crawford, R. L. and D. L. Crawford. 1996. *Bioremediation: Principles and Applications*. pp. 35-60. Cambridge University Press, NY.

Datta, A., A. Bettermann and T. K. Kirk. 1991. "Identification of a Specific Manganese Peroxidase Among Ligninolytic Enzymes Secreted by *Phanerochaete chrysosporium* During Wood Decay". *Appl. Environ. Microbiol.* 57:1453-1460.

Dosoretz C. G., N. Rothschild and Y. Hadar. 1993. "Overproduction of Lignin Peroxidase by *Phanerochaete chrysosporium* (BKM-F-1767) Under Nonlimiting Nutrient Conditions". *Appl. Environ. Microbiol.* 59:1919-1926.

Garbarini, D. R. and L. W. Lion. 1986. "Influence of the Nature of Soil Organics on the Sorption of Toluene and Trichloroethylene". *Environ. Sci. Technol.* 20:1263-1269.

Guerin, T. F. 2000."The Differential Removal of Aged Polycyclic Aromatic Hydrocarbons From Soil During Bioremediation". *Environ. Sci. Pollut. Res.* 7:19-26.

Herbes, S. E. and L. R. Schwall. 1978. "Microbial Transformation of Polycyclic Aromatic Hydrocarbons in Pristine and Petroleum Contaminated Sediments". *Appl. Environ. Microbiol.* 35:306-316.

Kieley, K. M., R. A. F. Matheson and P. A. Hennigar. 1986. *Polynuclear Aromatic Hydrocarbons in the Vicinity of Two Atlantic Region Wood Preserving Operations*. Environmental Protection, Environment Canada, Dartmouth, N.S. Canada. Report No. EPS-5-AR-86-3.

McBean, E. A. and F. A. Rovers. 1999. *Statistical Procedures for Analysis of Environmental Monitoring Data and Risk Assessmnet. Prentice Hall PTR Environmental Management and Engineering Series Vol #3.* Prentice Hall PTR, NJ.

Mueller, J. G., D. P. Middaugh, S. E. Lantz and P. J. Chapman. 1991. "Biodegradation of Creosote and Pentachlorophenol in Contaminated Groundwater: Chemical and Biological Assessment". *Appl. Environ. Microbiol.* 57:1277-1285.

Page, A. L., R. H. Miller and D. R. Keeney (Eds.). 1982. *Methods of Soil Analysis Part 2-Chemical and Microbiological Properties. 2nd ed.* American Society of Agronomy, Inc. and Soil Science Society of America, Inc. Madison, WI.

Reid, B. J., K. C. Jones and K. T. Semple. 2000. "Bioavailability of Persistent Organic Pollutants in Soil and Sediments-A Perspective on Mechanisms, Consequences and Assessment". *Environ. Pollut.* 108: 103-112.

Stelmack, P. L., M. R. Gray and M. A. Pickard. 1999. "Bacterial Adhesion to Soil Contaminants in the Presence of Surfactants". *Appl. Environ. Microbiol.* 65:163-168.

Wang, X., X. Yu and R. Bartha. 1990. "Effect of Bioremediation on Polycyclic Aromatic Hydrocarbon Residues in Soil. *Environ. Sci. Technol.* 24:1086-1089.

ANAEROBIC POLYCYCLIC AROMATIC HYDROCARBON (PAH)-DEGRADING ENRICHMENT CULTURES UNDER METHANOGENIC CONDITIONS

Wook Chang, Triana N. Jones, and *Tracey R. Pulliam Holoman*
(University of Maryland, College Park, Maryland, USA)

ABSTRACT: Baltimore Harbor (Baltimore, MD) sediments were utilized to initiate anaerobic enrichment cultures with polycyclic aromatic hydrocarbons (PAHs) under methanogenic conditions. Both naphthalene and phenanthrene showed complete degradation without a lag, but not pyrene. This is the first report confirming that methanogenic cultures are capable of complete PAH degradation.

INTRODUCTION

Polycyclic aromatic hydrocarbons (PAHs) are compounds which have two or more fused aromatic hydrocarbon rings and among the most common organic contaminants of aquatic sediments. The primary source of PAHs is from the combustion of fossil fuels, and these compounds have become widely distributed in the environment mainly through fuel spills, runoff, sewage treatment plants, waste incineration, and petrochemical industrial effluents. PAHs have been listed as priority pollutants by United States Environmental Protection Agency due to their carcinogenic and mutagenic properties (Keith *et al.*, 1979). In the stationary subsurface sediments, there is little direct human exposure to PAHs; however, the tendency to bioaccumulate in the aquatic food chain and the activities such as dredging operations can increase the potential for human contact (Menzie *et al.*, 1992).

Due to low aqueous solubilities and hydrophobicity, PAHs rapidly adsorb to particulates and ultimately settle in subsurface sediment, which is primarily anaerobic (Rockne and Strand, 1998). Past research generally led to the conclusion that PAHs could not be degraded under anaerobic conditions (10, 26). Only recently, studies began demonstrating PAH biodegradation under nitrate reducing (Mihelcic and Luthy, 1988; Rockne and Strand, 1998) and sulfate-reducing (Coates *et al.*, 1996; Zhang and Young, 1997) conditions. Methanogenic cultures have not been reported thus far with the ability to degrade PAHs completely. The depletion of electron acceptors such as nitrate and sulfate results in predominantly methanogenic conditions in petroleum-contaminated zone, especially, closet to the source of contamination (Lovley, 1997). Here we report on the first highly enriched anaerobic cultures capable of degrading both naphthalene and phenanthrene completely under methanogenic conditions.

MATERIALS AND METHODS

Enrichment cultures. The sediment samples were collected with a petite Ponar grab sampler from a subsurface depth of 9.1 m in the northwest branch of Baltimore Harbor (BH) (39°16.8'N, 76°36.1'W). Enrichment cultures were initiated by transferring Baltimore Harbor sediments (10% [wet wt/vol]) into sterile estuarine medium anaerobically in an atmosphere that contained N_2-CO_2 (4:1) (Holoman et al., 1998). Briefly, the medium contained the following constituents, in grams per liter of demineralized water: NaCl, 8.4; $MgCl_2 \cdot 6H_2O$, 3.95; KCl, 0.27; $CaCl_2 \cdot 2H_2O$, 3.0; NH_4Cl, 0.5; Na_2CO_3, 3.0; Na_2HPO_4, 1.12; cysteine-$HCl \cdot H_2O$, 0.25. In addition, resazurin (0.1% [vol/vol]), 0.1% of $Na_2S \cdot 9H_2O$ (2.5% [wt/vol]), vitamin, and trace solutions (1% [vol/vol] each) were also added. The medium was buffered at pH 6.8 by the addition of bicarbonate. Naphthalene (Sigma Chemical Co.), phenanthrene (Sigma Chemical Co.), and pyrene (Sigma Chemical Co.) were solubilized in acetone and then added to each culture to a final concentration of 200 µM. They were sealed with Teflon-coated rubber stoppers and aluminum crimp-seals. Sterile controls were autoclaved at 121°C for 30 min three times on consecutive days. All enrichment cultures were made in triplicate including the sterile controls. The cultures were incubated without shaking in the dark at 30°C.

Extraction and gas chromatography analysis. For PAH extraction, the cultures were well mixed and 0.75 ml of samples was withdrawn anaerobically. Each sample was then mixed with the same volume of hexane (Sigma Chemical Co.). PAHs were extracted by shaking overnight in 1.5 ml vials with Teflon-lined caps. Biphenyl was used as an internal standard. PAHs were identified and quantified using gas chromatography (GC) (Hewlett Packard 5890A) with flame ionization detector (FID) and an autosampler (Hewlett Packard 7673A). Methane production in the headspace was verified also by GC with FID. The column was 30 m long with a 0.25 mm inner diameter (J&W Scientific DB-5 capillary column, Folsom, CA). The analysis conditions were as follows: the injector and detector temperatures were 280°C and 300°C respectively, the column temperature initially started at 80°C, was increased to 300°C at 20°C/min and held at 300°C for 2 minutes.

RESULTS AND DISCUSSIONS

Anaerobic biodegradation of naphthalene and phenanthrene. Enrichment cultures were initiated by inoculating 10% of the BH sediment into estuarine media amended with 200 µM of naphthalene, phenanthrene. As shown in Fig 1 and 2, even though the same concentrations of naphthalene and phenanthrene were added to each culture, initial concentrations were different. This phenomenon can be explained by the effect of sorption into the sediment particles. The cultures were monitored weekly to verify that degradation was occurring. Degradation of naphthalene and phenanthrene started immediately without a lag after cultures were initiated, indicating that enrichment conditions

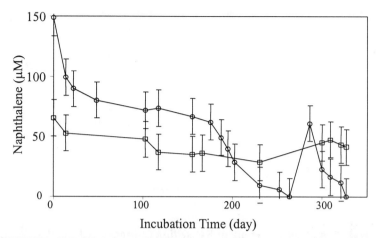

FIGURE 1. Initial degradation of naphthalene in BH sediment-inoculated, methanogenic enrichment cultures with refeeding on day 286 (active cultures [O], sterile controls [□]).

were already established. Concentrations in naphthalene and phenanthrene cultures dropped below detection limit (3 µM) after 286 and 168 days of incubation, respectively, while sterile controls showed no degradation. Methane production was detected throughout the degradation period in the naphthalene and phenanthrene cultures but not in the sterile controls. After the cultures were refed, the PAHs were readily degraded within 36 and 140 days in naphthalene and phenanthrene cultures without a lag (Fig. 1 and 2). Although only one refeeding is represented in Fig. 1 and Fig. 2, it was repeated several times in these enrichment cultures. After several successful refeedings, the actively degrading cultures were transferred into fresh medium to determine if activity could be maintained. Transferred cultures showed immediate degradation (data not shown).

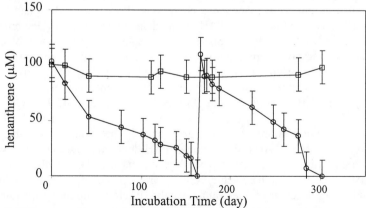

FIGURE 2. Initial degradation of naphthalene in BH sediment-inoculated, methanogenic enrichment cultures with refeeding on day 168 (active cultures [O], sterile controls [□]).

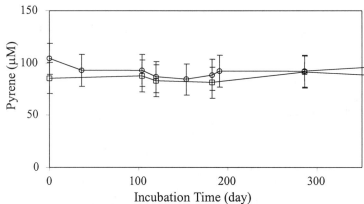

FIGURE 3. Degradation of pyrene in BH sediment-inoculated, methanogenic enrichment cultures (active cultures [O], sterile controls [□]).

Anaerobic biodegradation of pyrene. Even though enrichment cultures were initiated with pyrene as previously described, unlike naphthalene and phenanthrene cultures, no degradation of pyrene was observed over 350 days (Fig. 3). Methane was produced only at the initial period, but the amount was far less than in naphthalene and phenanthrene cultures. Pyrene is a highly recalcitrant compound due to its hydrophobicity, solubility, and stability. So far, there has been no report on pyrene degradation under anaerobic conditions.

CONCLUSIONS

To our knowledge, this is the first evidence that PAHs can be degraded completely under methanogenic conditions. Methanogenic *Archaea* typically predominate upgradient to the source of contamination (Lovley, 1997). Thus, these results have important implications for enhancing the potential benefit of PAH remediation under methanogenic conditions. Previous studies have not shown the complete degradation of PAHs in methanogenic cultures (Genthner *et al.*, 1997; Parker, 1995). Those results might be explained by the unestablished PAH-degrading microbial communities with long-term PAH exposure and appropriate electron acceptors (Coates *et al.*, 1996, Lovley, 1997). BH has suffered long-term exposure to high PAH concentrations and especially the northwest branch of BH has showed the highest concentrations (47,260 ng/g dry wt) of PAHs (Baker *et al.*, 1997).

It is becoming more evident that anaerobic bacteria are capable of mineralizing a variety of PAHs. Even though PAH-degrading pure cultures were not yet isolated in this study, the complete degradation of PAHs and the methane production suggest that methanogenic cultures are capable of degrading PAHs. Further studies are necessary to determine whether PAHs are directly mineralized under methanogenic conditions, and whether methanogens are responsible for PAH degradation.

ACKNOWLEDGEMENTS

This research was supported by National Science Foundation Faculty Early CAREER Award BES-9984285.

REFERENCES

Baker, J., R. Mason, J. Cornwell, and J. Ashley. 1997. *Spatial Mapping of Sedimentary Contaminants in the Baltimore Harbor/Patapsco River/Back River System- Final Report.* Maryland Geological Survey, Baltimore, MD

Coates, J. D., R. T. Anderson, and D. R. Lovley. 1996. "Oxidation of Polycyclic Aromatic Hydrocarbons under Sulfate-Reducing Conditions." *Appl. Environ. Microbiol.* 62:1099-1101

Genthner, B. R. S., G. T. Townsend, S. E. Lantz, and J. G. Mueller. 1997. "Persistence of Polycyclic Aromatic Hydrocarbon Components of Creosote Under Anaerobic Enrichment Conditions." *Arch. Environ. Contam. Toxicol.* 32:99-105

Holoman, T. R. P., M. A. Elberson, L. A. Cutter, H. D. May, and K. R. Sowers. 1998. "Characterization of a Defined 2,3,5,6-Tetrachlorobiphenyl-ortho-Dechlorinating Microbial Community by Comparative Sequence Analysis of Gene Coding for 16S rRNA." *Appl. Environ. Microbiol.* 64:3359-3367

Keith, L. H., and W. A. Telliard. 1979. "Priority Pollutants, I-a Perspective View." *Environ. Sci. Technol. Special Report.* 13:418-423

Lovley, D. R. 1997. "Potential for anaerobic bioremediation of BTEX in petroleum-contaminated aquifers." *J. Ind. Microbiol. Biotechnol.* 18:75-81

Menzie, C. A., B. B. Potocki, and J. Santodonato. 1992. "Exposure to carcinogenic PAHs in the environment." *Environ. Sci. Technol.* 26:1278-1284

Mihelcic, J. R., and R. F. Luthy. 1988. "Degradation of polycyclic aromatic hydrocarbon compounds under various redox conditions in soil-water systems." *Appl. Environ. Microbiol.* 54:1182-1187

Rockne, K. J., and S. E. Strand. 1998. "Biodegradation of Bicyclic and Polycyclic Aromatic Hydrocarbons in Anaerobic Enrichments." *Environ. Sci. Technol.* 32:3962-3967

Zhang, X., and L. Y. Young. 1997. "Carboxylation as an Initial Reaction in the Anaerobic Metabolism of Naphthalene and Phenanthrene by Sulfidogenic Consortia." *Appl. Environ. Microbiol.* 63:4759-4764

ACUTE PHYTOTOXICITY BIOASSAY (APB) APPLICATION ON PAH CONTAMINATED SOIL

C. Mollea, F. Bosco and *B. Ruggeri*
(Politecnico di Torino- C.so Duca degli Abruzzi, 24; 10129 TORINO, Italy)
e-mail: ruggeri@polito.it

ABSTRACT: This paper describes experiments for the standardization of a seed germination/root elongation as Acute Phytotoxicity Bioassay (APB) applied to PAH contaminated soil. Ten different seed species were tested on artificially contaminated soil: Carrot, Cucumber, Garden cress, Lettuce, Marrow squash, Radish, Soybean, Tomato (Dicotyledons), Onion, Perennial ryegrass (Monocotyledons). Among these species six seeds were selected using the following criteria: rapid growth, easy reading of the length, % of germination superior to 80% and good reproducibility of parameters. The evaluated parameters were: the number of germinated seeds in each dish, the length of hypocotyl and root and the Germination Index % (GI%). Anthracene, fluoranthene, naphthalene, phenanthrene and pyrene, at a concentration of 1000 ppm, were tested as PAH. Since all the seeds have shown a great sensitivity to naphthalene, a range-finding test (from 1000 to 0.1 ppm) was carried out. The most sensitive seed resulted to be Garden cress, while the least is the Cucumber, all the others presented almost the same sensitivity.

INTRODUCTION

Hazardous substances, when released in the environment, can have a deleterious impact on the ecosystem. The study of their potential toxicity is complex and require a range of different tests. Detailed chemical analyses are not sufficient because even if they give information about the occurrence and concentration of the toxic substances, they are not able to inform about the toxicity effect on the environment: only ecotoxicological bioassays respond to both the concentration of the chemicals and their interference on living organisms. A chemical investigation also is lacking when it needs to evaluate the reduction in toxicity of a parent compound after a bioremediation process, because the disappearance of one chemical does not necessarily mean the detoxification of the treated soil.

Using toxicity tests, it is possible to evaluate the combined effect of a complex of substances, the presence of an unknown chemical, and the products obtained during the detoxification or introduced with bioremediation. Toxicity assays are often used to check the bioremediation of contaminated sites or to evaluate the effectiveness of a new remedy. For all these reasons and owing to their simplicity and low cost, toxicity tests are useful and often used in the studies of hazardous aquatic and terrestrial waste sites. However they are not always sufficient, because even if it is possible to test a variety of substances on the same organism, it may not always be possible to determine which compound, or combination of compounds, is responsible for the observed toxicity (Wundram et al., 1997; Baudo et al., 1999).

There is a variety of toxicity tests that can be applied to different test organisms: algae test, luminescent bacteria test, crustacean bioassay, ittiotoxicity test, earthworm bioassay and phytotoxicity test. In particular, phytotoxicity bioassays have been recently included in environmental biomonitoring by an increasing number of researchers; the most common tests are algal growing and photosynthesis assay, *Lemna* test, vascular plant growth and seed germination/root elongation test. Algal assay cannot be applied to turbid solutions, cannot predict toxic effect on superior plants, and cannot be directly applied on solid samples (Wang, 1991). For these reasons the Environmental Protection Agency (US-EPA, 1996) and the American Society for Testing Materials (ASTM, 1994) have recently published a series of guidelines in which the use of superior plant seeds is recommended.

It is important to improve the use of phytotoxicity tests because they are useful to evaluate the potential toxicity of different toxic substances (such as heavy metals, mineral stress, chemical and allelopatic substances, herbicides, and salts) by observing their effect on plant germination. (Ratsch and Johndro, 1986). Those tests are biologically responsive, simple and low cost, they can be completed in a short time and statistically analyzed, they have a good reproducibility, and in most cases do not require artificial light; moreover they respect the ecology principles about the essential importance of primary producers (Baudo et al., 1999). These are acute toxicity tests (from 48 to 96 hours) based on germination and early seedling growth sensitivity: in fact, the initial growing phases are demonstrated to be the most sensitive to environmental changes (Wang, 1987). Plant seeds can stand long period of dryness, but in presence of water they undergo to rapid morphological changes during which they became much more sensitive to environmental stresses. Since this parameter can influence first growing phases, phytotoxicity tests may be used for testing the presence of toxic substances in culture media. They may be used to test sediment, wastewater, aqueous extracts and soil samples.

Different aspects should be considered in choosing plant species: sensitivity to toxic compounds of interest, commercial availability, costs, germination percentage and time, ecological relevance (presence in a specific habitat). Root length, number of germinated seeds and seedling dry weight are the most common measured endpoints. EPA guidelines (1996) and ASTM method (1994) suggest using both mono- and dicotyledons, and testing at least ten species each time, because sensitivity to different molecules is correlated to the species.

Different species have been tested on contaminated water, sediment and soil samples in ecotoxicological studies. Ronco et al. (1995) studied the sediment toxicity contaminated with organic compounds and heavy metals using a seed germination/root elongation test on Lettuce (*Lactuca sativa*) seeds. Common ryegrass (*Lepidium sativum*) is recommended as test species in toxicity test applied to compost (Barberis et al., 1996).

Monitoring toxicity reduction during and after remediation treatments is one of most important application of this kind of test. For example Meier *et al.* (1997) used a solvent extraction technique on a soil contaminated by PCB. Lettuce and oat (*Avena sativa*) have been applied in testing soil toxicity after solvent extraction: although PCB content was reduced by 99%, phytotoxicity remained essentially unchanged and was eliminated only by a post remediation water rinsing step . Chang *et al.* (1996) used a battery of bioassay to evaluate the remediation of a lead contaminated soil after a soil washing/ leaching treatment process: the increased toxicity after the treatment was caused by an increasing of salts introduced during the removal of heavy metals. Baud- Grasset et al. (1993) used phytotoxicity tests in order to observe the toxicity at the end of the bioremediation but also during a pilot scale treatability study.

This paper describes preliminary experiments for standardization of a seed germination/root elongation toxicity test used as an Acute Phytotoxicity Bioassay (APB) applied to PAH contaminated soils. Lettuce, cucumber, rye cress, radish, soybean, ryegrass, carrot, onion, tomato and zucchini seeds have been tested in order to identify the usefull species for PAH contaminated matrix.

MATERIALS AND METHODS

A natural soil, without contamination of PAH, was used for this study (pH 6.63, total moisture 12%). The soil was well mixed and sieved to remove particles greater than 1,5 cm, stored at 4°C in glass jars and dried at 105°C, in oven, for 24 hours, before starting the experiments.

Chemicals. Anthracene, fluoranthene, naphthalene, phenanthrene, pyrene, and dichloromethane were purchased from Fluka Chemical ; $K_2Cr_2O_7$ was purchased from Analyticals Carlo Erba.

Test organisms. Ten species were tested in order to select the seeds for the phytotoxicity experiments: Onion and Perennial ryegrass for the Monocotyledons; Carrot, Cucumber, Garden cress, Lettuce, Marrow squash, Soybean, and Tomato for the Dicotyledons. These species are recommended from APHA, AWWA, WEF, and U.S. EPA. The seeds were obtained from Ingegnoli SPA (Milano) and stored at room temperature before use. Ninety percent of all seeds were guaranteed to germinate and also not to be treated with fungicides, insecticides, and pelletizing agents. The seeds used were from the same lot of the individual species and of uniform size to reduce variability. Both seed germination and root elongation tests were performed in the same Petri dishes on test soil directly. Ten seeds for each species were grown in closed Petri dishes filled with 10 g of dried soil and 7 ml of deionized water, at 25 °C ±1, in the dark; three replicates were made; those parameters were maintained in all the experiments, including the positive and negative controls, and the artificial spiking of the soil. The incubation continued for the time after 80 % of the plants has germinated and the roots reached the length of 20 mm. The species were evaluated for short time of germination, easy measurements, high germination rate, and reproducibility.

Negative and positive control. In order to test the stability of the seeds, a positive control was used together with a negative one. The last one (dried soil additioned with water) is a seed growth rate control. The positive control (dried soil additioned with a CrVI solution) is a seed growth rate control in presence of a known toxic substance. The compound selected was CrVI because it is stable and non volatile; each seed was grown at a scale of concentrations of CrVI in order to find the one that cause the reduction of the 50% of root length in comparison with the negative control. The different concentrations used for each seed were 100, 50, 25, 12.5 6.25, 3.125, 1.5625 mg Cr L^{-1}. The stock solution was prepared dissolving 0.28281g of $K_2Cr_2O_7$ in 1l of deionized water (corresponding to 100 mg Cr L^{-1}).

Artificial spiking of the soil. First, all seeds were tested with 1000 ppm of five different PAH: anthracene, fluoranthene, naphthalene, phenanthrene, and pyrene. Next, a range-finding test with naphthalene at a scale of concentrations (1000, 100, 10, 1, 0.1 ppm) was conducted. Finally, this PAH was tested at 500 ppm. Two ml of solution of PAH dissolved in dichloromethane (DCM) were added to 10 g of dried soil. After five minutes to allow the DCM to evaporate, the soil was wetted with 7 ml of deionized water. Two ml of pure DCM were added to the positive and negative control in order to maintain the same starting conditions. Before starting to use the DCM, the same preliminary test was performed in order to be sure that the solvent had no negative effect on the growing seeds compared with a control without DCM.

Petri dishes were placed with their cover into different glass boxes to separate different PAH incubations and concentrations from each other and from the controls; a layer of blotting paper was positioned at the bottom of the boxes to maintain moisture.

Calculations. At the end of the incubation the number of germinated seeds in each dish was counted, and the lengths for separated hypocotyl and root were measured. Mean, standard deviation, and Germination Index % (GI%) were determined for each measurement, where:

$$GI\% = (Gs* Ls)/(Gc* Lc)* 100$$

Gs= mean number of germinated seed in the sample, Ls= mean length of the sample, Gc= mean number of germinated seed in the negative control, Lc= mean length of the negative control.

RESULTS AND DISCUSSION

Test organisms. The six seeds selected for the test represent both the groups of the Dicotyledons and the Monocotyledons. The species have a rapid growth, 72 or 96 hours, their lengths are easy to read because the roots are not delicate, they guarantee good reproducibility and all overtake 80 % of germination (Table 1). Lettuce and Carrot seeds were reject because they grow too slowly; Marrow squash seeds were not used because they are too large and it was not possible to place ten seeds in the dishes; soybean was rejected because it grows with molds.

TABLE 1. Selected seeds.

DICOTYLEDONS			
Species	Common name	Time of growth	% of germination
Lepidium sativum	Garden cress	72	100
Raphanus sativus	Radish	72	92
Cucumis sativus	Cucumber	72	90
Lycopersicon esculentum	Tomato	96	85
MONOCOTYLEDONS			
Species	Common name	Time of growth	% of germination
Allium cepa	Onion	96	98
Lolium perenne	Perennial ryegrass	96	94

Positive control. For each seed we selected different CrVI concentrations: the most sensitive is Garden cress (25 ppm), the least is Cucumber (100 ppm) while the others showed the same sensitivity to CrVI (50 ppm). As it's possible to notice (Figure 1) the concentration selected is not always the one that give the 50% of inhibition of root length, but it's an arrangement between the inhibition of the root length and the GI%.

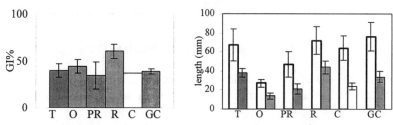

FIGURE 1. GI% for the positive control (Left). Comparison between length of negative control (empty bars) and positive control (Right). (T= Tomato, O= Onion, PR= Perennial ryegrass, R= Radish, C= Cucumber, GC= Garden cress).

Artificial spiking of the soil. First, we evaluated the effect of scale concentrations on root and hypocotyl separately. We abandoned this approach because the trend of the two parts is truly represented from that of the total length (Figure 2).

The preliminary test with 1000 ppm of the five PAH showed that naphthalene was the most toxic: the growth of all the seeds was inhibited by naphthalene (Figure 3). The least sensitive seeds were Onion, Radish, and Cucumber; the most sensitive seeds were Tomato, Garden cress, and Perennial ryegrass. In particular the Perennial ryegrass showed total inhibition of the germination process.

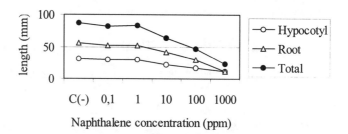

FIGURE 2. Trend of root and hypocotyl for Cucumber seeds in the presence of naphthalene.

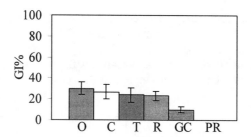

FIGURE 3. GI% of all the seeds on naphthalene (1000 ppm).

The species tested were not sensible to the other PAH, with the only exception being Cucumber, which had a similar reaction to anthracene, fluoranthene, phenanthrene, and pyrene.
We observed an inhibition effect caused by the phenanthrene for Onion, Tomato and Perennial ryegrass (Figure 4).

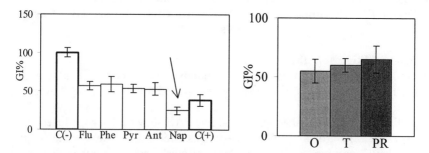

FIGURE 4. Cucumber on the five PAH (1000 ppm) →naphthalene (Left). Onion, Tomato and Perennial ryegrass on phenanthrene (Right).

After these results, a range-finding test was conducted: all the seeds were exposed at a scale of concentration of naphthalene, the most toxic PAH. According to the

GI% values, it is possible, for all the species, to see a significant inhibition only between 100 and 1000 ppm of naphthalene (Figure 5).

For this reason the seeds were tested at a concentration of 500 ppm in order to examine the trend between 100 and 1000 ppm. The growth is significantly proportional to this intermediate value (Figure 6).

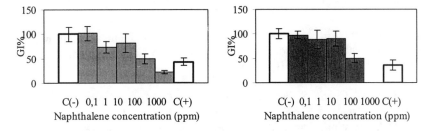

Figure 5. GI% for Onion (Left) and Perennial ryegrass (Right).

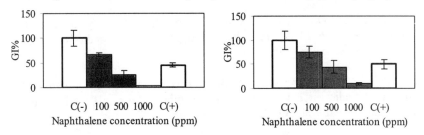

Figure 6. GI% for Garden cress (Left) and Tomato (Right).

CONCLUSIONS

The APB on artificially spiked soil gives reproducible results, so they can be considered useful instrument in the monitoring of contaminated sites. The results obtained at the lower concentrations (0.1, 10 ppm) are not statistically significant because of the high volatility of the naphthalene. From the observations of the Germination Index % versus the contaminant concentrations tested (naphthalene) (Figure 6.), it is possible to argue that the GI% is inversely proportional to the concentration. Interesting is the behaviour of Cucumber seeds which gives similar response for the five PAH tested: it could be used in real conditions because contaminated soils contain different PAH.

REFERENCES

American Society for Testing and Materials. 1994. Standard Practice for Conducting Early Seedling Growth Tests. E 1598- 94.

Barberis, R., P. Nappi, P. Boschetti, and E. Vincenzino. 1996. "Valutazione della Qualità del Compost". Acqua & Aria. 3: 297- 303.

Baud-Grasset, F., S. Baud-Grasset, and S. I. Safferman. 1993. "Evaluation of the Bioremediation of a Contaminated Soil with Phytotoxicity Tests". Chemosphere. 26(7): 1365-1374.

Baudo, R., M. Beltrami, P. Barbero, and D. Rossi. 1999. "Test di Germinazione e Allungamento Radicale". Acqua & Aria. 2: 69-85.

Chang, L. W., and M. K. Meier. 1997. "Application of Plant and Earthworm Bioassays to Evaluate Remediation of a Lead-Contaminated Soil". Arch. Environ. Contam. Toxicol. 32: 166-171.

Meier, J. R., L. W. Chang, S. Jacobs, J. Torsella, M. C. Meckes, and M. K. Smith. 1997. "Use of Plant and Earthworm Bioassays to Evaluate Remediation of soil from a Site Contaminated with Polychlorinated Biphenyls". Environ. Toxicol. and Chem. 16(5): 928-938.

Ratsch, H. C., and D. Johndro. 1986. "Comparative Toxicity of Six Test Chemicals to Lettuce Using Two Root Elongation Test Methods. Environ. Monit". And Assess. 6: 267-276.

Ronco, A. E., M. C. Sobrero, G. D. Bulus Rossini, and P. R. Alzuet. 1995. "Screening for Sediment Toxicity in the Rio Santiago Basin: a Baseline Study". Environ. Toxicol. and Water Qual. 10: 35-39.

U.S.Environmental Protection Agency. 1996. Ecological Effects Test Guidelines. OPPTS850.4200. Seed Germination/ Root Elongation Toxicity Test. EPA 712-C-96-154.

Wang, W. 1987. "Root Elongation Method for Toxicity Testing of Organic and Inorganic Pollutants". Environ. Toxicol. and Chem. 6: 409-414.

Wang, W. 1991. "Literature Review on Higher Plants for Toxicity Testing". Water, Air and Soil Pollution. 59: 381-400.

Wundram, M., D. Selmar, and M. Bahadir. 1997. "Representative Evaluation of Phytotoxicity- Reliability and Peculiarities." Angew. Bot. 71: 139-143.

BIOLOGICAL REMOVAL OF IRON AND PAHS FROM MGP-IMPACTED GROUNDWATER

Don E. Richard (University of Minnesota, Minneapolis, MN)
Daryl F. Dwyer (University of Minnesota, Minneapolis, MN)

ABSTRACT: Dissolved iron has been viewed primarily as a nuisance in the design and operation of biological treatment systems for organic contaminants in groundwater. However, iron is an essential micro-nutrient, can have an important role in attachment of bacteria to solid surfaces, and may confer other benefits to biological systems designed to treat recalcitrant organic compounds. Many European communities use biological filters to remove iron from drinking water and the operating concepts from these filters were used to develop a new biofilter system. In this system, iron is retained while polycyclic aromatic hydrocarbons (PAHs) are degraded by biological activity. In this paper we present a summary of the findings from a laboratory-scale aerated biofilter, and discuss the application of these results for remediation of impacted groundwater at a former manufactured gas plant (MGP) site.

INTRODUCTION

Biological treatment is a commonly used remedial technology for removing organic contaminants from groundwater. However, groundwater from reduced environments also contains dissolved, reduced iron (Fe^{2+}), which can easily be oxidized into insoluble ferric hydroxides in aerobic biological treatment systems (Martineau et al., 1999). The conventional approach to this problem has been to pre-treat the water to precipitate and remove the iron prior to biological treatment (Hickey et al., 1995; Aherns et al., 1996). While this two-stage process is effective, the iron removal process adds significantly to the system operating requirements and the cost of groundwater treatment.

Many European communities use biologically-based iron removal filters (Mouchet, 1992), and these systems have seen limited use in the United States (Smith and Smith, 1994). In a biological filter, iron is retained by precipitation on the surfaces of microorganisms. This approach has several advantages over conventional iron removal systems, including increased iron removal efficiency, and greater iron retention capacity (Mouchet, 1992). Recent studies have even shown that the operational performance of a biological iron filter can be maintained indefinitely, without back washing, provided the iron loading rate is controlled (Michalakos et al., 1997).

In limited examples, the degradation of organic compounds such as dissolved organic matter from algae (Terauchi et al., 1995) and phenoxy acids from landfill leachate (Pedersen and la Cour Jansen, 1992) has been observed in iron filters. These findings provided the basis for our work to investigate the degradation of PAHs in MGP-impacted groundwater using an iron retaining biofilter.

EXPERIMENTAL PROCEDURES

This section contains a summary of the experimental procedures that were used to evaluate simultaneous removal of iron and PAHs from MGP-impacted groundwater. A detailed description of the experimental materials and methods is presented in Richard and Dwyer (2001).

MGP-Site Groundwater: A complex mixture of organic and inorganic contaminants often impacts MGP groundwater. PAHs are present at MGP sites because they are make up a significant percentage of coal tar, a common by-product of MGP operations (Hayes et al., 1996). Iron is one of the most common inorganic constituents in MGP-impacted groundwater, due to the low oxidation-reduction potential in most groundwater environments, and at MGP sites in particular due to on-site disposal of oxide box wastes. For this study, reduced groundwater containing both dissolved iron and PAHs was collected from the former Waterloo Coal Gasification Plant Site in Waterloo, Iowa. Field measurements during collection of the water showed that the water had a pH of 6.9, dissolved oxygen concentration of 0.2 mg/L, temperature of 11°C, and an oxidation-reduction potential of −144 mV.

Experimental Set-up: The groundwater treatment experiment was conducted using duplicate laboratory columns (50 mm by 600 mm glass cylinders with Teflon end-caps). The columns were filled with autoclaved gravel media (4.5 mm). Each of the treatment columns was operated with hydraulic loading rate of approximately 15 cm/day for 60 days. After 60 days one column was removed and the second was operated for an additional ten days with a loading rate of 30 cm/day. The column study was conducted in the dark and in a temperature-controlled room (11°C). This operational scheme was selected to model potential subsurface construction of a full-scale system. A small volume of filtered air (4 ml/min), the only biostimulant added to the groundwater, was added in a countercurrent flow to the groundwater at the base of the columns through a porous diffuesr stone.

Sampling and Analysis: Throughout the study, influent and effluent groundwater samples were collected from the columns for analysis of dissolved oxygen, pH, PAHs, and total iron. Dissolved oxygen and pH were measured with direct-reading probes. A mass balance approach was used to quantify PAH degradation and iron retention within the column. The mass of PAHs removed was calculated as the difference between the applied mass and the total mass of PAHs measured in the column effluent, in the column exhaust gas, or retained within the columns. The iron removal efficiency was calculated as the difference between the influent and effluent mass. PAHs in water samples were analyzed using high-pressure liquid chromatography (HPLC), while total iron was analyzed by flame atomic absorption. To quantify the mass of PAHs that volatilized from the columns two carbon filters (Supelco) were placed in series on the air exhaust line. The air-carbon traps were removed monthly and extracted for analysis of retained PAHs.

PAHs retained within the exhaust traps or within the column were solvent-extracted and quantified by HPLC.

RESULTS AND CONCLUSIONS

Influent Groundwater Characteristics: The influent groundwater applied to the column had the following average characteristics:

- pH → 6.9
- dissolved oxygen → 1.5 mg/l
- total iron → 3.8 mg/L
- total PAHs → 3.0 mg/L

The influent dissolved oxygen in the water applied to the columns was slightly higher than the value measured in the field, indicating minimal oxidation of the water occurred during collection, transfer, and storage. However, special techniques used in the field and in the laboratory (Tedlar bags or glass bottles with nitrogen headspace) effectively limited oxidation of the water before it was applied to the columns. The majority of the influent PAH was naphthalene. Seven other PAHs were quantified consistently including two-ring (acenaphthlyene, acenaphthene, fluorene), three-ring (phenanthrene, anthracene), and four-ring (fluoranthene, pyrene) PAHs. Benzo(a)anthracene and indeno(1,2,3-cd)pyrene were not detected in the influent, and chrysene, benzo(k)fluoranthene, dibenzo(ah)anthracene, benzo(ghi)perylene and benzo(a)pyrene were detected infrequently. Only the results of the eight PAHs commonly identified in this study were included in the removal efficiency calculations.

Column Observations and Effluent Results: Microbial growth and iron retention occurred primarily within the column. This conclusion is based on the observations of iron precipitation (orange-brown film on the gravel media) in the upper half of the columns (50 mm to 350 mm as measured from the top). The absence of flooding at the top of the column during the study, and visual observation of the flow diffuser from the top of each column when the system was disassembled at the end of the study confirmed that neither biofilm or iron precipitate was present at the top of the column, which may have indicated overloading or over-aeration of the influent.

The effluent water from both columns had an average pH of 8.3 and an average dissolved oxygen concentration of 9.2 mg/L. The concentration of iron in the effluent during the first three weeks of operation was generally less than 1 mg/L. After three weeks of operation, the effluent iron concentration had decreased to less than the detection limit (0.1 mg/L), and remained near the detection limit throughout the study. Approximately 97% of the applied iron was retained within the columns. The PAH removal efficiency for the two- and three-ring PAHs ranged from 83% to 99% as shown in Figure 1. This efficiency was observed after the first three days of operation, and was maintained throughout the study, including the final ten days when the loading to the remaining column was doubled. The four-ring PAHs, fluoranthene and pyrene were conserved

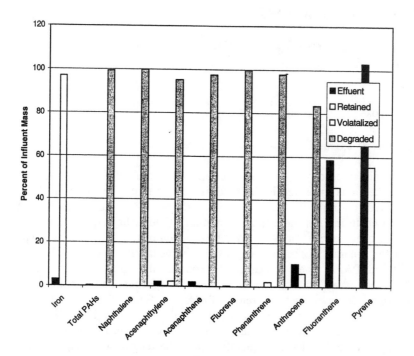

FIGURE 1. Percentage of Influent Iron and PAHs in the Effluent, Retained, Volatilized, or Degraded.

APPLICATION

We believe that a single-stage treatment system in which PAHs are biodegraded and iron is retained may have several advantages over two-stage treatment systems including elimination of the iron-sludge waste stream, reduction in operating procedures, and reduction in operating costs. The loading rate used in this study suggests that an area of approximately 400 square meters will be needed to treat a flowrate of 40 liters per minute. If constructed aboveground, the footprint for an iron/PAH biofilter is larger than other biological treatment systems. However, since the laboratory system we have evaluated was operated at ambient subsurface conditions, the entire system could be constructed below-ground – for example, within the limits of an excavation to remove MGP-impacted soils from the vadose zone – thus facilitating redevelopment or reuse of the property. In addition, because our experimental procedures did not use any biostimulants other than air, we believe that our findings may provide insight into the natural attenuation processes that occur at oxic/anoxic boundaries in the groundwater environment.

Use in Engineered Systems: We believe the results of this work can be used to design an engineered biofilter that could be constructed on-site using conventional drain-field technology. One issue that our study has not addressed is the long-term hydraulic performance for an engineered system. This issue would be addressed during field scale testing or full-scale operation. However, the results of our work suggest that iron precipitation will not be detrimental to the overall operation of a biofilter. Iron was retained primarily in the top half of our columns, but not near the inlet. This suggests that the hydraulic loading rate relative to the rate of oxidation was sufficient to distribute the iron as well as the biologically available carbon into main body of the column before it was precipitated or consumed, respectively. More importantly, we did not observe an increase in resistance to flow through the column, for either the air or water pathways, or any other significant changes in hydraulic performance that would suggest the potential for a long-term decrease in the treatment capacity of the system. Experience with other iron removal systems suggests that iron precipitation will not adversely affect the hydraulic performance of the proposed biofilter within the operating life for typical groundwater remediation systems (Rott and Lamberth, 1993; Hallberg and Martinell, 1976).

Natural Attenuation Considerations: In the groundwater environment, iron precipitation at the oxic/anoxic boundary may provide an improved region for bacterial adhesion to aquifer sediments, and may help to maintain iron and other important micro-nutrients within a biologically active zone at the margin of a contaminant plume. In laboratory studies, iron coated sand particles have been shown to facilitate significantly higher bacterial adhesion than uncoated quartz particles (Mills et al., 1994; Truesdail et al., 1998), while bacterial surfaces have been shown to have a positive influence on iron precipitation (Warren and Ferris, 1998). Retaining an easily reducible form of iron at the margin of the plume may also facilitate increased iron cycling as a plume is expanding, thus increasing the natural retardation of the plume (Lovley, 1997).

At a smaller scale, retention of iron by bacteria may affect the initial uptake or transport of non-polar compounds across the bacterial cell wall. Accumulation of iron along the bacterial cell membrane will tend to neutralize the surface charge on the cell membrane (Ferris et al., 1989; Lunsdorf et al., 1997), which has been suggested as a mechanism to improve the uptake of non-polar hydrocarbons, such as PAHs (Beveridge, 1989). This potential mechanism for enhancing natural attenuation of PAHs by bacteria is currently the focus of additional study.

ACKNOWLEDGEMENTS

Funding for this work was provided through a research grant from MidAmerican Energy Company, Sioux City, Iowa. The authors thank Robert Buschbom and Sam Nelson for their assistance during this study.

REFERENCES

Aherns, B. W., Leahy, M. C., and Blazicek, T. L. 1996. "Field Demonstration of Comparative Control Technologies for the Treatment of Groundwater at a Former MGP Facility." In: *Hazardous and Industrial Wastes: Proceedings of the Twenty-Eighth Mid-Atlantic Industrial and Hazardous Waste Conference*. Technomic Publishing Company, Inc., Lancaster, PA.

Beveridge T. J. 1989. "Interactions of Metal Ions with Components of Bacterial Cellwalls and their Biomineralization." In Poole and Gadd, (Eds.) *Metal-Microbe Interactions*. pp. 65-83. Oxford University Press, Oxford, UK.

Ferris, F. G., S. Schultze, T. C. Witten, W. S. Fyfe, and T. J. Beveridge. 1989. "Metal Interactions with Microbial Biofilms in Acidic and Neutral pH Environments." *Appl. Environ. Microbiol.* 55(5):1249.

Hallberg, R. O., and R. Martinell. 1976. "Vyredox - In Situ Purification of Ground Water." *Ground Water.* 14(2):88.

Hayes, T. D., D. G. Linz, D. V. Nakles, and A. P. Leuschner 1996. *Management of Manufactured Gas Plant Sites.* Amherst Scientific Publishers, Amherst, MA.

Hickey, B., R. Rajan, A. Sunday, D. Wagner, C. Cooper, V. Groshko, M. Benovska, A. Leuschner, P. Rice, G. Gromicko, and others. 1995. "GAC-FBR for Treating Water from MGP Sites." In: *Eighth International Symposium on Gas, Oil, and Environmental Biotechnology.* Institute of Gas Technology, Des Plaines, IL.

Lovley, D. R. 1997. "Microbial Fe(III) Reduction in Subsurface Environments." *FEMS Microbiological Reviews.* 20:305.

Lunsdorf H., I. Brummer, K. N. Timmis, and I. Wagner-Dobler. 1997. "Metal Selectivity of In Situ Microcolonies in Biofilms of the Elbe River." *J. Bacteriol.* 179(1):31.

Martineau, G., A. Tetreault, J. Gagnon, L. Deschenes, and R. Samson. 1999. "Development of a Biofiltration System for In Situ Remediation of BTEX-Impacted Groundwater." In B. C. Alleman and A. Leeson, (Eds.) *In Situ Remediation of Petroleum Hydrocarbon and Other Organic Compounds.* pp. 451-456 Battelle Memorial Institute, Columbus, OH.

Michalakos, G. D., J. M. Nieva, D. V. Vayenas, and G. Lyberatos. 1997. "Removal of Iron from Potable Water Using a Trickling Filter." *Water Res.* 31(5):991.

Mills, A. L., J. S. Herman, G. M. Hornberger, and T. H. DeJesus. 1994. "Effect of Solution Ionic Strength and Iron Coatings on Mineral Grains on the Sorption of Bacterial Cells to Quartz Sand." *Appl. Environ. Microbiol.* 60(9):3300.

Solution Ionic Strength and Iron Coatings on Mineral Grains on the Sorption of Bacterial Cells to Quartz Sand." *Appl. Environ. Microbiol.* 60(9):3300.

Mouchet, P. 1992. "From Conventional to Biological: Removal of Iron and Manganese in France." *Journal of the American Water Works Association.* 84(4):158.

Pedersen, B. M., and J. la Cour Jansen. 1992. "Treatment of Leachate Polluted Groundwater in an Aerobic Biological Filter." *European Water Pollution Control.* 2(4):40.

Richard, D. E., and D. F. Dwyer. 2001. "Aerated Biofiltration for Simultaneous Removal of Iron and Polycyclic Aromatic Hydrocarbons (PAHs) from Groundwater." *Water Environment Research. (submitted for publication).*

Rott, U., and B. Lamberth. 1993. "Groundwater Clean Up by In-Situ Treatment of Nitrate, Iron, and Manganese." *Water Supply.* 11(3/4):143.

Smith, C. D., J. F. Smith. 1994. "Comparison of Biological and Chemical/ Physical Iron Removal." In J. N. Ryan and M. Edwards (Eds.) *Critical Issues in Water and Wastewater Treatment: Proceedings of the 1994 National Conference on Environmental Engineering.* ASCE, New York, NY.

Terauchi, N., T. Ohtani, K. Yamanaka, T. Tsuji, T. Sudou, and K. Ito. 1995. "Studies on a Biological Filter for Musty Odor Removal in Drinking Water Treatment Processes." *Water Sci. Technol.* 31(11):229.

Truesdail, S. E., J. Lukasik, S. R. Farrah, D. O. Shah, and R. B. Dickinson. 1998. "Analysis of Bacterial Deposition on Metal (Hydr)oxide-Coated Sand Filter Media." *J. Colloid Interface Sci.* 203:369.

Warren, L. A., and F. G. Ferris. 1998. "Continuum Between Sorption and Precipitation of Fe(III) on Microbial Surfaces." *Environ. Sci. Technol.* 32(15):2331.

Ecotoxicological characterization of metabolites produced during PAH biodegradation in contaminated soils.

Frank Haeseler, Denis Blanchet (Institut Français du Pétrole,
Rueil Malmaison, France)
Peter Werner (Technische Universität, Dresden, Germany)
Jean-Paul Vandecasteele (Institut Français du Pétrole, Rueil Malmaison, France)

ABSTRACT

The variations of the ecotoxicological properties resulting from microbiological PAH degradation in industrial soils polluted by products of coal pyrolysis were studied in the laboratory. Two bacterial tests have been used, inhibition of bacterial luminescence in *Vibrio fischerii* for the determination of acute toxicity, and SOS-Chromotest with *Escherichia coli PQ 37* for evaluation of genotoxicity. The tests were performed on leachates prepared from untreated soils and from biologically-treated soils. Toxicity changes in the aqueous phases during PAH biodegradation in soil-suspension reactors were also studied, as well as during the biodegradation of a model PAH, phenanthrene.

In leachates from untreated soils from former manufactured gas plants (MGP), positive acute toxicity responses could be observed and correlated to the concentrations of the most abundant PAH, naphthalene and phenanthrene. These leachates did not exhibit genotoxicity, in accordance with their low concentrations in genotoxic PAH.

In the course of biodegradation, transitory positive responses to the acute toxicity and genotoxicity tests were observed. Acute toxicity was quite strongly reduced at the end of biodegradation. A positive response for direct genotoxicity (without induction by liver microsome extract S9), was observed in an early phase of PAH biodegradation and disappeared when degradation was complete.

Although toxic biodegradation intermediates do not appear to be long-lived, the question of their possible accumulation in PAH-contaminated soils in sites presenting environmental limitations to biodegradation such as oxygen availability, deserves further studies.

INTRODUCTION

Polycyclic aromatic hydrocarbons (PAH) are widespread genotoxic pollutants (Kramers and van der Heijden, 1990). They are present as components of coal tar at coal pyrolysis sites such as the former manufactured gas plant (MGP) sites where they constitute persistent pollutants. Several authors have shown that only partial degradation of PAH could be achieved in laboratory experiments even in optimized conditions (Weissenfels et al., 1991; Tiehm et al., 1997). Haeseler et al., (1999b) presented evidence relating these results to the fact that PAH pollutants remaining in soils after extensive biological treatment were unavailable for further biodegradation and for leaching. Besides PAH, water-soluble metabolites arising from PAH microbial degradation constitute another concern because of their possible toxicity. In particular, biodegradation involving oxygenases and enzymatic oxygenation by cytochrome P-450 of pro-genotoxic PAH into genotoxic metabolites constitute the basis of their mutagenicity and of their genotoxicity to mammals (DePierre and Ernster, 1978). In the present work, we studied the ecotoxicological properties of PAH (acute toxicity and genotoxicity) during their microbiological degradation in MGP soils in laboratory reactors.

MATERIALS AND METHODS

Soil samples. The soil samples originated from different former MGP sites (SA, M, LH and G) and have been previously described from analytical and biological points of view (Haeseler et al., 1999a, b, c).

Biodegradation. Experiments were performed in stirred flask reactors in a Sapromat respirometer (type D, Voith, Ravensburg, Germany) as described by Bouchez et al., (1997). Phenanthrene was added as an acetone solution, which was evaporated from the flask, before mineral salt medium and inoculum addition. One gram of contaminated soil from MGP site SA was used as inoculum after being characterized for PAH degraders as described by Haeseler et al., (1999c).

For the experiments with soil suspensions, the flask reactors contained 40 g of contaminated soil and 900 mL of mineral salt medium. Sampling during incubation was done by stopping flasks at time intervals indicated by O_2 consumption, and filtering their content through a regenerated cellulose filter (Schleicher & Schuell, Dassel, Germany).

Leaching Experiments. For soil leaching experiments, 100 g (dry weight) of soil were shaken with 1 L of distilled water at 20°C in 2-liter glass flasks for 24 hours on a rotary shaker at 10 rpm.

Analyses. PAH analysis was performed after solvent extraction by HPLC with fluorescence detection for water samples and by GC with flame ionization detection for soil samples, as described by Haeseler et al., (1999b). The metabolites produced during PAH degradation were determined as dissolved organic carbon (DOC) with a Carbon Analyser DC 80 (Dohrmann, NL).

Ecotoxicological Characterization. Ecotoxicity tests were carried out on soil leachates or on aqueous phases obtained during biodegradation experiments. Acute toxicity was determined with the Microtox test according to standard methods. The results are expressed as the maximal inhibition of bioluminescence for the undiluted sample (referred to as maximal inhibition in %) and as toxicity units (TU_{50}) calculated as ($100/EC_{50}$), with EC_{50} being the effective concentration of the tested solution, which induced a bioluminescence inhibition of 50%.

Genotoxicity was evaluated using the SOS-Chromotest with the tester strain *Escherichia coli PQ37 (sulA::lacZ),* kindly provided by Philippe Quillardet (Institut Pasteur, Paris, France), according to Quillardet and Hoffnung (1982). This test was adapted to a protocol using microtiter plates as described by Haeseler et al., (1999b). Results were expressed in equivalents of 4-nitroquinoline 1-oxide (4NQO) response for direct genotoxicity (assays without enzymatic activation) and as equivalent of benzo(a)pyrene (BaP) response for pro-genotoxicity (assays with rat liver microsome-preparation S9 mix).

RESULTS AND DISCUSSION

Toxicity of soil leachates. As shown in Table 1, the highest PAH concentrations in leachates were observed for the most polluted soils. The leachates of soils SA and M had substantial PAH concentrations and also presented significant responses for acute toxicity (maximal inhibition was 85% for soil SA and 50% for soil M). These toxicity values could be correlated to the concentrations of the most abundant PAH, naphthalene (for soil SA) and phenanthrene (for soil M) according to the acute toxicity values measured for each of them with the same test by Stieber (1995). These leachates did not exhibit genotoxicity, in accordance with their low concentrations in genotoxic PAH. Leachates from biologically-treated soils presented no acute toxicity and no genotoxicity, which reflected the negligible concentrations of PAH and also showed the absence of genotoxic dead-end metabolites of PAH biodegradation in these soils.

Table 1 : PAH concentration and toxicity values measured in leachates of soils originating from different former MGP sites (untreated soils and biologically-treated soils).

	soil SA		soil M		soil LH		soil G	
	untreated soil	treated soil	untreated soil	treated soil	untreated soil	treated soil	untreated soil	treated soil
Concentration of the sum of the 16 EPA PAH [g.kg^{-1}]	6.7	3.9	4.3	1.0	0.87	0.53	0.31	0.14
Naphthalene	5 059	0.09	7	0.05	2.12	0.10	0.08	0.09
Acenaphthylene	236	<0.01	99	<0.01	<0.01	<0.01	<0.01	<0.01
Acenaphthene	37	0.01	28	<0.01	0.11	0.02	0.07	0.01
Fluorene	112	0.05	124	<0.01	0.33	0.09	0.17	0.10
Phenanthrene	134	0.05	213	0.03	0.60	0.07	0.19	0.07
Anthracene	27	0.01	38	0.01	0.26	0.01	0.31	0.01
Fluoranthene	24	0.08	35	0.03	0.47	0.06	0.09	0.03
Pyrene	17	0.05	22	<0.01	0.30	0.08	0.10	0.05
Benz(a)anthracene	0.26	0.00	0.22	0.02	0.01	<0.01	0.01	<0.01
Chrysene	0.10	0.02	0.08	<0.01	0.03	0.02	0.02	0.02
Benzo(b)fluoranthene	0.25	<0.01	0.01	<0.01	0.03	0.01	0.03	<0.01
Benzo(k)fluoranthene	0.10	<0.01	<0.01	<0.01	0.01	<0.01	0.01	<0.01
Benzo(a)pyrene	0.15	<0.01	<0.01	<0.01	0.02	0.01	0.02	<0.01
Indeno(c,d)pyrene	0.03	<0.01	<0.01	<0.01	0.01	<0.01	0.01	<0.01
Dibenz(a,h)anthracene	0.04	<0.01	<0.01	<0.01	0.04	<0.01	0.03	<0.01
Benzo(g,h,i)perylene	0.12	<0.01	<0.01	<0.01	0.04	<0.01	0.04	<0.01
Sum of the 16 EPA PAH	5 647	0.36	566	0.13	4.38	0.47	1.18	0.38
Toxicity Units (100/EC$_{50}$)	905	nd	200	nd	nd	nd	nd	nd
Genotoxicity (+S9 mix)$^{(1)}$ [μgL^{-1}]	< 130	< 130	< 130	< 130	< 130	< 130	< 130	< 130
Genotoxicity (-S9 mix)$^{(2)}$ [μgL^{-1}]	< 13	< 13	< 13	< 13	< 13	< 13	< 13	< 13

nd : not detected (below calculation limit when EC$_{50}$<50%)

1 : expressed in μg.L^{-1} BaP equivalents (the lowest significant concentration for BaP was 178 μg.L^{-1})

2 : expressed in μg.L^{-1} 4NQO equivalents (the lowest significant concentration for 4NQO was 17 μg.L^{-1})

Toxicity during phenanthrene biodegradation. The variations of responses to the acute toxicity and genotoxicity tests in the course of biodegradation were studied. The results obtained in the case of a pure PAH, phenanthrene, are presented in figures 1 and 2. Figure 1 illustrates the time course of phenanthrene biodegradation, followed by oxygen consumption. For this poorly soluble compound (1.5 mg.L^{-1}) which is degraded with a high rate of mineralization (O_2/C about 0.73 mol/mol), continuous monitoring of O_2 consumption is a precise indicator of degradation progress (Bouchez et al., 1997). As represented in Figure 2 by the variations in bioluminescence inhibition (% inhibition and TU_{50} values), a large increase in the acute toxicity response of the incubation medium was observed, starting at the onset of phenanthrene degradation. The curves presented two successive peaks, at 3.5 days and 6 days respectively. The variations in overall concentrations of metabolites (measured by DOC) presented a clearly different profile with a maximum at 4 days (Figure 1), showing that changes in the nature of the degradation intermediates of phenanthrene were occurring. Residual acute toxicity was very low at the end of biodegradation, clearly below that of initial phenanthrene. Concerning genotoxicity, a positive response for direct genotoxicity (without induction by liver microsome extract S9), was observed at the beginning of phenanthrene biodegradation and disappeared when degradation was complete (Figure 2). No pro-genotoxicity was observed in tests with induction by rat liver microsome extracts (data not shown) in accordance with the low mutagenicity of phenanthrene (IARC, 1986). These genotoxicity results thus indicated the transitory production of genotoxic bacterial intermediates of biodegradation that were probably different from those known to be produced by enzymatic epoxidation with S9 of mutagenic PAH.

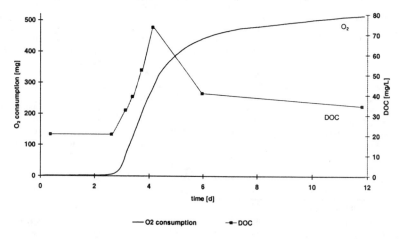

Figure 1 : Oxygen consumption and metabolite production (DOC) in the aqueous phase during the biodegradation of phenanthrene.

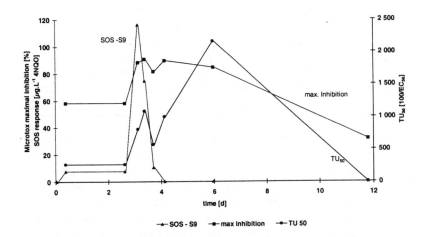

Figure 2 : Acute toxicity (Microtox) and genotoxicity (SOS Chromotest) of the aqueous phase during the biodegradation of phenanthrene.

Toxicity during soil PAH biodegradation. Compared to the evolution of PAH concentrations in soil suspensions during biodegradation in stirred reactors (Table 2), the diauxie observed in the corresponding O_2 consumption curve (Figure 3) appeared to be related to the rapid biodegradation of naphthalene (day one) followed by that of the 3-ring PAH (days 1 to 4). During 7 days the DOC increased significantly and reflected the production of metabolites, which were further degraded.

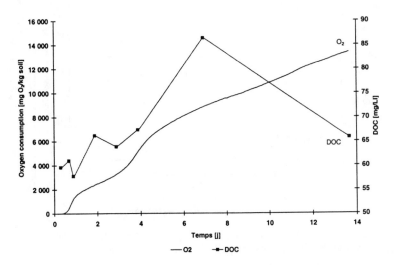

Figure 3 : Oxygen consumption and metabolite production (DOC) in the aqueous phase during the biodegradation of the soil SA. .

Table 2 : PAH concentrations in soil SA during biological treatment.

time [d]	Concentration [mg/kg]						
	0	1	2	3	4	7	14
Naphthalene	802	121	48	36	34	27	21
3 ring PAH	1 565	1 177	971	782	441	254	145
4 ring PAH	3 099	2 429	2 149	2 008	1 966	1 701	1 431
5 and 6 ring PAH	2 134	1 988	1 742	1 630	1 570	1 588	1 671
Sum of 16 EPA PAH	7 600	5 716	4 911	4 456	4 011	3 569	3 269
Total PAH	18 013	13 141	11 093	10 253	9 768	8 331	7 303

The results of the ecotoxicological analyses performed on aqueous phases during biodegradation are presented in Figure 4. A strong increase of the acute toxicity and a significant genotoxicity response can be seen during the biodegradation of naphthalene. During this step the aqueous concentration of naphthalene was reduced from 4,500 to 80 μg.L^{-1} (Figure 5). The direct genotoxicity measured without addition of rat liver microsomes rose up to an equivalent response of 212 μg.L^{-1} 4NQO and decreased rapidly. The evolution of genotoxicity was parallel to that of the acute toxicity. The latter was strongly reduced at the end of the biodegradation of the 3-ring PAH which decreased in the aqueous phase from 650 to 3.5 μg.L^{-1}.

In contrast to the observation made during phenanthrene biodegradation, no increase of acute toxicity or genotoxicity was noticed during the biodegradation of the three-ring PAH. An explanation could be a lower bioavailability of PAH when they were bound in the heavy coal tar matrix contaminating the soil (Haeseler et al., 1999a).

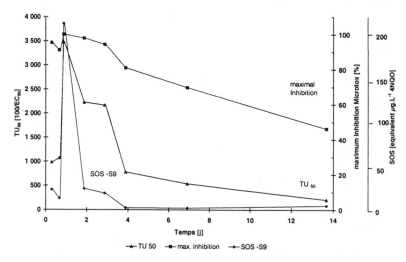

Figure 4 : Acute toxicity (Microtox) and genotoxicity (SOS Chromotest) of the aqueous phase during the biodegradation of the soil SA.

Like in the case of phenanthrene biodegradation, no genotoxicity was observed in tests with induction by rat liver microsome extracts (data not shown) in accordance with the low concentrations of the genotoxic PAH like benzo(a)pyrene (Figure 5).

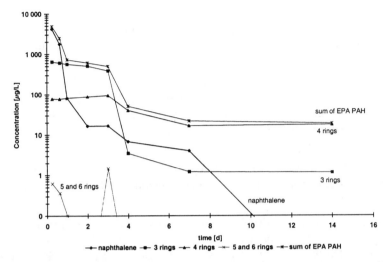

Figure 5 : PAH concentrations in the aqueous phase during biological treatment of soil SA.

CONCLUSIONS

These results show the relevance of ecotoxicological tools to assess and to follow the biodegradation progress of PAH. The SOS-Chromotest was found efficient for the evaluation of genotoxic metabolites generated during biodegradation of phenanthrene and naphthalene by bacterial consortia. These results are in accordance with the reports of Belkin et al. (1994) with the Mutatox test. It is interesting to note that genotoxic metabolites were produced from PAH with low mutagenicity. Although such intermediates did not appear to be long-lived in our experimental conditions, the question of their possible accumulation in PAH-contaminated soils in sites presenting environmental limitations to biodegradation such as oxygen availability, deserves further study.

REFERENCES

Belkin, S., Stieber, M., Tiehm, A., Frimmel, F.H., Abeliovich, A., Werner, P., Ulitzur, S., 1994 "Toxicity and Genotoxicity Enhancement During Polycyclic Aromatic Hydrocarbons Biodegradation", *Environmental Toxicology and Water Quality*, 9: 303-309.

Bouchez, M., Blanchet, D., Besnainou, B., Leveau, J.-Y., Vandecasteele, J.-P. 1997 "Kinetic studies of biodegradation of insoluble compounds by continuous determination of oxygen consumption", *J. Appl. Bacteriol.*, 82: 310-316.

DePierre, J.W., Ernster, L., 1978, "The metabolism of polycyclic hydrocarbons and its relationship to cancer", *Biochim Biophys Acta*, 473: 149-196.

Haeseler, F., Blanchet, D., Druelle, V., Werner, P., Vandecasteele, J.-P., (a) 1999, "Analytical characterization of soils from former manufactured gas plants" *Environ. Sci. Technol.*, 33: 825-830.

Haeseler, F., Blanchet, D., Druelle, V., Werner, P., Vandecasteele, J.-P., (b) 1999, "Ecotoxicological Assessment of Former Manufactured Gas Plants Soils: Bioremediation Potential and Pollutant Mobility." *Environ. Sci. Technol.*, 33: 4379-4384.

Haeseler, F., Blanchet, D., Druelle, V., Werner, P., Vandecasteele, J.-P., (c) 1999, *In The Fifth International Symposium on In Situ and On-Site Bioremediation*, Leeson A. and Alleman B.C. (eds), Battelle Press, Columbus Richland, April 19-22 San Diego CA, 5(8): pp 117-122.

IARC, 1986, "PAH as occupational carcinogens", *In: Bjørseth A, Becker G (eds)* PAH in work atmospheres: occurrence and determination. CRC Press, Boca Raton FL, USA.

Kramers, P.G.N., van der Heijden C.A., 1990, "Polycyclic aromatic hydrocarbons (PAH): Carcinogenicity data and risk extrapolations", *In: Rose J. ed. Environmental health. The impact of pollutants.* Gordon and Breach Science Publishers, New York.

Quillardet, P., Hoffnung, M., 1988. "The screening, diagnosis and evaluation of genotoxic agents with batteries of bacterial tests", *Mutation Research*, 205: 107-118.

Stieber, M., 1995, "Untersuchungen zum mikrobiellen Abbau von polycyclischen aromatischen Kohlenwasserstoffen. Monitoring, limitierende Faktoren, Ökotoxizität". Doctoral thesis at the Fakultät für Forst-, Geo- und Hydrowissenschaften der Technischen Universität Dresden, Germany.

Tiehm, A., Stieber, M., Werner, P., Frimmel, F.H., 1997, "Surfactant-Enhanced Mobilization and Biodegradation of Polycyclic Aromatic Hydrocarbons in Manufacctured Gas Plant Soil", *Environ. Sci. Technol.*, 31: 2570-2576.

Weissenfels, W.D., Beyer, M. Klein, J., Rehm, H.J., 1991, "Microbial metabolism of fluoranthene: isolation and identification of ring fission products", *Appl. Microbiol. Biotechnol.*, 34: 528-535.

EFFECTS OF AGING, BACTERIAL SOURCE AND DESORPTION ON PAH BIODEGRADATION

Sangchul Hwang and *Teresa J. Cutright*
(The University of Akron, Akron, OH 44224, USA)

Abstract: The objective of this study was to investigate the effects of compound aging in soil, bacterial source inoculated, and the number of desorption steps on the biodegradation of phenanthrene (PHE) and pyrene (PYR). A batch biodegradation study was conducted in the 3×3 Latin Square design to achieve this objective. To statistically compare the results, the Tukey's HSD test with an $\alpha = 0.05$ was performed. The results indicated that the number of desorption steps had a statistical difference. Among desorption steps, 0-time desorption yielded a more significant increase ($P<0.05$) in biodegradation than two and six time desorption which yielded similar results. Furthermore, the presence of a recalcitrant cosolute (PYR) had a detrimental effect on PHE degradation with bacteria that had been acclimated with PHE as the sole carbon source. An increase in desorption steps contributed to the decrease in toxicity of the PYR to the PHE-degraders. More conclusive evaluations will be made in the upcoming full-paper publication.

INTRODUCTION

The physicochemical interactions of polycyclic aromatic hydrocarbons (PAHs) with soils result in a strong binding and slow release rates that significantly influence their bioavailability [Luthy et al., 1997; Wilcke, 2000]. Hatzinger and Alexander [1995] and Yeom and Ghosh [1998] reported that PAH mineralization rate in soil was significantly retarded with an increase in contamination age due to elevated mass transfer limitations. Guerin and Boyd [1992] found that some bacteria species were able to degrade sorbed naphthalene, whereas other species required naphthalene to be in the solution phase. Laor et al. [1996] also demonstrated that some organisms were able to use phenanthrene directly from the bound phase at the same rate as from the solution phase.

These studies showed degradation rate was compound-soil-bacteria specific. However, bioremediation is still commonly believed to be a sequential process whereby a contaminant must first be desorbed and then degraded. Thus, desorption has been identified as the rate limiting step for soil systems [Hatzinger and Alexander, 1995; Zhang et al., 1995]. An emphasis must be placed on the desorption steps used when evaluating the data from these studies. For example, Burgos et al. [1999] reported that successive batch desorption was dependent on how often the water refilled, but not the time-scale of the water desorption.

Objective. Although it is well known that the PAH biodegradation is dependent on aging, bacteria and/or desorption, no study has been conducted that considers all three factors simultaneously. Therefore, the objective of this study was to investigate the effects of the three factors on the biodegradation of PAHs. Two model PAHs were chosen for use in the experiment. Phenanthrene (PHE) was selected because it is commonly found in petroleum-contaminated sites, is widely used as a representative PAH, and is readily degradable. Pyrene (PYR) was used as the other model compound because of its ubiquitous presence in the environment, carcinogenic property, simplicity of analysis and its resistance to biodegradation. To meet the objectives, a batch biodegradation study was conducted in the 3×3 Latin Square Design. Due to space limitations, only the experimental methods used and results of the statistical analysis will be made at this time.

MATERIALS AND METHODS

Soil Source and Characterization. Clean silty sand soil was collected from a depth of 3 to 6 m from a pristine area in Colombia, South America. The characteristics and preparation of soil were provided in detail in an early study [Hwang et al., 2000]

Compound Aging. Compound aging in soil was initiated by spiking ≥96% purity PHE and PYR in hexane at 100 mg of each compound per kg of soil. The samples were "homogenized" by vigorous shaking, and then hexane was immediately evaporated. The spiked soils were aged in the dark at room temperature for the allocated time, 50 and 200 days. The 0-day aging experiment was initiated immediately after the hexane was evaporated.

Bacterial Source. PHE-degrading, PYR-degrading, and a mixture of (PHE+PYR)-degrading bacteria were isolated from a petroleum-contaminated soil. Each culture was maintained over eight months via a standard transfer method. Each of the bacterial consortiums was harvested by centrifugation at 1070 g for 20 minutes and resuspended in 5.5 % NaCl solution. The microbial growth was monitored via agar plates containing 0.3 %(w/v) tryptic soy broth, 1.5 % agar, and an inorganic medium.

Desorption and Analysis Methods. Four grams of soil and 40 mL of 5 mM $CaCl_2$ solution were placed into the reactor. After being vortexed for 20 seconds, the reactor was centrifuged at 1070 g for 20 minutes. The successive desorption was repeated via a decant-and-refill manner, in which the decanted supernatant (87.5 %) was replaced with the fresh 5 mM $CaCl_2$ solution. The two and six-time successive desorptions were utilized to obtain the labile fraction and the desorption-resistant fraction, respectively [Lahlou and Ortega-Calvo, 1999].

The synchronous scan method was applied to detect PHE and PYR simultaneously via a Perkin-Elmer LS-50B fluorescence spectrophotometer. A constant wavelength interval, $\Delta\lambda = 112$ nm (365 nm ($\lambda_{emission}$) - 253 nm ($\lambda_{excitation}$)) was used for PHE detection and $\Delta\lambda = 44$ nm (379 nm - 335 nm) was employed for

PYR detection. The soil-phase concentration of the compound was determined by gas chromatography (GC) after Soxhlet extraction with methylene chloride for 15 hours. A Shimadzu 14A GC equipped with an Rtx-5 capillary column (30 m, 0.32 mm ID, 0.25 μm, cross bond of 5 % diphenyl-95 % dimethyl polysiloxane) and flame ionization detector was used. The specific method utilized was comprised of an initial temperature of 70 °C for 5 min, and 6 °C/min ramp to 310 °C, which was held for 1 min. Helium was used for a carrier gas.

Biodegradation Experiment. A destructive batch biodegradation test was carried out in duplicate for each treatment. After the allocated desorption steps, 25 mL of the mineral medium was added. The mineral medium for biodegradation was comprised of 5 mM of $CaCl_2$ with the following salts (in g/L): 0.053 KH_2PO_4, 0.1068 K_2HPO_4, 2.0 NH_4Cl, 2.0 Na_2SO_4, 1.0 KNO_3 and 0.2 $MgSO_4 7H_2O$. The reactors were pre-equalized on an orbital shaker at 125 rpm and 28±2°C for 24 hours. Following this, 10 mL of the washed bacteria solution was used to inoculate each reactor. The initial bacteria number was determined to be 2×10^6/mL for the PYR-degraders, 3×10^7/mL for PHE-degraders, and 4×10^7/mL for (PHE+PYR)-degraders inoculation. The total volume of the liquid phase was 40 mL. Hydrogen peroxide (H_2O_2) was added at 75 mg/L to ensure the presence of an aerobic environment. The reactors were removed in the time course and both bacterial number and compound concentrations were monitored.

Table 1 shows the 3×3 Latin Square experimental design containing three independent factors; the compounds aging time in soil, the bacteria sources inoculated, and the number of successive desorption steps. The response variables were the compound degradation in the solution. Since biodegradation rates were very fast (particularly for the 0-day aged sample), biodegradation data interpretation was based on the phenomena exhibited at the three-day results. The growth trend of total heterotrophic bacteria was phenomenogically evaluated based on the results over a 14-day experiment since three-day analysis was not enough to consider the lag and growth-phase in this study.

Table 1. Design matrix for 3×3 Latin Square to investigate PAH biodegradation.

Treatment	Aging (days)	Bacteria Source [a]	Number of Desorption (times)
1	0	A	0
2	0	B	2
3	0	C	6
4	50	B	0
5	50	C	2
6	50	A	6
7	200	C	0
8	200	A	2
9	200	B	6

Note) (a): A, B and C represent PYR, PHE and (PHE+PYR) degraders, respectively.

RESULTS AND DISCUSSION

Biodegradation of PHE and PYR. Table 2 shows the analysis of variance for the data. As shown, only the desorption frequency made a statistical difference for both PYR and PHE degradation. Other two factors, aging time and bacterial source did not yield a significant difference in this experiment.

Table 2. Analysis of variance for the 3×3 Latin Square experiment.

Source of Variation	DF[a]	Sum of Square		Mean of Square		F_o[b]	
		PYR	PHE	PYR	PHE	PYR	PHE
Aging	2	2510	402	1255	201		
Bacteria	2	1147	71878	574	35939		
Desorption	2	91913	297913	45956	1489565	29.325	67.971
Error	2	3131	43829	1566	21915		
Total	8	98702	3095240				

Note) (a) degree of freedom; (b) test statistic

Since it was found that either aging and bacterial type didn't yield a statistical difference in the compound biodegradation, the difference in desorption steps in conjunction with either aging time or bacterial source was used for the further statistical data analysis. The Tukey's Honestly Significant Difference (HSD) test with an $\alpha = 0.05$ was performed in order to statistically compare the desorption effects. The results indicated that the effects of 0-time desorption for both PYR and PHE was statistically different from that of either two or six-time desorption. However, two-time desorption didn't yield a substantial difference from 6-time desorption (Figure 1). This indicates that the two-step desorption removed the loosely bound PAH which would be easily bioavailable. Exceptions were found for PYR when aged for 50 days (Figure 1a) and inoculated with bacteria A (Figure 1b). These exceptions are partly indicative of interactive effects among the factors which could not be evaluated in 3×3 Latin Square design. For a better understanding these phenomena, a three-factor factorial design is required.

Bacterial Growth. Figure 2 exhibits a different trend of the bacterial growth for the 14-day experiment. For PYR-degraders, the growth was affected by aging and desorption which would significantly influence bioavailability. As with the increase in compound aging, the growth was substantially inhibited even with two-time desorption. In freshly contaminated soil, the bacterial growth was negligible with inoculation of PHE-degraders and two-time desorption. These were attributed to the toxicity of PYR to these organisms and to the resistant fraction to desorption, respectively. The similar trend was found for PHE-degraders in 200-day aged soil with six-time desorption. For (PHE+PYR)-degraders, the growth peak appeared later with an early population decrease regardless of aging and desorption. Unlike for the freshly aged soil, PYR toxicity to PHE-degraders was not found (treatment 4) in 50-day aged soil. This can be explained by the phenomena that PYR was more tightly bound and therefore not available as a result of the aging time. Blank reactors were also prepared to

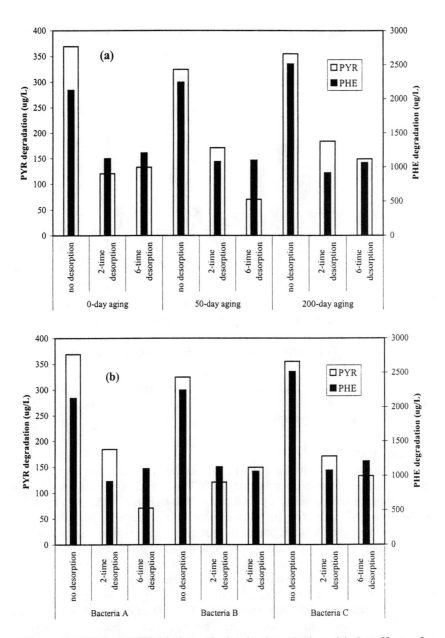

Figure 1. PYR and PHE biodegradation in the solution: (a) the effects of desorption frequency given the aging period when bacteria source was not considered; and (b) the effects of desorption frequency in accordance with bacteria source inoculated when aging period was not taken account.

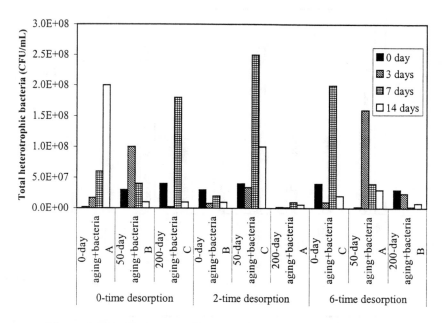

Figure 2. Total heterotrophic bacteria growth in treatments.

determine any abiotic loss and degradation by the indigenous bacteria. The results (data not shown) didn't show a significant impact on the biodegradation of compounds during the experiment.

CONCLUSIONS

Based on the statistical analysis, the number of desorption steps made a significant difference in biodegradation of PYR and PHE. The other two factors, compound aging and bacterial source didn't yield a statistical difference ($P>0.05$). The Tukey's HSD indicated that 0-time desorption produced a more substantial increase in compounds biodegradation than two and six-time desorption; i.e., more compound was present for the subsequent degradation. The biodegradation difference between two and six-time desorption was not statistically significant. From the phenomenogical analysis on the bacteria growth, PYR was toxic to PHE-degraders, which diminished with compounds aging and desorption frequency. Considering all three factors, (PYR+PHE)-degraders showed a better degradation efficiency and growth trend. This was because bacterial acclimation can provide for an increased biodegradation rate if the correct carbon source(s) is employed.

In this experiment, the response variable was the biodegradation of PAHs after three days. If three-day time constraint removed, results will be different. For this case, it is anticipated that compound aging or bacterial source used (or both) would become a more dominant factor. Thus, it is recommended that a three factor factorial design be used to assess a detailed contribution of factors as well as interactive effects between them. More comprehensive evaluations will be made in the upcoming full-paper publication.

REFERENCES

Burgos, W.D., Munson, C.M., and Duffy, C.J. 1999. "PHE Adsorption-desorption Hysteresis in Soil Described using Discrete-interval Equilibrium Models." *Water Resour. Res.* 35(7):2043-2051.

Guerin, W.F., Boyd, S.A. 1992. "Differential Bioavailability of Soil-sorbed Naphthalene to Two Bacterial Species." *Appl. Environ. Microbiol.* 58(4):1142-1152.

Hatzinger, P.B., and Alexander, M. 1995. "Effect of Aging of Chemicals in Soil on their Biodegradability and Extractability." *Environ. Sci. Technol.* 29(2):537-545.

Hwang, S., Ramirez, N, and Cutright, T. 2000. "Sequestration of Pyrene by Clay Minerals in a Natural Soil." *The 220th American Chemical Society National Meeting.* pp. 158-160. Washington DC.

Lahlou, M., and Ortega-Calvo, J.J. 1999. "Bioavailability of Labile and Desorption-resistant Phenanthrene Sorbed to Montmorillonite Clay Containing Humic Fractions." *Environ. Toxicol. Chem.* 18(12):2729-2735.

Laor, Y., Strom, P.F., and Farmer, W.J. 1996. "The Effect of Sorption on Phenanthrene Bioavailability." *J. Biotechnol.* 51(3):227-234.

Luthy, R.G., Aiken, G.R., Brusseau, M.L., Cunningham, S.D., Gschwend, P.M., Pignatello, J.J., Reinhard, M., Traina, S., Weber, W.J., and Westall, J.C. 1997. "Sequestration of Hydrophobic Organic Contaminants by Geosorbents." *Environ. Sci. Technol.* 31(12):3341-3347.

Wilcke, W. 2000. "Polycyclic Aromatic Hydrocarbons (PAHs) in Soil - A Review." *J. Plant Nutr. Soil Sci.* 163:229-248.

Yeom, I.-T., and Ghosh, M.M. 1998. "Mass Transfer Limitation in PAH-contaminated Soil Remediation." *Wat. Sci. Tech.* 37(8):111-118.

Zhang, W., Bouwer, E., Wilson, L., and Durant, N. 1995. "Biotransformation of Aromatic Hydrocarbons in Substrate Biofilms." *Water Sci. Technol.* 31(1):1-14.

EFFECT OF A RHAMNOLIPID BIOSURFACTANT ON THE PHENANTHRENE DESORPTION FROM A CLAY-LOAM SOIL

C. Amézcua-Vega; R. Rodríguez-Vázquez, H. M. Poggi-Varaldo and E. Ríos-Leal
(CINVESTAV-IPN., México, D.F., México)

ABSTRACT:. It is known that phenanthrene (Phe) and other polynuclear aromatic hydrocarbons (PAH's) show a great affinity to clay and organic matter of soils. Thus, adsorption-desorption processes play an important role in PAH's transport, availability and degradation. A laboratory study was performed to investigate the adsorption-desorption behavior of Phe onto a non-contaminated clay-loam soil. This work also examined the effect of the addition of a rhamnolipid biosurfactant on the desorption pattern of Phe. The soil was characterized as an acid, clay-loam soil with high organic matter content and exchange capacity. The first series of tests showed that approximately 25% of Phe was rapidly retained in the soil; which was followed by a slow adsorption process with an average rate of 0.29 % /day for the following 60 days. Further isotherm adsorption experiments indicated that a linear model was capable of describing the Phe adsorption onto soil, with an equilibrium adsorption constant (K_a) of 777 mL/g (coefficient of determination $r^2 = 0.995$). According to the K_d value and the desorption equilibria, the system seemed to be strongly irreversible. Semi-purified rhamnolipid biosurfactants added to soil enhanced the desorption of Phe. It was concluded that our clay-loam soil shows an important retention of Phe and the weathering significantly increased the Phe retention. The latter was confirmed by the low desorption determined in the second series of experiments. However, the addition of a semi-purified rhamnolipid surfactant significantly improved the Phe desorption from soil. This suggests that rhamnolipid biosurfactants could have an important beneficial effect on Phe availability in polluted soil.

INTRODUCTION

Phenanthrene (Phe) is one of the most abundant polycyclic aromatic hydrocarbon (PAH's) present in oil-polluted Mexican soils. It is known that the desorption from these persistent toxic compounds from soil is often very slow, with an estimated half-life on the order of months to many years (Ryan, *et al.* 1988). The complex nature and heterogeneity of soil are factors that makes more difficult the hydrocarbon release. Karickhoff, *et al.* (1981) reported that the desorption of pesticides and PAH's from soils showed strong irreversibilities. One of the strategies used to enhance the removal rates of soil pollutants is the addition of synthetic surfactants (Laha, *et al.* 1992, Tiemh, *et al.* 1997) or biosurfactants (Van Dyke, *et al.* 1993; Zhang, *et al.* 1995). Synthetic surfactants have contributed significantly to the understanding of the mechanisms that improve apparent solubility of PAH's, but, their low biodegradability, relative toxicity and limited efficiency at low concentrations has reduced their potential for application on soil (Laha, et al. 1992). On the other hand, biosurfactants seem to overcome several of the disadvantages of synthetic surfactants (Zhang, et al. 1995; Jordan, et al. 1999). The objectives of this study were i) to determine the fraction of Phenanthrene retention on a clay-loam soil and the effect of time (aging), and ii) to evaluate the hysteretic behavior of Phenanthrene on such soil and the influence of the addition of a biosurfactant on pollutant desorption. The biosurfactant used was a rhamnolipid produced by *Pseudomonas putida*.

MATERIALS AND METHODS

Sampling and characterization of soil. The soil was collected from an uncontaminated site in the Central Mexico Valley. The soil characteristics determined were; pH, cation exchange capacity, organic matter and texture, according to methods described in Alef and Nannipieri (1995).

Production of biosurfactant. Rhamnolipid was obtained from the cell-free extracelullar fluid of *Pseudomonas putida* CDBB-100 (Amézcua, 2001) and recovered with ethyl acetate (Schenk, et al. 1995).

Sorption test. The effect of time on Phe sorption was examined by Phe extraction tests at room temperature following the Weissenfels, et al. (1992) protocol. The sieved soil was contaminated with the initial Phe concentration of 400 mg/kg. Sorption was investigated for six periods of incubation (0 to 60 day range), 150 rpm at 25°C. The entire contents of vials were extracted three times with 10 mL of acetone. The organic extracts were analyzed by HPLC at 240 nm.

Adsorption-desorption isotherms of Phenanthrene. The experiments of adsorption-desorption equilibria were carried out according to procedures described by Fall, et al. (2000), in batch tests. Five g of soil were put in contact with 50 mL of water varying concentrations of Phe (0.2-1.0 mg/L) with $CaCl_2$ and NaN_3 and mixed at 25°C, 120 rpm for 3 days. At the end of the adsorption, soils from flasks that had 0.6 mg/L initial concentration of Phe were desorbed without or with a solution containing 250 mg/L of partially-purified rhamnolipid biosurfactant. The solid-desorption liquid ratio was 1 g to 50 mL, as recommended by Fall et al. (2000).

RESULTS AND DISCUSSION

Physico-chemical properties of the soil. Table 1 summarizes the properties of the soil. The soil was characterized as an acid, clay-loam soil with high organic matter content and exchange capacity.

Effect of phenanthrene sorption. The clay-loam soil showed an important retention of Phe and the further aging increased the Phe retention. Approximately 25% of Phe was rapidly retained in the soil, and this was followed by a slow sorption process with an average rate of 0.29% /day for the following 60 days (Fig. 1). It has been suggested (Alexander, 1995) that the initial fast process is related to be an adsorption of the pollutant onto the hydrophobic areas of soil surfaces, whereas the following adsorption is associated to migration and adsorption to less accesible sites within the soil matrix, this fraction of incorporated pollutant seems to represent its non-bioavailable portion.

TABLE 1. Physico-chemical characteristics of soil.

Parameters	Soil
o.m (%) [a]	5.0
pH	5.4
CEC (meq/100g) [b]	21.9
Clay (%)	29
Silt (%)	32
Sand (%)	39

Notes: a: o.m:organic matter; b: CEC: cation exchange capacity

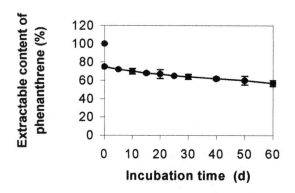

FIGURE 1. Influence to sorption processes on the recovery of phenanthrene added to uncontaminated soil.

Adsorption-desorption isotherms. Isotherm adsorption experiments indicated that a linear model was capable of describing the Phe adsorption onto clay-loam soil (Fig.2) with a high value equilibrium adsorption constant (K_a) of 777 mL/g (Table 2). Results of isotherm desorption of Phe, using only water, were also linear and showed a low value of the corresponding desorption coefficient (K_d: 139 mL/g). Desorption isotherm tended to be almost horizontal. Poggi-Varaldo et al. (2000) presented a general concept of hysteresis coefficient (HC) based on the ratio of the derivatives of the adsorption and desorption curves. They discussed the usefulness of such coefficient for the quantitative determination of the effect of sorption irreversibilities. They also demonstrated that, in the particular case of linear isotherms, the HC can be expressed as follows;

$$HC = \left(\frac{K_a}{K_d}\right) - 1 \qquad [1]$$

It is worth emphasizing that the larger the irreversibility, the larger the value of HC. The HC was 4.6 for the adsorption-desorption of Phe in soil-water system, suggesting a strong hysteretic behavior. It has been suggested (Biswas, et al. 1992). that, systems showing strong irreversibilities are associated to high adsorption constants. Fall, et al. (2000) observed that irreversibility occurs in soil with a large adsoption capacity that occurs when K_a is approximately 50 mL/g or higher. Cox, et al. (1998) also reported a hysteretic behavior in adsorption-desorption of imidacloprid on soils.

TABLE 2. Adsorption-desorption parameters of Phenanthrene on a clay-loam soil.

Isotherm	K_a [a] (mL/g)	K_d [b] (mL/g)	r^2 [c]	HC [d]
Adsorption	777	---	0.99	---
Desorption without biosurfactant	---	139	0.90	4.6
Desorption with biosurfactant	----	268	0.88	2.0

Notes: a: K_a is the linear adsorption coefficient; b: K_d is the linear desorption coefficient; c: r^2: Coefficient of determination of the linear regression model; d: HC: Hysteresis coefficient.

Effect of rhamnolipid on Phenanthrene desorption. The ability or partially purified *P.putida* CDBB-100 biosurfactant to enhance Phe desorption from clay-loam soil was investigated (Fig. 2). It was apparent that the slope of the desorption curve with biosurfactant was larger than the corresponding to the desorption without biosurfactant. At 250 mg/L of biosurfactant concentration, the K_d was 268 mL/g, (Table 2), that is, a 100% increase over the K_d of the desorption without biosurfactant. Correspondingly, the hysteresis coefficient in the presence of biosurfactant was lower (2.0) than that of the system without rhamnolipid. Moreover, if we interpret the interception of the desorption curves to the origin as the amount of pollutant irreversibly attached to soil (obtained by extrapolating the isotherms), this fraction decreased from 86.4% [(5.7/6.6)*100] to 40.9% [(2.7/6.6)*100] for non-biosurfactant and biosurfactant conditions, respectively.

The difference between the values of the retained fraction of pollutant obtained by the Weissenfels *et al.* (1992) protocol and that reported for the adsorption-desorption protocol is notable. In effect, the first is in the order of 25% whereas the second is 86.4% for desorption without biosurfactant. This discrepancy could be ascribed to the differences between the systems examined by the two protocols. In the first protocol, the Phe is spiked at a relatively high concentration, and most of the pollutant is probably in solid form. In the second protocol, Phe is present as a soluble species in the test solution. It can be speculated that the Weissenfels protocol measures the retention of both adsorbed and enmeshed solid pollutant onto soil, whereas the second protocol measures the retention of only the adsorbed Phe.

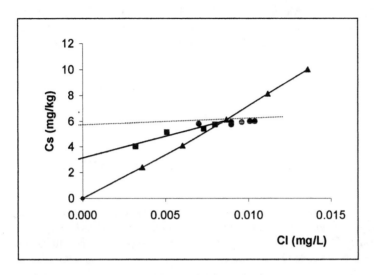

**FIGURE 2. Adsorption and desorption isotherms of phenanthrene.
(▲) Adsorption, (●) desorption without rhamnolipid, and (■) desorption
with ramnolipid.**

CONCLUSIONS

The analysis of the results presented in this work leads to the following conclusions:
-The soil was characterized as an acid, clay-loam soil with high organic matter content and exchange capacity.
-The first series of tests showed that approximately 25% of an initial Phe concentration of 400 mg/kg soil was rapidly retained in the soil; which was followed by a slow adsorption process with an average rate of 0.29 % /day for the following 60 days. The aging significantly increased the Phe retention.
-Further isotherm adsorption experiments indicated that a linear model was capable of describing the Phe adsorption onto soil, with an equilibrium adsorption constant (K_a) of 777 mL/g (coefficient of determination r^2= 0.995). According to the low K_d value and the desorption equilibria (139 mL/g) and the high value of the hysteresis coefficient (4.6), the desorptive behavior Phe-soil seemed to be strongly irreversible.
-Semi-purified rhamnolipid biosurfactant added to soil at 250 mg/L enhanced the desorption of Phe. The desorption coefficient was 100% higher (268 mL/g) and consequently, the hysteresis coefficient was lower (2.0) than those values obtained without biosurfactant. These results indicated that the rhamnolipid effected an important decrease on the irreversible behavior of the Phe in the model soil.

Overall, results of this work suggest that spiking rhamnolipids to Phe-soil systems would facilitate the Phe desorption from soil and could have an important beneficial effect on Phe availability.

ACKNOWLEDGEMENTS

Financial support of IMP, México (Project FIES-97-09-VI) and CONACyT, México (graduate scholarship) is gratefully acknowledged. The authors wish to thank Cirino Rojas, Professor of Analytical Central, Biotechnology Department, and Alfredo Medina for their valuable assistance on this project.

REFERENCES

Alef, K and Nannipieri. 1995. *Methods in Applied Soil Microbiology and Biochemistry*. Academic Press. USA.

Alexander, M. 1995. "How Toxic are Toxic Chemicals in Soil?". *Environ. Sci. Technol.* 29:2713-2717.

Amézcua V, C. 2001. *Relationship between biosurfactant production and phenanthrene removal from soil by stimulation of isolated rhizosphere microorganisms.* Doctoral Thesis. Dept. Biotechnology and Bioengineering, CINVESTAV-IPN, México D.F.; México.

Biswas, N.; Zytner, R.G. and Beutra, J.K. 1992."Model for Predicting PCE Desorption from Contaminated Soils". *Water Environ. Res.* 64:170-176

Cox, L.; Koskinen, W.C. and Yen, P.Y. 1998. "Influence of Soil Properties on Sorption-Desorption of Imidacloprid". *J. Environ. Sci. Health.* B33, 2:123.

Fall, C. Chaouki, J. and Chavarie, C. 2000."Desorptive Behavior of Pentachlorophenol (PCP) and Phenanthrene in Soil-Water Systems". *Water Environ. Res.* 72:162-169.

Jordan, N.R., Nichols, P.E. and Cunningham, B.A. 1999. "The role of (Bio) Surfactant Sorption in Promoting the Bioavailability of Nutrients Localized at the Solid-Water Interface". *Water. Sci.Technol.* 39:91-98.

Karickhoff, S.W. 1981. "Semi–Empirical Estimation of Sorption of Hydrophobic Pollutants on Natural Sediments and Soils". *Chemosphere*.10:833.

Laha, S. and Luthy, R. 1992. "Effects of Nonionic Surfactants on the Solubilization and Mineralization of Phenanthrene in Soil-Water Systems". *Biotech Bioeng.* 40: 1367-1380.

Poggi, V. H.M.; Caffarel, M. S. and Rinderknecht, S. N. 2000. *A new differential hysteresis coefficient for characterizing the adsorption-desorption behavior of pollutants on solid matrices*. Cuadernos de Tecnología No. 1, iii + 35 pp. TESE. Ecatepec, Edo. de México, México. In Spanish. Extended abstract in English.

Ryan, J.A., Bell, R.M., Davidson, J.M. and O'Connors, G.A. 1988. "Plant Uptake of Nonionic Organic Chemicals from Soils". *Chemosphere*, 17:2299-2323.

Schenk, T.; Schuphan, I. and Schimidt, B. 1995. "High-performance Liquid Chromatographic Determination of the Rhamnolipids Produced by *Pseudomonas aeruginosa*." *J. Chrom*. A 693:7-13.

Tiehm, A., Stieber, M., Werner, P. and Frimmerl, H.F. 1997. "Surfactant-Enhanced Mobilization and Biodegradation of Polycyclic Aromatic Hydrocarbons in Manufactured Gas Plant Soil". *Environ. Sci. Technol.* 31:2570-2576.

Van Dyke, I. M; Couture, P.; Bravuer, M; Lee, H and Trevors, J.T. 1993. " *Pseudomonas aeruginosa* UG2 Rhamnolipid Biosurfactants: Structural Characterization and their Use in Removing Hydrophobic Compounds from Soil". *Can.J. Microbiol.* 39: 1071-1078.

Weissenfels, W.D., Klewer, H. and Langhoff, J. 1992. "Adsorption of Polycyclic Aromatic Hydrocarbons (PAH's) by Soil Particles: Influence on Biodegradability and Biotoxicity. *Appl. Environ. Biotechnol.* 36:689-696.

Zhang,Y. and Miller, M.R. 1995. "Effect of Rhamnolipid (Biosurfactant) Structure on solubilization and Biodegradation of n-Alkanes". *Appl. Environ. Microbiol.* 61:2247-2251.

BACTERIAL STRATEGIES TO OPTIMIZE THE BIOAVAILABILITY OF SOLID PAH

Lukas Y. Wick and Hauke Harms
(Swiss Federal Institute of Technology, Lausanne, Switzerland)

ABSTRACT: Experimental and theoretical considerations show that microbial degradation of polycyclic aromatic hydrocarbons (PAH) in soil requires bioavailability enhancing strategies of the bacteria such as i) direct adhesion to the substrate source, ii) reduction of the cell surface concentration due to enhanced substrate affinity, or iii) dispersion or solubilization of the substrate by means of biosurfactant excretion. By using poorly water-soluble solid anthracene as sole carbon source in batch cultures, we examined the possible role of bioavailability-promoting strategies in anthracene degradation by *Mycobacterium sp.* LB501T. *M. sp.* LB501T exhibited a high specific affinity for but no production of surface-active compounds. When solid anthracene served as sole carbon source, *M. sp.* LB501T grew as a confluent biofilm on the solid substrate. No biofilm formation on the anthracene was observed, when glucose was added as co-substrate. This difference was attributed to a surface modification of the bacteria. In batch adhesion experiments, anthracene-grown cells exhibited significantly better adhesion efficiency to hydrophobic surfaces than glucose-grown cells. This observation indicates that attachment may be an actively regulated strategy to optimize substrate bioavailability.

INTRODUCTION

Many hydrophobic organic compounds (HOC) are priority soil contaminants because of their toxicity and their tendency to persist in soils and to escape biological degradation even in engineered bioremediation, such as mixing, slurrying, sonication, or heating of soil. As bacteria appear to degrade chemicals only when they are dissolved in water, physically retarded HOC-transfer to the aqueous phase combined with an unequal spatial distribution of microorganisms and HOC often results in limited efficiency of HOC-degradation *in-situ* (Luthy et al., 1994, Zhang et al., 1998). Consequently, HOC availability for biodegradation has been defined as the extent of mass transfer to microbial cells relative to the intrinsic catabolic potential of these organisms (Bosma et al., 1997, Harms and Bosma, 1997). This definition accounts for the dynamics of biodegradation and points at the importance of the substrate supply for a consumptive process. Although limited bioavailability appears to be primarily a physically controlled phenomenon, recent observations indicate that bioavailability-promoting, organism-specific, biological strategies may exist and that generalizations about the bioavailability of HOC, such as polycyclic aromatic hydrocarbons (PAH) may be inappropriate. Some bacteria, for instance, degraded HOC, either sorbed to a solid matrix or dissolved in an organic phase, faster than desorption or partitioning rates in the absence of bacteria would suggest. Moreover, different isolates degraded sorbed HOC at different rates indicating organism-specific bioavailability. It has been further shown, that sorption-limited bioavailability of PAH plays an important role in the selection of PAH-degrading bacteria, and that different bacteria inhabiting the same soil may be adapted to different degrees of PAH bioavailability (Bastiaens et al., 2000, Friedrich et al., 2000). Members of the genera *Sphingomonas* and *Mycobacterium* and related actinomycetes have been identified as efficient degraders of PAH (Balkwil et al., 1997, Kleespies et al., 1996, Linos et al., 2000) and thus may be adapted to the utilization of poorly bioavailable compounds. These bacteria may apply strategies to improve the bioavailability of their potential substrates. Theoretically, HOC-degrading bacteria may enhance the substrate transfer by (i) shortening their distance to a substrate source, for instance by attaching to it, (ii) by use of biosurfactants to disperse and to transfer the substrate into the bacteria's aqueous surrounding and (iii) by developing high-affinity uptake systems that efficiently reduce the aqueous phase substrate concentration (Figure 1).

Objective. Here we present a study that examined whether *Mycobacterium* sp. LB501T makes use of the before-mentioned possible strategies. This strain was isolated from PAH-contaminated soil by applying a new Teflon membrane based extraction that promotes hydrophobic bacteria likely to adhere to hydrophobic surfaces (Bastiaens, et al., 2000). *M.* sp LB501T grows readily on poorly water-soluble anthracene as sole carbon source. The provision of low amounts of solid anthracene has been shown to cause considerable mass transfer limitation of this strain that led to pseudo-linear growth, whereas provision of >30 g L^{-1} resulted in exponential growth, respectively (Wick et al., 2001). Thus, the possibility to exert limited bioavailability in a controlled way by adding only little anthracene made this bacterium a good candidate for the examination of possible bacterial bioavailability-enhancing mechanisms.

MATERIAL AND METHODS

Bacterium and Culture Conditions. *Mycobacterium* sp. LB501T is an aerobic, rod-shaped bacterium capable of degrading and growing on anthracene (Bastiaens, et al., 2000). *M.* sp. LB501T was grown in 300-mL Erlenmeyer flasks on a gyratory shaker at 130 rpm at 25°C in a minimal medium (Wick, et al., 2001) containing of 2-3 g L^{-1} of solid anthracene or 2 gL^{-1} glucose, resp. Cells for adhesion experiments were harvested in the late exponential stage (glucose-grown bacteria). Linear growing cultures on solid PAHs were used after about 15 days growth.

Analytical methods. Surface tensions of bacterial cultures were measured with a plate device according to the Wilhelmy technique (Adamson, 1990). The electrophoretic mobility and the cell surface hydrophobicity of the bacterial cells were analyzed according to a modified method (Jucker et al., 1996) of van

Loosdrecht et al. (Loosdrecht et al., 1987). Batch adhesion experiments were performed according to a method described by Rijnaarts et al. (Rijnaarts et al., 1993). The adhesion efficiency of bacteria α_t upon collision (Elimelech and O'Melia, 1990) is commonly quantified by the ratio of the rate of attachment η_t to the rate of bacterial transport to the surfaces η_{trans} ($\alpha_t = \eta_t / \eta_{trans}$)..

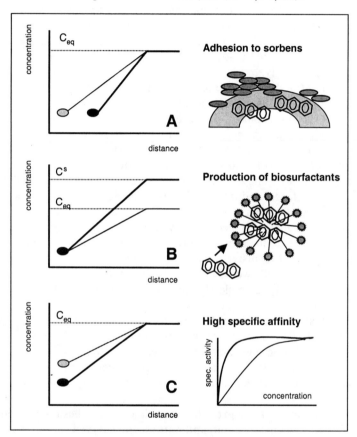

FIGURE 1. Bacterial strategies to enhance the diffusive flux of poorly available organic pollutants. Optimization of the diffusive flux (symbolized by bold lines), i.e. steepening of the concentration gradient can be obtained by shortening of the diffusion distance by attachment to the substrate (A), enhancing the dispersion and the (pseudo)solubility (C_s) of the contaminant by biosurfactant production (B), and reducing the cell surface concentration due to high substrate affinity (C).

RESULTS AND DISCUSSION

Adhesion to solid PAH. When small amounts of anthracene crystals (<2 g L^{-1}) were supplied as the sole carbon source in shaken batch cultures, biofilm forming on the crystals contributed up to 40% to the total biomass during all growth phases. The ratio of suspended to crystal-bound biomass depended on the culture conditions, such as shaking velocity. *M*. sp. LB501T biofilms were visualized by scanning electron microscopy (SEM) (Figure 2). Cryo-SEM images exhibited direct contact of extracellular polymeric substances (EPS) with the anthracene surface (Wick, et al., 2001). The distance of the nearest bacteria from the surface was ≤ 1 μm. *M*. sp. LB501T did not colonize anthracene crystals provided in small amounts when dissolved glucose as additional carbon source was present. The observation that biofilm formation did generally not occur under high substrate regimes suggested that it might allow growth under conditions of low substrate availability only.

The impact of the substrate on the attachment was further studied in batch adhesion experiments. Glucose- and anthracene-grown *M*. sp. LB501T cells were tested for their (dimensionless) adhesion efficiency α_t to glass, Teflon and anthracene- or phenanthrene-covered surfaces, respectively. Glucose-grown bacteria did not discriminate between the different surfaces, exhibiting adhesion efficiencies α_t of about 0.1 to 0.2 to all surfaces studied (Table 1). Anthracene-grown cells, however, showed a clear adhesion-preference for hydrophobic surfaces: For instance, adhesion to hydrophobic Teflon, anthracene-and phenanthrene-covered surfaces was 40-, 270-, and 240-fold higher than to hydrophilic glass. It must be noted that α_t values for anthracene-grown cells were significantly above 1. In the absence of repulsive forces α_t reaches unity and falls below 1 in presence of repulsive forces, respectively. As anthracene and glucose-grown *M*. sp. LB501T cells were mobile and calculation of α_t relies on knowledge of the transport rate (η_{trans}), we clearly have underestimated η_{trans} by assuming that diffusion-driven processes exclusively drove transport of *M*. sp. LB501T. α_t values thus should be considered as relative adhesion efficiencies only. Our results of the adhesion experiments thus may explain why *M*. sp. LB501T was isolated when a Teflon membrane was used to enrich anthracene-degrading bacteria from soil, yet not when conventional enrichment protocols were applied to the same soil (Bastiaens, et al., 2000).

TABLE 1. Influence of the growth substrate on the adhesion efficiency, α_t of *M*. sp. LB501T to glass, Teflon, and anthracene-covered surfaces. Because of enhanced contact probabilities due to the mobility of bacteria some α_t exceed 1 and thus should be considered as relative adhesion efficiencies only.

substrate	α_t glass	α_t Teflon	α_t anthracene	α_t phenanthrene
glucose	0.20	0.12	0.11	0.11
anthracene	0.03	1.14	8.10	7.20

Lack of biosurfactant production. No drop of the surface tension was observed within 250 h of growth, indicating no release of surface-active compounds into the medium. This does not exclude the presence of bioemulsifiers, as this class of molecules does not reduce the surface activity. However, the effect of a bioemulsifier on a crystalline compound would be restricted to the dispersion of the crystals and not increase the aqueous concentration.

Substrate Affinity. The specific affinity $a°_A$ (L $g_{protein}^{-1}$ h^{-1}) of *M*. sp. LB501T was calculated according to ₁Button (Button, 1985) based on the previously published (Wick, et al., 2001) maximal rate of substrate uptake, q_{max} (mol $mg_{protein}^{-1}$ h^{-1}) and the half saturation constant K_m (mol m^{-3}) using the following definition: $a°_A = q_{max} \cdot K_m^{-1}$. $a°_A$ expresses the efficiency of a bacterium to reduce the substrate concentration at its surface in relation to the bulk aqueous substrate concentration. High specific affinities create steep concentration gradients and fast diffusive substrate transfer (Figure 1C). The specific affinity of *M* sp. LB501T was 32500 L $g_{protein}^{-1}$ h^{-1}. This is in the range of other efficient HOC-degraders, and significantly above the $a°_A$-values of bacteria degrading much better soluble substrates (Button, 1985). This indicates that strain *M*. sp. LB501T is adapted to the naturally occurring low range of aqueous anthracene concentrations ($c^w_{sat} \approx 62$ μgL^{-1}).

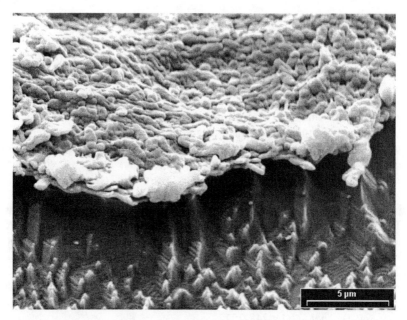

FIGURE 2. Scanning electron micrograph of an anthracene crystal covered by a confluent *Mycobacterium* sp. LB501T biofilm. (enlargement 4100x). *M.* sp. LB501T grows in close contact to solid anthracene. The gap between the anthracene surface and the biofilm is an artifact due to the air drying of the sample

CONCLUSIONS
Due to its very low aqueous solubility, anthracene is slowly transferred to bacteria even under most favorable conditions, for instance when it is supplied as accessible crystals in liquid medium. The heterogeneity of soil further restricts its

bioavailability and anthracene mass transfer rates will always be far below those that most bacteria would need to grow and survive. From our results, it appears that *M.* sp. LB501T may be specialized in the utilization of HOC compounds by being extremely oligotrophic. This means that *M.* sp. LB501T cells make use of high affinity uptake systems and that their substrate requirements for cell maintenance are very low. The fact that *M.* sp. LB501T tends to attach to HOC crystals only when it is beneficial, because e.g. the substrate flux is low, indicates that (a) attachment is an actively regulated strategy to optimize substrate availability, and (b) that *M.* sp. LB501T may optimize the anthracene flux within the physical constraints of its habitat.

ACKNOLEDGMENTS

This study has been financially supported by EC Biotech programs (contracts BIO4-CT97-2015 and QLK3-CT-1999-00326) and the Swiss Federal Office for Education and Science (contracts 96.0404 and 99.0366). The support is greatly acknowledged.

REFERENCES

Adamson, A. W. 1990. *Physical Chemistry of Surfaces*. John Wiley, New York, NY.

Balkwil, D. L., G. R. Drake, R. H. Reeves, J. K. Frederickson, D. C. White, D. B. Ringelberg, D. P. Chandler, M. F. Romine, D. W. Kennedy and C. M. Spadoni. 1997. "Taxonomic study of aromatic-degrading bacteria from deep-terrestrial-subsurface sediments and description of *Sphingomonas aromaticivorans* sp. nov., *Sphingomonas subterranea* sp. nov., and *Sphingomonas stygia* sp. nov." *Int. J. Syst. Bacteriol. 47*: 191-201.

Bastiaens, L., D. Springael, P. Wattiau, H. Harms, R. d. Wachter, H. Verachtert and L. Diels. 2000. "Isolation of new polycyclic aromatic hydrocarbon (PAH) degrading bacteria using PAH sorbing carriers" *Appl. Environ. Microbiol.* 66(5): 1834-1843.

Bosma, T. N. P., P. J. M. Middeldorp, G. Schraa and A. J. B. Zehnder. 1997. "Mass Transfer Limitation of Biotransformation: Quantifying Bioavailability" *Environ. Sci. Technol.* 31(1): 248-252.

Button, K. D. 1985. "Kinetics of Nutrient-Limited Transport and Microbial Growth" *Microbiol. Rev.* 49(3): 270-297.

Elimelech, M. and C. R. O'Melia. 1990. "Kinetics of deposition of colloidal particles in porous media" *Eviron. Sci. Technol.* 24: 1528-1536.

Friedrich, M., R. J. Grosser, A. Kern, W. P. Inskeep and D. M. Ward. 2000. "Effect of model sorptive phases on phenanthrene degradation: Molecular analysis of enrichments and isolates suggests selection based on bioavailability" *Appl. Microbiol. Biotechnol. 66*: 2703-2710.

Harms, H. and T. N. P. Bosma. 1997. "Mass transfer limitation of microbial growth and pollutant degradation" *J. Ind. Microbiol. Biotechnol. 18*: 97-105.

Jucker, B. A., H. Harms and A. J. B. Zehnder. 1996. "Adhesion of the positively Charged Bacterium Stenotrophomonas (Xanthomonas) maltophilia 70401 to Glass and Teflon" *J. Bact. 178*(18): 5472-5479.

Kleespies, M., R. M. Kroppenstedt, F. A. Rainey, L. E. Webb and E. Stackebrandt. 1996. "*Mycobacterium hodleri* sp. nov., a new member of the fast growing mycobacteria capable of degrading polycyclic aromatic hydrocarbons." *Int. J. Syst. Bacteriol. 46*: 683-687.

Linos, A., M. M. Berekaa, R. Reichelt, U. Keller, J. Schmitt, H. C. Flemming, R. M. Kroppenstedt and A. Steinbüchel. 2000. "Biodegradation of cis-1,4-Polyisoprene Rubbers by Distinct Actinomycetes: Microbial Strategies and Detailed Surface Analysis" *Appl. Environ. Microbiol. 66*(4): 1639-1645.

van Loosdrecht, M. C. M., J. Lyklema, W. Norde, G. Schraa and A. J. B. Zehnder. 1987. "The role of bacterial cell wall hydrophobicity in adhesion" *Appl. Microbiol. Biotechnol. 53*: 1893-1897.

Luthy, R. G., D. A. Dzombak, C. A. Peters, S. B. Roy, A. Ramaswami, D. V. Nakles and B. R. Nott. 1994. "Remediating Tar-Contaminated Soils at Manufactured Gas Plant Sites" *Environ. Sci. Technol. 28*(6): 266A-277A.

Rijnaarts, H. H. M., W. Norde, E. J. Bouwer, J. Lyklema and A. J. B. Zehnder. 1993. "Bacterial adhesion under static and dynamic conditions" *Appl. Eviron. Microbiol. 59*: 3255-3265.

Wick, L. Y., T. Colangelo and H. Harms. 2001. "Kinetics of Mass Transfer-Limited Bacterial Growth on Solid PAHs" *Environ. Sci. Technol. 35*(2): 354-361.

Zhang, W. X., E. J. Bouwer and W. P. Ball. 1998. "Bioavailability of Hydrophobic Organic Contaminants: Effects and Implications of Sorption-Related Mass Transfer on Bioremediation" *Ground Water Monit. R. 18*(1): 126-138.

FIELD-SCALE STUDIES ON REMOVAL OF CREOSOTE FROM CONTAMINATED SOIL

Harry L. Allen (U.S. EPA/ERTC, Edison, NJ USA)
Leo Francendese (U.S. EPA, Atlanta, GA USA)
Greg Harper (Tetra Tech EM, Inc., Duluth, GA USA)
T. Ferrell Miller (Lockheed Martin/REAC, Inc., Edison, NJ USA)

Abstract: The Walker Springs Wood Treating site is an abandoned wood treating facility located near the city of Jackson, Alabama. Site soil was found to be extensively contaminated with creosote oil. Bench-scale studies were conducted to screen soils for creosote-degradative activity and to develop a nutrient recipe that would promote rapid degradation of creosote in pilot-scale and field-scale studies. Screening studies identified a site soil sample which contained microbial populations capable of removing almost 89% of target creosote components in 28 days. Recipes were subsequently developed which promoted the removal of 87% to 90% of target creosote components in 84 days in bench-scale solid-phase studies. In 1997, four pilot-scale land farm cells were constructed with individual capacities of 4.7 yd^3 (3.6 m^3) to evaluate recipes developed from bench-scale studies. Initial creosote concentrations in solid-phase studies ranged from 3,300 to 5,500 mg/kg. One of the recipes promoted the removal of over 94% of target creosote components in 90 days. A field-scale unit was constructed at the site with a capacity of 4,800 yd^3 (3,700 m^3). Almost two years (699 days) were required before creosote component removal achieved the end results obtained in bench-scale and pilot-scale studies.

INTRODUCTION

Creosote is a commercial product widely used in the wood preservation industry. This product, along with pentachlorophenol (PCP), and copper, chromium, and arsenic (CCA), is one of the most widely used pesticides on a volume basis (Mueller et al., 1989). Creosote is a complex mixture of over 200 chemicals with less than 20 present in amounts greater than one percent of the total. Coal tar creosote is reported to contain 85% polycyclic aromatic hydrocarbons (PAHs), 10% phenolic, and 5% heterocyclic compounds (Mueller et al., 1989). The creosote found in the environment may differ widely in composition and may also include petroleum hydrocarbons and PCP residues.

There is an ample amount of literature indicating that creosote components can be degraded by purified or mixed microbial cultures (Gibson and Subramanian, 1984; Cerniglia, 1984; Guerin and Jones, 1988; Heitkamp et al., 1988; Mueller et al., 1990; Mueller et al., 1991). Biodegradation is an accepted technology for removing creosote from contaminated soil.

In September, 1996, soil samples were collected at the Walker Springs Wood Treating site, shipped to the U.S. Environmental Protection Agency Environmental Response Team Center (U.S. EPA/ERTC) bioremediation laboratory, and screened for creosote-degradative activity. The screening study identified potent creosote-degradative activity in one of the samples tested. Bench-scale solid-phase studies were subsequently conducted to identify a suitable nutrient recipe which promoted the rapid degradation of target creosote components. Results from this study indicated that approximately 90% of the target components evaluated were degraded by indigenous microbial populations in test recipes.

Bioremediation was chosen by the U.S. EPA On-Scene Coordinator to treat the creosote-contaminated soil at this site. In July, 1997, pilot-scale cells and a field-scale land farm unit were constructed at the site to initiate the cleanup action and monitored over a two year period. Results from the bench-scale, pilot-scale, and field-scale studies are presented.

Objective. The objective of these studies was to assess the performance of solid-phase bioremediation technology in removing creosote from site soil. Bench-scale studies were conducted to screen site soils for creosote-degradative activity and to develop a nutrient recipe which promoted the rapid degradation of creosote components in site soil. Developed recipes were tested in pilot-scale and field-scale studies.

Site Description. The Walker Springs Wood Treating site is an abandoned wood treating facility located near the town of Jackson, Alabama. The site, covering an area of approximately one acre (0.4 h), was the location of a small family-owned company, which produced wood products treated with creosote. Very little information is available about the operating history of this site. Site soil was found to be extensively contaminated with creosote, presumably due to poor housekeeping and improper handling procedures. Soil analysis showed that PCP contamination was quite localized and found only in sediment samples collected from a site lagoon.

MATERIALS AND METHODS

Soil Screening Studies. Four soil samples were screened for biodegradative activity in a nutrient-amended creosote-saturated screening medium. The medium was prepared by washing creosote-contaminated soil with a 0.50% volume/volume (v/v) solution of Tween 80® surfactant solution in 0.10 Normal (N) sodium hydroxide. Sodium hydroxide was included in the solution to sterilize the wash water as well as to dissolve any residual PCP from the contaminated soil. Before use, the wash water was allowed to stand at room temperature until spread plates showed no evidence of viable cells. The wash water was then neutralized with sterile dilute sulfuric acid and amended with sterile inorganic nutrient solutions using a modified recipe of Bushnell and Haas (Mueller et al., 1991).

Two-grams of each site soil test sample were added to 200 mL aliquots of the screening medium to make a test slurry. Twenty-five mL aliquots of each test slurry were dispensed into six 125 mL screwcapped Erlenmeyer flasks. Each flask cap was fitted with a Teflon liner to minimize abiotic losses due to volatilization. Flasks were placed on a gyratory shaker set at a temperature of 30°C and at an agitation rate of

200 rpm and incubated for 28 days. Control flasks, sterilized with formaldehyde, were harvested at Days 0 and 28 to evaluate abiotic losses of creosote components. All test and control flasks were prepared in duplicate.

Twenty-five creosote components, selected for their abundance or their environmental significance, were monitored during the study. These components are listed in Table 1. PAHs were grouped according to the number of fused benzene rings as proposed by Mueller et al. (Mueller et al., 1991). Relative losses of individual creosote components and target analyte groups were monitored over the experimental time period.

TABLE 1. Creosote components evaluated in bioremediation studies

Analyte Group	Creosote Compound
Group 1 PAHs	Naphthalene, 1-Methylnaphthalene, 2-Methylnaphthalene, 2,6-Dimethylnaphthalene, Biphenyl
Group 2 PAHs	Acenaphthylene, Acenaphthene, Fluorene, Phenanthrene, Anthracene, 2-Methylanthracene
Group 3 PAHs	Fluoranthene, Pyrene, Chrysene, Benz(a)anthracene, Benzo(a)pyrene, Benzo(b)fluoranthene, Benzo(k)fluoranthene, Indeno(1,2,3-c,d)pyrene
Phenolics	o-Cresol, m-Cresol, p-Cresol, Pentachlorophenol (PCP)
Heterocycles	Dibenzofuran, Carbazole

Bench-Scale Solid-Phase Bioremediation Studies. Studies were conducted to identify a recipe promoting rapid removal of creosote components. These recipes are summarized in Table 2. Biometer flasks (Bellco Glass) were used as the test reactor in all bench-scale studies. Procedures for the use of the flasks were according to those recommended by Bartha and Pramer (Bartha and Pramer, 1965) and Bartha (personal communication).

TABLE 2. Recipes used in bench-scale solid-phase studies

Recipe #	Limestone	Sawdust	NH_4NO_3-N	Na_2HPO_4-P
	(g/kg)	(g/kg)	(g/kg TOC)	(g/kg TOC)
1	10	None	None	None
2	10	None	16.7	1.25
3	10	50	None	None
4	10	50	16.7	1.25

TOC - Total organic carbon

Screening studies showed that the sample collected from the Old Wood Chip Pile (OWCP) location not only exhibited potent creosote-degradative activity but contained only trace amounts of creosote. Due to these properties, this sample could be used both as an inoculum soil and as a diluent to reduce the creosote concentration in highly contaminated soil to a more desirable starting concentration.

Each recipe was dispensed in 30 gram (dry weight) aliquots into each of six biometer flasks. The total creosote concentration ranged from 2,803 to 3,098 mg/kg in the four recipes. Sufficient water was added to achieve 70% of the water holding capacity of each recipe. In recipes using inorganic nitrogen and phosphorus sources, nutrients were prepared as concentrated stock solutions and added to deionized water used to hydrate the test soil. Nutrients were added at a carbon to nitrogen (C:N) ratio of 60:1 and carbon to phosphorus (C:P) ratio of 800:1 (Dibble and Bartha 1979). After mixing the soil, the flasks were assembled and incubated in the dark for a total of 84 days. Carbon dioxide production was measured over the entire 84 day period. Samples were collected for creosote analysis from duplicate flasks harvested at Days 0, 42, and 84.

Pilot-Scale Solid-Phase Studies. Recipes 3 and 4 were tested in pilot-scale land farm cells. Four land farm cells were prepared with dimensions of 8 ft (2.4 m) by 8 ft (2.4 m) by 2 ft (0.6 m) in depth and a capacity of 4.7 yd^3 (3.6 m^3). Two sheets of 1/4 in (0.64 cm) plywood, four feet (1.2 m) by 8 ft (2.4 m), were placed on the plastic liner used to line the cell to protect it during tilling. Approximately 5 to 6 tons (4,536 to 5,443 kg) of contaminated soil were added to each cell and amended with 500 to 600 lbs (227 to 272 kg) of sawdust and 50 lbs (22.7 kg) of limestone. Seventy-five lbs (34.1 kg) of granular ammonium nitrate and 5 lbs (2.3 kg) of neutralized triple super phosphate were added to Cells 2 and 4. These levels approximated the nitrogen and phosphorus concentrations used in bench-scale solid-phase studies. Five-gallon (0.02 m^3) buckets of OWCP soil were added to Cells 1 and 2 to determine the effects of biologically active soil on creosote removal. The total creosote concentration ranged from 3,313 to 5,451 mg/kg in the four cells. Each cell was then hydrated with tap water, tilled, and then covered with a black tarpaulin. Samples were collected monthly for a total of 6 months and analyzed for creosote content.

Field-Scale Solid-Phase Studies. Contaminated soil was stockpiled on-site pending determination of the optimum nutrient recipe from the pilot results. The field-scale land farm unit was 175 ft (53.3 m) by 115 ft (35.1 m) and 5 ft (1.5 m) deep. Due to the limited land area available for on-site treatment, the final depth of the unit was much greater than desired. Contaminated soil was mixed with sawdust (5% on a weight basis) and active soil as in Recipe 3, and then uniformly dispensed throughout the unit. The initial total creosote concentration was 4,812 mg/kg. The mixture was covered with a geotextile sheet, which allowed air exchange, while resisting water infiltration. The cover could be removed for sampling and tilling. The unit was set up in July, 1997, and was sampled monthly for about 8 months. The poor performance of Recipe 3 in pilot-scale studies indicated the need for inorganic nutrients in the treatment unit. Therefore, granular ammonium nitrate and dibasic sodium phosphate, respectively, were added as the nitrogen and phosphorus sources as in Recipe 4, beginning in March, 1998. The nutrients were dispensed in three monthly applications with ammonium nitrate and dibasic sodium phosphate added at rates of 7.6 tons (6,895 kg) and 0.90 tons (816 kg) per application.

Sampling. In bench-scale screening and solid-phase studies, the entire sample was collected and analyzed for creosote content. In pilot-scale studies, the test cells were divided into nine member grids. At Day 0, triplicate soil cores were collected within each grid and mixed together to make up a grid composite sample. Thus, nine samples were generated for each cell at Day 0 and averaged to obtain a starting creosote value for the cell. At other sampling times, the same procedures were followed except that the composite samples prepared from each of the nine grids were further combined. Composite samples were prepared from samples collected from three grids which yielded a net of three samples at each sampling time. The results from the three composite samples were then averaged.

In the field-scale treatment unit, a nine member grid was established, and five soil cores were taken from each grid. The cores were composited to yield a total of nine samples at each sampling time. The results of the nine samples were then averaged.

RESULTS AND DISCUSSION

Soil Screening Studies. Results from screening the four site soil samples showed that the OWCP sample exhibited potent creosote-degradative activity. Total creosote and PAH removal were 89.3% and 88.5%, respectively. Group 1 PAH, Group 2 PAH, Group 3 PAH, and heterocycle-degradative activity was found with removal levels of 100% (limit of detection - 10 µg/mL), 100%, 67.4%, and 100%, respectively. The remaining three samples exhibited excellent Group 1, Group 2, and heterocycle-degradative activity but exhibited no activity against Group 3 PAHs. Test soil samples were also screened for PCP-degradative activity, but activity was not detected.

Bench-Scale Solid-Phase Studies. Of the four recipes tested, Recipes 2 and 4 (inorganic nutrients added) promoted the highest extent of removal of target creosote components. By Day 84, total creosote and PAH removal was 87.1% and 88.7%, respectively, in Recipe 2 and 90.1% and 90.7%, respectively, in Recipe 4. Recipes 1 and 3 (no added nutrients) promoted the removal of approximately two-thirds of the total creosote and total PAH content over the 84 day test period. The extent of heterocycle removal ranged from 74% to 86% for all recipes. The results from Recipe 4 are shown in Figure 1.

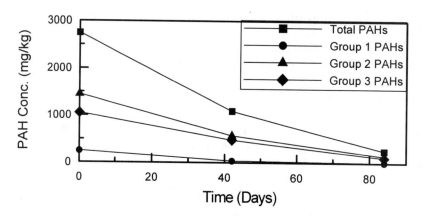

FIGURE 1. PAH Degradation in Bench-Scale Solid-Phase Studies Using Recipe 4

Excellent removal of Group 1 and Group 2 PAHs was observed in all recipes tested. Group 1 PAH removal levels ranged from 94% to 97% while Group 2 PAH removal levels ranged from 76% to 92%. Losses of these analytes were due to biodegradative as well as abiotic processes (volatilization). The main difference in the performance of the test recipes was the extent of removal of Group 3 PAHs. Group 3 PAH removal levels were approximately 81% and 90% in Recipes 2 and 4, respectively, but only approximately 34% and 48% in Recipes 1 and 3, respectively. It is clear that the addition of inorganic nutrients facilitated the removal of components of this analyte group.

Pilot-Scale Solid-Phase Studies. Recipes 3 and 4 were compared in the field in pilot-scale studies. Results showed that Recipe 4 was superior to Recipe 3 in total creosote removal, especially in the removal of Group 3 PAHs. The degradation profile for Recipe 4 (Cell 4) is shown in Figure 2.

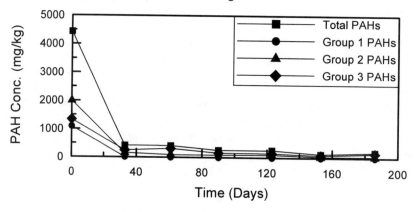

FIGURE 2. PAH degradation in Cell 4 using Recipe 4

Field-Scale Solid-Phase Treatment. The degradation profile for the field-scale unit is shown in Figure 3.

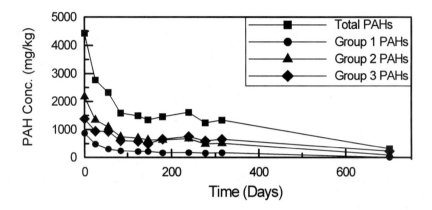

FIGURE 3. PAH degradation in the field-scale land farm treatment unit

There was a precipitous drop in creosote content for the first 84 days, but the rate of reduction was considerably less thereafter. Total creosote and total PAHs were reduced by 64.7% and 64.2%, respectively. After the addition of nutrients at Day 239, there was no immediate improvement in the extent of creosote removal. By Day 315, total creosote and total PAHs had only been reduced by 70.7% and 69.8%, respectively. After extending the treatment time for an additional 384 days, however, there was a profound decrease in total creosote and total PAH concentration. Results showed that by Day 699, the total creosote content and total PAH content had decreased by 93.2% and 92.8%, respectively. The extent of heterocycle removal was greater than 97%.

The lack of sustained degradative activity was probably due to the lack of addition of inorganic nutrients at Day 0. Once the nutrients from sawdust were used up, PAH biodegradation essentially came to a standstill. The added nutrients eventually stimulated the growth of indigenous microbial populations to degrade creosote components, but the rate of removal never approached the rates observed in pilot studies using Recipe 4. Another contributing factor affecting process performance was the depth of the treatment unit. The depth of the treatment unit made it difficult to till the soil effectively, and limited the availability of oxygen for optimal degradative activity to occur.

CONCLUSIONS

The following conclusions can be drawn from this study:

1. Nutrient recipes were developed that promoted the removal of high levels of creosote in bench-scale, pilot-scale, and field-scale solid-phase studies. Recipes containing inorganic nitrogen (ammonium nitrate) and phosphorus

[sodium phosphate (dibasic) or triple super phosphate] sources gave the best performance.
2. In bench-scale and pilot-scale solid-phase studies, total creosote levels, ranging from 2,803 to 5,451 mg/kg, were reduced by approximately 90.1% and 94.1% in approximately 3 months.
3. In field-scale studies, the total creosote content (4,812 mg/kg) was reduced by similar amounts (93.2%), although it took considerably longer (699 days) to achieve the same extent of removal. The lack of sustained degradative activity in the field-scale unit was probably due to the lack of addition of inorganic nutrients at Day 0, inadequate mixing, and reduced oxygen availability due to the depth of the treatment unit.

REFERENCES

Bartha, R., and D. Pramer. 1965. "Features of a Flask and Method for Measuring the Persistence and Biological Effects of Pesticides in Soil." *Soil Sci.* 100(1): 68-70.

Cerniglia, C. E. 1984. "Microbial Metabolism of Polycyclic Aromatic Hydrocarbons." *Advan. Appl. Microbiol.* 30: 31-71.

Dibble, J. T., and R. Bartha. 1979. "Effect of Environmental Parameters on the Biodegradation of Oil Sludge." *Appl. Environ. Microbiol.* 37(4): 729-739.

Gibson, D. T., and V. Subramanian. 1984. Microbial Degradation of Aromatic Hydrocarbons, p. 181-252. *In* D. T. Gibson (ed.). Microbial Degradation of Organic Compounds. Marcel Dekker, Inc., New York.

Guerin, W. F., and G. E. Jones. 1988. "Two-Stage Mineralization of Phenanthrene by Estuarine Enrichment Cultures." *Appl. Environ. Microbiol.* 54(4): 929-936.

Heitkamp, M. A., W. Franklin, and C. E. Cerniglia. 1988. "Microbial Metabolism of Polycyclic Aromatic Hydrocarbons: Isolation and Characterization of a Pyrene-Degrading Bacterium." *Appl. Environ. Microbiol.* 54(10): 2549-2555.

Mueller, J. G., P. J. Chapman, and P. H. Pritchard. 1989. "Creosote Contaminated Sites: Their Potential for Bioremediation." *Environ. Sci. Technol.* 23(10): 1197-1201.

Mueller, J. G., P. J. Chapman, B. O. Blattmann, and P. H. Pritchard. 1990. "Isolation and Characterization of a Fluoranthene-Utilizing Strain of *Pseudomonas paucimobilis*." *Appl. Environ. Microbiol.* 56(4): 1079-1086.

Mueller, J. G., S. E. Lantz, B. O. Blattmann, and P. J. Chapman. 1991. "Bench-Scale Evaluation of Alternative Biological Treatment Processes for the Remediation of Pentachlorophenol- and Creosote-Contaminated Materials: Solid-Phase Bioremediation." *Environ. Sci. Technol.* 25(6): 1045-1055.

ISOLATION OF A SOIL BACTERIAL STRAIN CAPABLE OF DIBENZOTHIOPHENE DEGRADATION

Simona Di Gregorio, Daniela Lizzari, Oscar Massella, Chiara Zocca and Giovanni Vallini [LA]

University of Verona, Department of Science & Technology, Verona, Italy

ABSTRACT: Biocatalytic degradation of organo-sulphur aromatic heterocycles is still not a commercial technology. However, exploitation of this biological process through molecular engineering as well as analysis of biotransformation rates have revealed that this approach is very promising for environmental applications. Dibenzothiophene (DBT) has been selected as a model for the chemical architecture of the organic sulphur in the fossil fuel matrices.

A bacterium belonging to *Burkholderia* sp., capable of oxidising DBT, has been isolated from toxic sediments contaminated by oil refinery wastewater. The corresponding genetic elements responsible for DBT oxidation has been cloned. The cloned genetic determinants can be classified as a new group of catabolic genes dealing with biodegradation of polycyclic aromatic sulphur-containing heterocycles (PASHs). They are organised in two separate transcriptional units. To the best of our knowledge the data reported here are the first showing a divergent organisation of genes involved in DBT oxidation, and they represent a step forward to understand the prevalence and distribution of the different groups of PASHs catabolic genes that occur in a contaminated soil.

INTRODUCTION

Although aerobic catabolism of polycyclic aromatic hydrocarbons (PAHs) by bacteria has been extensively studied, dynamics of microbial degradation of PASHs such as benzothiophenes is not yet completely understood. Sulphur-containing hydrocarbons are generally integrated in the matrix of fossil fuels, namely crude oil and coal. They enter the atmosphere either as release from industrial processes such as coal gasification and liquefaction, and fuel refinement, or as a consequence of carbon black dispersion and crude oil spills. Among these xenobiotics, the condensed thiophenes represent a class of compounds of prominent environmental concern, because of their high potential for biological magnification and toxicity. Condensed thiophenes account for about 50% of fossil fuel content in organic sulphur compounds and, therefore, they contribute significantly to the emission of sulphur oxides into the atmosphere during combustion processes.

The present study focuses on a microorganism isolated from toxic sediments contaminated by oil refinery discharges. This bacterium belongs to the genus *Burkholderia*; it has been shown to be capable of growing on DBT as sole carbon and energy source. Furthermore, *in vitro* DBT biodegradation by the

---

[LA] Lead Author (vallini@sci.univr.it)

isolate (*Burkholderia* sp. strain DBT1) is not repressed in the presence of an alternative carbon source. Actually, either in minimal or in rich media containing DBT, the microorganism produces five degradation intermediates. By means of analyses with UV-visible and GC-MS spectrometry, as well as nuclear magnetic resonance spectroscopy, three of these products were identified as 3-hydroxy-2-formyl-benzothiophene, benzothienopyran-2-one, and dibenzothiophene-5-oxide. This suggests that the bacterium metabolises DBT via the oxidative cleavage of the aromatic ring with a mechanism analogous to that described for naphthalene degradation in *Pseudomonas* (Denome et al., 1993). Nevertheless, at molecular level, the phenotype cannot be completely referred to that characterised in *Pseudomonas*. The genetic elements responsible for the biodegradative phenotype are organised in two distinct clusters of genes which have been cloned by insertional mutagenesis. The study has evidenced a phylogenetically cluster of catabolic genes divergent from those of microorganisms previously characterised.

Objective. The enlargement of the current information on the genetics governing bacterial degradation of sulphur-containing aromatic hydrocarbons should be regarded to as a prominent step to understand the prevalence and distribution of new groups of PASHs catabolic genes to be possibly applied in environmental biotechnology.

MATERIAL AND METHODS

Substrate and reagents. DBT (analytical grade) was purchased from Sigma. Growth media for cultivation of *Burkholderia* sp. DBT1 were as follows: Minimal Defined Medium (DM) - Na_2HPO_4 2.2 g, KH_2PO_4 0.8 g, NH_4NO_3 3.0 g, yeast extract 0.05 g, deionised H_2O (MilliQ) to 1000 mL; Yeast Mannitol Agar (YMA) - K_2HPO_4 0.5 g, $MgSO_4 \cdot 7H_2O$ 0.2 g, NaCl 0.1 g, yeast extract 1.0 g, mannitol 10 g, noble agar (Difco) 15 g, deionised H_2O (MilliQ) to 1000 mL. After sterilisation, a trace mineral solution and 500 mg/L DBT were added to both substrates as an *N-N*-dimethylformamide solution. Recombinant constructs in *E. coli* were routinely grown and maintained on Luria Bertani (LB) solid medium (1.5 % agar) or broth containing ampicillin (100 µg/mL) or spectinomycin-streptomycin (50 µg/mL).

Bacterial strain. *Burkholderia* sp. strain DBT1 was routinely maintained on solid mineral salt medium DM containing 500 mg/L DBT.

Enrichment cultures and isolation of DBT-degrading microbial strain. Three 500 mL Erlenmeyer flasks with 100 mL of DM containing 500 mg/L DBT were inoculated with 10 mL of aerobic sediments collected from a canal draining the wastewater of an oil refinery operating in Central Italy. The cultures were incubated at 28°C on a rotary shaker (100 rpm). After 7 days, 10 mL from each flask were transferred to fresh medium and incubated under the same conditions. Altogether, five consecutive transfers were performed within a span of 35 days. Detection of a putative DBT degraders was carried out by serial dilution plating on agar from enrichment cultures. Growth medium was DM containing 15 g/L of noble agar (Difco) and DBT (500 mg/L). Colonies producing orange diffusible pigment on solidified DM were repeatedly streaked onto nutrient agar (Difco) plates containing DBT (500 mg/L), then streaked on slant of Yeast Mannitol Agar (YMA) containing DBT (500 mg/L) and finally maintained at 4 °C.

DBT degradation assay. *Burkholderia* sp. strain DBT1 was incubated in yeast extract free DM broth containing 0.5 mL/L of a vitamin solution (biotin 2.0 mg, folic acid 2.0 mg, thiamine hydrochloride 5.0 mg, riboflavin 5.0 mg, pyridoxine hydrochloride 10 mg, cyanocobalamine 0.1 mg, nicotinic acid 5.0 mg, calcium pantothenate 5.0 mg, *p*-aminobenzoic acid 5.0 mg, lipoic acid 5.0 mg, deionised H_2O to 1000 mL) and DBT (100 and 500 mg/L). The growth of the isolate on DBT as a sole carbon and energy source was evaluated by means of CFU count on YMA plates. The amount of residual DBT was determined by spectrophotometric analysis with a Beckman DU-7 apparatus (Setti et al., 1995).

Molecular techniques. Standard procedures were used for plasmid DNA preparation, manipulation and agarose gel electrophoresis (Sambrook et al., 1989). Plasmid DNA was extracted using the Qiaprep Spin Plasmid Kit (Qiagen), bacterial genomic DNA was isolated using the Nucleospin Tissues Kit (Clontech) following the manufacturer instructions. Cloned nucleotide sequences were determined by the PRIMM DNA sequencing facilities with a Prism Ready Reaction DNA terminator cycle sequencing kit (Perkin-Elmer). The reactions were resolved with an Applied Biosystems Inc. (ABI) model 377 sequencer. Nucleotide sequence data was assembled by the ABI Fractura and Assembler computer packages and analysed by Clustal W (Thompson et al., 1994).

Cloning of genes in *Burkholderia* sp. strain DBT1. Insertional mutants of *Burkholderia* sp. strain DBT1 were generated (Wilson et al., 1995) and screened for the loss of the capacity to produce the orange coloured intermediates in a DBT containing medium. The suicide vector (kindly provided by Stephan Sidler of UPM, Spain) was a *mTn5gusAoriV* transposon located on pSS240, a pCAM140-based plasmid (Wilson et al., 1995) harbouring the *mTn5gusAoriV* transposon. The transposon contains an oriV origin of replication (ColE1-type), a spectinomycin-streptomycin resistance and a promoterless reporter gene *gusA*. Putative insertional mutants (>10,000 total) were selected on YMA containing DBT (500 mg/L) and spectinomycin-streptomycin (50 µg/mL), and tested for the capacity to develop the antibiotic resistance along with the loss of capacity to release the orange coloured intermediate. Genomic DNA of putative transposon recipients (~ 0.1%) was extracted and digested with *Pst*I, a restriction enzyme which does not cut inside the transposon sequence. The digested DNA was circularised by using T4 ligase and transformed by electroporation in *E. coli* DH5α competent cells. These were successively streaked on LB agar added with spectinomycin-streptomycin (50 µg/ml). Different replicative units were obtained as plasmid preparation from several insertional mutants selected. The mentioned replicative units were sequenced by inverse PCR on both strands to gain information on the nucleotide sequence of the transposon flanking regions. At this scope, two primers *gus55*: TTTGATTTCACGGGTTGG and *aada1846*: GCTGGCTTTTTCTTGTTATCG, annealing respectively on the two very far ends of the transposon, were designed. Sequencing reactions have been programmed on each replicative units derived from each insertional mutants which lost the capacity to release the orange coloured intermediate.

Preparation of genomic DNA, Southern blot DNA transfer and hybridisation. Genomic DNA was obtained from overnight cultures and digested with *EcoRV*. Restriction endonuclease digestion, DNA separation on agarose gel electrophoresis and Southern blotting analysis were carried out as described by Sambrook et al., 1989. The probe was obtained by *EcoRV* digestion of the p46 replicative unit. It was labelled by random primer extension with digoxigenin-11-dUTP (DIG), according to the manufacturer specification (Boheringer). Hybridisation and detection conditions were also done following the producer recommendations.

RT-PCR. Samples (10^8 cells) from strain DBT1 cultures, grown until late exponential phase on DBT (500 mg/L), citrate (250 mg/L) or DBT (500 mg/L) and 1% glucose were pelletted. RNA was isolated from cells by using Rneasy total RNA kit (Qiagen). Appropriate primers were designed for RT and PCR reactions. Positive and negative controls were included. Photos of the southern, SDS-PAGE, and RT-PCR analysis are not reported in this presentation.

Nucleotide sequences. The nucleotide sequences of cloned genes has been deposited in GenBank under accession number AF380367.

RESULTS

Six replicative units derived from insertional mutants were partially sequenced by inverse PCR in order to find the transposon flanking regions which encode for the structural and regulative genes responsible for DBT oxidation. Based on sequence similarities, putative gene identifications were assigned to each mutated sequences (Table 1).

As shown in Table 1, the identified putative genes are responsible for the oxidation of low molecular PAHs in *Pseudomonas* and *Burkholderia* species (Denome et al., 1993; Laurie and Lloyd-Jones, 1999).

To determine whether the insertional event in each mutant occurred in a single ORF, Southern analysis on genomic DNA of either the strain DBT1 or selected mutants (M46, M51, and M61) was performed. The genomic DNA was digested with *EcoRV*. On the other hand, the labelled probe encoded for the 5' end of the transposon reporter gene (GUS), flanked by a portion of the ORF matching with the iron-sulphur protein (ISP) large (α) subunit. These analyses showed that only a single copy of the mutating transposon was present either in each mutant or in distinct ORFs.

The complete nucleotide sequences of p46 (replicative unit obtained from M46) and p51 (replicative unit obtained from M51) were determined in order to find the different ORFs involved in the oxidation of DBT. The analysis revealed the presence of five ORFs on p46 and two ORFs on p51. The function of individual ORFs was determined on the basis of nucleotide homology to previously described genes and their encoded amino acid sequences (Table 2).

TABLE 1. Identification of interrupted open reading frames (ORFs) in insertional mutants of *Burkholderia* sp., strain DBT1

Mutant	Best gene match	BLAST score E value[a]
M46	ISP[b] large (α) subunit (structural)	8×10^{-43}
M51	Extradiol dioxygenase (structural)	5×10^{-22}
M61	*rpoN* (regulative)	6×10^{-17}
M11	NA[c]	
M115	NA	
M119	NA	

[a]E value represents the probability that the query sequences matches the database sequence by chance. The closest the value to "0" the highest the probability of matching.
[b]ISP, iron-sulphur protein.
[c]NA, not applicable. The transposon flanking regions match unknown sequences.

The different ORFs were subcloned in a modified pBS plasmid and *in vitro* expressed in order to verify that they encode for peptides of the predicted size. Recombinant peptides were successfully expressed in *E. coli* (DH5α) and visualised by SDS-PAGE. The results confirmed that all the putative ORFs encode for enzymes involved in the oxidation of low molecular PAHs, as observed for the *dox* genes (DBT oxidising) in *Pseudomonas* sp. (Denome et al., 1993). As a consequence, the DBT1 cloned genes were indicated as *dbt* from DBT oxidation and their designation is consistent with those given to known genes involved in the same pathway.

The microbial capability of PAH oxidation has always been related to the highly conserved *nah* genotypes cloned in PAH-degrading pseudomonads. Nevertheless, deduced amino acid sequences of Dbts show a low homology to the Nah-like class of proteins (encoded by the *nah*-like gene family). In the meanwhile, Dbts show a low homology also to Phn amino acid sequences (encoded by the *phn*-like genes), a divergent phenotype capable of oxidising low molecular PAHs, recently cloned in *Burkholderia* sp. strain RP007 (Laurie and Lloyd-Jones, 1999). The results show that ORF 1 encodes for DbtD, a putative isomerase. DbtD shows a 63% similarity with the Nah-like and a 60% to the Phn-like isofunctional aminoacid sequences. ORF 2 and 3 encode for DbtAc and DbtAd, respectively the large (α) and small (β) subunit of the iron-sulphur protein (ISP) of a DBT initial dioxygenase. DbtAc shows 68% homology to PhnAc and Nah-like isofunctional proteins. DbtAd shows a similar arrangement with 60% amino acid homology to PhnAd and 54% to Nah-like. ORF 4 encodes for DbtB a putative dihydrodiol dehydrogenase. As in PhnB, the predicted amino acid sequence shows closest similarity to analogous dehydrogenases from biphenyl catabolic pathway (65%); nevertheless, it is quite divergent from PhnB (76%).

ORF 5 encodes for a reductase, ORF 6 encodes for an extradiol dioxygenase, and ORF 7 encodes for a ferredoxin. The phylogenetic analysis of ORF 5, 6 and 7 is still in progress.

TABLE 2. Features of the *dbt* genes cloned

ORF	Gene	Replicative Unit	Protein feature
1	*dbt*D	p46	isomerase
2	*dbt*Ac	p46	ISP (α) subunit
3	*dbt*Ad	p46	ISP (β) subunit
4	*dbt*B	p46	dihydrodiol dehydrogenase
5	*dbt*Aa	p46	reductase
6	*dbt*Ab	p51	ferredoxin
7	*dbt*C	p51	extradiol dioxygenase

As previously mentioned, *Burkholderia* sp. strain DBT1 grows at the expense of DBT as a sole carbon and energy source. In order to determine the transcription of the cloned genes, DNA-free RNA was isolated from DBT1 induced (DM+DBT), not induced (DM+citrate), and putatively repressed (DM+1% glucose+DBT). The extracts were used as the target for RT-PCR amplification of partial gene sequences. ssDNA derived from induced cells showed amplified products of the anticipated size. ssDNA derived from not induced and repressed cells failed to give a product. Thus *dbt*DAcAdBAa and *dbt*CAb result to be co-transcribed in two distinct policistronic messengers, regulated at the transcriptional level, induced by DBT, and repressed by glucose. The close arrangement of the *dbt*DAcAdBAa genes suggests that these latter are co-transcribed and may constitute an operon. The expected initiation site should be located inside the 801 bp that precedes *dbt*D, where no putative ORF has been recognised. Also, no ORF has been recognised either upstream from *dbt*C or downstream from *dbt*Aa and *dbt*Ab.

Sequencing results, Southern analysis and RT-PCR reactions show that catabolic cistrons cloned in DBT1 appear to be organised in two distinct *loci* (Figure 1).

Most catabolic pathways in bacteria that have been studied so far have a genetic organisation characterised by clustering of the genes of entire pathways, or independently functioning pathways segments, in single transcriptional units transcribed from highly regulated promoters (Franklin et al., 1981). To the best of our knowledge the data reported are the first showing a divergent organisation of genes involved in DBT oxidation. We have identified genetic elements for the DBT oxidative pathway which are organised in two separate transcriptional units.

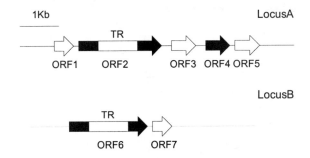

FIGURE 1. Genetic organisation of *loci* encompassing *dbts* genes
TR: *mTn5GusAoriV*

The organisation of the DBT oxidation pathway in *Burkholderia* sp. strain DBT1 is thus clearly different from that of other isofunctional pathways so far characterised in several bacteria. Whether this atypical organisation has a physiological or evolutionary significance remains unclear but the finding may be consistent with a more dynamic genetic organisation in relation to the better studied pathways in microorganisms such as *Pseudomonas* sp. It is worth noting that the genetic organisation of the same catabolic pathway changes also within *Burkholderia* sp.. The gene clustering of isofunctional genes in DBT1 is different in respect to *Burkholderia* sp. strain RP007, which resembles the one of *Pseudomonas* (Laurie and Lloyd-Jones, 1999). This finding suggests for a catabolic versatility of the members of *Burkholderia* sp., that is actually becoming more and more evident, and may indicate a dynamic genetic organisation of this microorganism suitable for possible interesting relapses in bioremediation.

CONCLUSIONS

Understanding the functions of diverse genetic determinants that encode enzymes for PAH catabolism is important for the evaluation of the ecological significance of such genes in the context of environmental pollution. Likewise, sequencing of genes which encode for distinct, although isofunctional, catabolic enzymatic pools allows to perfect of specific probes and PCR amplification experiments. This is essential to determine the prevalence of distinct microbial genotypes in different aromatic hydrocarbon contaminated soils. Moreover, it is made possible individuation of specific microorganisms which might show improved biodegradative potential in response to selective pressure exerted by different PAHs in a range of soils. The biochemical characterisation of the DBT degradative pathway in *Burkholderia* sp. strain DBT1 is consistent with these goals. Particular attention will be given to the activity and specificity of the catabolic enzymes for the application of this specific microbial isolate in environmental biotechnology.

ACKNOWLEDGEMENTS

Funding for this study was provided by a grant from the Italian National Environmental Protection Agency (ANPA).

REFERENCES

Denome, S. A., D. C. Stanley, E. S.Olson, and K. D. Young. 1993. "Metabolism of dibenzothiophene and naphthalene in *Pseudomonas* sp. strain C18: complete DNA sequence of upper naphthalene catabolic pathway". *J. Bacteriol.* 176: 2158-2164.

Franklin, F. C., M. Bagdasarian, M. M. Bagdasarian, and K. N. Timmis. 1981. "Molecular and functional analysis of the TOL plasmid pWWO from *Psuedomonas putida* and cloning of genes for the entire regulated aromatic ring *meta* cleavage pathway". *Proc. Natl.Acad. Sci.* USA 78: 7458-7462.

Laurie, A.D., and G. Lloyd-Jones. 1999. "The *phn* genes of *Burkholderia sp.* strain RP007 constitute a divergent gene cluster of polycyclic aromatic hydrocarbon catabolism" *J. Bacteriol.* 181: 531-540.

Sambrook, J., E. F. Fritsch, and T. Maniatis. 1989. "Molecular cloning: a laboratory manual" 2nd ed. Cold Spring Harbour Laboratory Press, Cold Spring Harbor, N.Y..

Setti, L., G. Lanzarini, and P. G. Pifferi. 1995. "Dibenzothiophene biodegradation by a *Pseudomonas sp.* in model solutions". *Process. Biochem.* 30: 721-728.

Thompson, J. D., H. Y. Tang, C. L. Joannou, N. H. Abdel-Wahab, and J. R. Mason. 1994. "CLUSTALW: improving the sensitivity of progressive multiple sequence alignment through sequence weighting, position-specific gap penalities and weight matrix choice". *Nucleic Acid Res.* 22: 4673-4680.

Wilson, K. J., A. Sessitch, J. C. Corbo, K. Giller, A. D. L. Akkermans, and R. A. Jefferson. 1995. "β-glucoronidase (GUS) transposons for ecological and genetic studies of *Rhizobia* and Gram-negative bacteria". *Microbiology* 141: 1691-1705.

2001 AUTHOR INDEX

This index contains names, affiliations, and volume/page citations for all authors who contributed to the ten-volume proceedings of the Sixth International In Situ and On-Site Bioremediation Symposium (San Diego, California, June 4-7, 2001). Ordering information is provided on the back cover of this book. The citations reference the ten volumes as follows:

6(1): Magar, V.S., J.T. Gibbs, K.T. O'Reilly, M.R. Hyman, and A. Leeson (Eds.), *Bioremediation of MTBE, Alcohols, and Ethers*. Battelle Press, Columbus, OH, 2001. 249 pp.
6(2): Leeson, A., M.E. Kelley, H.S. Rifai, and V.S. Magar (Eds.), *Natural Attenuation of Environmental Contaminants*. Battelle Press, Columbus, OH, 2001. 307 pp.
6(3): Magar, V.S., G. Johnson, S.K. Ong, and A. Leeson (Eds.), *Bioremediation of Energetics, Phenolics, and Polycyclic Aromatic Hydrocarbons*. Battelle Press, Columbus, OH, 2001. 313 pp.
6(4): Magar, V.S., T.M. Vogel, C.M. Aelion, and A. Leeson (Eds.), *Innovative Methods in Support of Bioremediation*. Battelle Press, Columbus, OH, 2001. 197 pp.
6(5): Leeson, A., E.A. Foote, M.K. Banks, and V.S. Magar (Eds.), *Phytoremediation, Wetlands, and Sediments*. Battelle Press, Columbus, OH, 2001. 383 pp.
6(6): Magar, V.S., F.M. von Fahnestock, and A. Leeson (Eds.), *Ex Situ Biological Treatment Technologies*. Battelle Press, Columbus, OH, 2001. 423 pp.
6(7): Magar, V.S., D.E. Fennell, J.J. Morse, B.C. Alleman, and A. Leeson (Eds.), *Anaerobic Degradation of Chlorinated Solvents*. Battelle Press, Columbus, OH, 2001. 387 pp.
6(8): Leeson, A., B.C. Alleman, P.J. Alvarez, and V.S. Magar (Eds.), *Bioaugmentation, Biobarriers, and Biogeochemistry*. Battelle Press, Columbus, OH, 2001. 255 pp.
6(9): Leeson, A., B.M. Peyton, J.L. Means, and V.S. Magar (Eds.), *Bioremediation of Inorganic Compounds*. Battelle Press, Columbus, OH, 2001. 377 pp.
6(10): Leeson, A., P.C. Johnson, R.E. Hinchee, L. Semprini, and V.S. Magar (Eds.), *In Situ Aeration and Aerobic Remediation*. Battelle Press, Columbus, OH, 2001. 391 pp.

Aagaard, Per (University of Oslo/NORWAY) 6(2):181
Aarnink, Pedro J.P. (Tauw BV/THE NETHERLANDS) 6(10):253
Abbott, James E. (Battelle/USA) 6(5):231, 237
Accashian, John V. (Camp Dresser & McKee, Inc./USA) 6(7):133
Adams, Daniel J. (Camp Dresser & McKee, Inc./USA) 6(8):53
Adams, Jack (Applied Biosciences Corporation/USA) 6(9):331
Adriaens, Peter (University of Michigan/USA) 6(8):19, 193
Adrian, Neal R. (U.S. Army Corps of Engineers/USA) 6(6):133
Agrawal, Abinash (Wright State University/USA) 6(5):95
Aiken, Brian S. (Parsons Engineering Science/USA) 6(2): 65, 189

Aitchison, Eric (Ecolotree, Inc./USA) 6(5):121
Al-Awadhi, Nader (Kuwait Institute for Scientific Research/KUWAIT) 6(6):249
Alblas, B. (Logisticon Water Treatment/THE NETHERLANDS) 6(8):11
Albores, A. (CINVESTAV-IPN/MEXICO) 6(6):219
Al-Daher, Reyad (Kuwait Institute for Scientific Research/KUWAIT) 6(6):249
Al-Fayyomi, Ihsan A. (Metcalf & Eddy, Inc./USA) 6(7):173
Al-Hakak, A. (McGill University/CANADA) 6(9):139
Allen, Harry L. (U.S. EPA/USA) 6(3):259
Allen, Jeffrey (University of Cincinnati/USA) 6(9):9
Allen, Mark H. (Dames & Moore/USA) 6(10):95
Allende, J.L. (Universidad Complutense/SPAIN) 6(4):29
Alonso, R. (Universidad Politecnica/SPAIN) 6(6):377
Alphenaar, Arne (TAUW bv/THE NETHERLANDS) 6(7):297
Alvarez, Pedro J. J. (University of Iowa/USA) 6(1):195; 6(3):1; 6(8):147, 175
Alvestad, Kimberly R. (Earth Tech/USA) 6(3):17
Ambert, Jack (Battelle Europe/SWITZERLAND) 6(6):241
Amezcua-Vega, Claudia (CINVESTAV-IPN/MEXICO) 6(3):243
Amy, Penny (University of Nevada Las Vegas/USA) 6(9):257
Andersen, Peter F. (GeoTrans, Inc./USA) 6(10):163
Anderson, Bruce (Plan Real AG/AUSTRALIA) 6(2):223
Anderson, Jack W. (RMT, Inc./USA) 6(10):201
Anderson, Todd (Texas Tech University/USA) 6(9):273
Andreotti, Giorgio (ENI Sop.A.) 6(5):41

Andretta, Massimo (Centro Ricerche Ambientali Montecatini/ITALY) 6(4):131
Andrews, Eric (Environmental Management, Inc./USA) 6(10):23
Andrews, John (SHN Consulting Engineers & Geologists, Inc./USA) 6(3):83
Archibald, Brent B. (Exxon Mobil Environmental Remediation/USA) 6(8):87
Archibold, Errol (Spelman College/USA) 6(9):53
Aresta, Michele (Universita di Catania/ITALY) 6(3):149
Arias, Marianela (PDVSA Intevep/VENEZUELA) 6(6):257
Atagana, Harrison I. (Mangosuthu Technikon/REP OF SOUTH AFRICA) 6(6):101
Atta, Amena (U.S. Air Force/USA) 6(2):73
Ausma, Sandra (University of Guelph/CANADA) 6(6):185
Autenrieth, Robin L. (Texas A&M University/USA) 6(5): 17, 25
Aziz, Carol E. (Groundwater Services, Inc./USA) 6(7):19; 6(8):73
Azizian, Mohammad (Oregon State University/USA) 6(10): 145, 155

Babel, Wolfgang (UFZ Center for Environmental Research/GERMANY) 6(4):81
Bae, Bumhan (Kyungwon University/REPUBLIC OF KOREA) 6(6):51
Baek, Seung S. (Kyonggi University/REPUBLIC OF KOREA) 6(1):161
Bagchi, Rajesh (University of Cincinnati/USA) 6(5):243, 253, 261
Baiden, Laurin (Clemson University/USA) 6(7):109
Bakker, C. (IWACO/THE NETHERLANDS) 6(7):141
Balasoiu, Cristina (École Polytechnique de Montreal/CANADA) 6(9):129
Balba, M. Talaat (Conestoga-Rovers & Associates/USA) 6(1):99; 6(6):249; 6(10):131

Banerjee, Pinaki (Harza Engineering Company, Inc./USA) 6(7):157
Bankston, Jamie L. (Camp Dresser and McKee Inc./USA) 6(5):33
Barbé, Pascal (Centre National de Recherche sur les Sites et Sols Pollués/FRANCE) 6(2):129
Barcelona, Michael J. (University of Michigan/USA) 6(8):19, 193
Barczewski, Baldur (Universitat Stuttgart/GERMANY) 6(2):137
Barker, James F. (University of Waterloo/CANADA) 6(8):95
Barnes, Paul W. (Earth Tech, Inc./USA) 6(3): 17, 25
Basel, Michael D. (Montgomery Watson Harza/USA) 6(10):41
Baskunov, Boris B. (Russian Academy of Sciences/RUSSIA) 6(3):75
Bastiaens, Leen (VITO/BELGIUM) 6(4):35; 6(9):87
Batista, Jacimaria (University of Nevada Las Vegas/USA) 6(9): 257, 265
Bautista-Margulis, Raul G. (Centro de Investigacion en Materiales Avanzados/MEXICO) 6(6):361
Becker, Paul W. (Exxon Mobil Refining & Supply/USA) 6(8):87
Beckett, Ronald (Monash University/AUSTRALIA) 6(4):1
Beckwith, Walt (Solutions Industrial & Environmental Services/USA) 6(7):249
Beguin, Pierre (Institut Pasteur/FRANCE) 6(1):153
Behera, N. (Sambalpur University/INDIA) 6(9):173
Bell, Nigel (Imperial College London/UK) 6(10):123
Bell, Mike (Coats North America/USA) 6(7):213
Beller, Harry R. (Lawrence Livermore National Laboratory/USA) 6(1):195
Belloso, Claudio (Facultad Catolica de Quimica e Ingenieria/ARGENTINA) 6(6): 235, 303
Benner, S. G. (Stanford University/USA) 6(9):71
Bensch, Jeffrey C. (GeoTrans, Inc/USA) 6(7):221

Béron, Patrick (Université du Québec à Montréal/CANADA) 6(3):165
Berry, Duane F. (Virginia Polytechnic Institute & State University/USA) 6(2):105
Betts, W. Bernard (Cell Analysis Ltd./UK) 6(6):27
Billings, Bradford G. (Brad) (Billings & Associates, Inc./USA) 6(1):115
Bingler, Linda (Battelle Sequim/USA) 6(5):231, 237
Birkle, M. (Fraunhofer Institute/GERMANY) 6(2):137
Bitter, Paul (URS Corporation./USA) 6(2):261
Bittoni, A. (EniTecnologie/ITALY) 6(6):173
Bjerg, Poul L (Technical University of Denmark/DENMARK) 6(2):11
Blanchet, Denis (Institut Français du Pétrole/FRANCE) 6(3):227
Bleckmann, Charles A. (Air Force Institute of Technology/USA) 6(2):173
Blokzijl, R. (DHV Environment and Infrastructure/THE NETHERLANDS) 6(8):11
Blowes, David (University of Waterloo/CANADA) 6(9):71
Bluestone, Simon (Montgomery Watson/ITALY) 6(10):41
Boben, Carolyn (Williams/USA) 6(1):175
Böckle, Karin (Technologiezentrum Wasser/GERMANY) 6(8):105
Boender, H. (Logisticon Water Treatment/THE NETHERLANDS) 6(8):11
Böhler, Anja (BioPlanta GmbH/GERMANY) 6(3):67
Bonner, James S. (Texas A&M University/USA) 6(5):17, 25
Bononi, Vera Lucia Ramos (Instituto de Botânica/BRAZIL) 6(3):99
Bonsack, Laurence T. (Aerojet/USA) 6(9):297
Borazjani, Abdolhamid (Mississippi State University/USA) 6(5):329; 6(6):279

Borden, Robert C. (Solutions Industrial & Environmental Services/USA) 6(7):249

Bornholm, Jon (U.S. EPA/USA) 6(6):81

Bosco, Francesca (Politecnico di Torino/ITALY) 6(3):211

Bosma, Tom N.P. (TNO Environment/THE NETHERLANDS) 6(7):61

Bourquin, Al W. (Camp Dresser & McKee Inc./USA) 6(5):33; 6(6):81; 6(7):133,

Bouwer, Edward J. (Johns Hopkins University/USA) 6(2):19

Bowman, Robert S. (New Mexico Institute of Mining & Technology/USA) 6(8):131

Boyd, Sian (CEFAS Laboratory/UK) 6(10):337

Boyd-Kaygi, Patricia (Harding ESE/USA) 6(10):231

Boyle, Susan L. (Haley & Aldrich, Inc./USA) 6(7):27, 281

Brady, Warren D. (IT Corporation/USA) 6(9):215

Breedveld, Gijs (University of Oslo/NORWAY) 6(2):181

Bregante, M. (Istituto di Cibernetica e Biofisica/ITALY) 6(5):157

Brenner, Richard C. (U.S. EPA/USA) 6(5):231, 237

Breteler, Hans (Oostwaardhoeve Co./THE NETHERLANDS) 6(6):59

Bricka, Mark R. (U.S. Army Corps of Engineers/USA) 6(9):241

Brickell, James L. (Earth Tech, Inc./USA) 6(10):65

Brigmon, Robin L. (Westinghouse Savannah River Co/USA) 6(7):109

Britto, Ronnie (EnSafe, Inc./USA) 6(9):315

Brossmer, Christoph (Degussa Corporation/USA) 6(10):73

Brown, Bill (Dunham Environmental Services/USA) 6(8):35

Brown, Kandi L. (IT Corporation/USA) 6(1):51

Brown, Richard A. (ERM, Inc./USA) 6(7):45, 213

Brown, Stephen (Queen's University/CANADA) 6(2):121

Brown, Susan (National Water Research Institute/CANADA) 6(7):321, 333, 341

Brubaker, Gaylen (ThermoRetec North Carolina Corp./USA) 6(7):1

Bruce, Cristin (Arizona State University/USA) 6(8):61

Bruce, Neil C. (University of Cambridge/UK) 6(5):69

Buchanan, Gregory (Tait Environmental Management, Inc./USA) 6(10):267

Bucke, Christopher (University of Westminster/UK) 6(3):75

Bulloch, Gordon (BAE Systems Properties Ltd./UK) 6(6):119

Burckle, John (U.S. EPA/USA) 6(9):9

Burden, David S. (U.S. EPA/USA) 6(2):163

Burdick, Jeffrey S. (ARCADIS Geraghty & Mills/USA) 6(7):53

Burgos, William (The Pennsylvania State University/USA) 6(8):201

Burken, Joel G. (University of Missouri-Rolla/USA) 6(5):113, 199

Burkett, Sharon E. (ENVIRON International Corp./USA) 6(7):189

Burnell, Daniel K. (GeoTrans, Inc./USA) 6(2):163

Burns, David A. (ERM, Inc./USA) 6(7):213

Burton, Christy D. (Battelle/USA) 6(1):137; 6(10):193

Buscheck, Timothy E. (Chevron Research & Technology Co/USA) 6(1): 35, 203

Buss, James A. (RMT, Inc./USA) 6(2):97

Butler, Adrian P. (Imperial College London/UK) 6(10):123

Butler, Jenny (Battelle/USA) 6(7):13

Büyüksönmez, Fatih (San Diego State University/USA) 6(10):301

Caccavo, Frank (Whitworth College/USA) 6(8):1

Callender, James S. (Rockwell Automation/USA) 6(7):133

Calva-Calva, G. (CINVESTAV-IPN/MEXICO) 6(6):219

Camper, Anne K. (Montana State University/USA) 6(7):117

Camrud, Doug (Terracon/USA) 6(10):15
Canty, Marietta C. (MSE Technology Applications/USA) 6(9):35
Carman, Kevin R. (Louisiana State University/USA) 6(5):305
Carrera, Paolo (Ambiente S.p.A./ITALY) 6(6):227
Carson, David A. (U.S. EPA/USA) 6(2):247
Carvalho, Cristina (Clemson University/USA) 6(7):109
Case, Nichole L. (Haley & Aldrich, Inc./USA) 6(7):27, 281
Castelli, Francesco (Universita di Catania/ITALY) 6(3):149
Cha, Daniel K. (University of Delaware/USA) 6(6):149
Chaney, Rufus L. (U.S. Department of Agriculture/USA) 6(5):77
Chang, Ching-Chia (National Chung Hsing University/TAIWAN) 6(10):217
Chang, Soon-Woong (Kyonggi University/REPUBLIC OF KOREA) 6(1):161
Chang, Wook (University of Maryland/USA) 6(3):205
Chapuis, R. P. (École Polytechnique de Montréal/CANADA) 6(4):139
Charrois, Jeffrey W.A. (Komex International, Ltd./CANADA) 6(4):7
Chatham, James (BP Exploration/USA) 6(2):261
Chekol, Tesema (University of Maryland/USA) 6(5):77
Chen, Abraham S.C. (Battelle/USA) 6(10):245
Chen, Chi-Ruey (Florida International University/USA) 6(10):187
Chen, Zhu (The University of New Mexico/USA) 6(9):155
Cherry, Jonathan C. (Kennecott Utah Copper Corp/USA) 6(9):323
Child, Peter (Investigative Science Inc./CANADA) 6(2):27
Chino, Hiroyuki (Obayashi Corporation/JAPAN) 6(6):249
Chirnside, Anastasia E.M. (University of Delaware/USA) 6(6):9
Chiu, Pei C. (University of Delaware/USA) 6(6):149
Cho, Kyung-Suk (Ewha University/REPUBLIC OF KOREA) 6(6):51
Choung, Youn-kyoo (Yonsei University/REPUBLIC OF KOREA) 6(6):51
Clement, Bernard (École Polytechnique de Montréal/CANADA) 6(9):27
Clemons, Gary (CDM Federal Programs Corp./USA) 6(6):81
Cocos, Ioana A. (École Polytechnique de Montréal/CANADA) 6(9):27
Cocucci, M. (Universita' degli Studi di Milano/ITALY) 6(5):157
Coelho, Rodrigo O. (CSD-GEOLOCK/BRAZIL) 6(1):27
Collet, Berto (TAUW bv/THE NETHERLANDS) 6(10):253
Compton, Joanne C. (REACT Environmental Engineers/USA) 6(3):25
Connell, Doug (Barr Engineering Company/USA) 6(5):105
Connor, Michael A. (University of Melbourne/AUSTRALIA) 6(10):329
Cook, Jim (Beazer East, Inc./USA) 6(2):239
Cooke, Larry (NOVA Chemicals Corporation/USA) 6(4):117
Coons, Darlene (Conestoga-Rovers & Associates/USA) 6(1):99; 6(10):131
Costley, Shauna C. (University of Natal/REP OF SOUTH AFRICA) 6(9):79
Cota, Jennine L. (ARCADIS Geraghty & Miller, Inc./USA) 6(7):149
Covell, James R. (EG&G Technical Services, Inc./USA) 6(10):49
Cowan, James D. (Ensafe Inc./USA) 6(9):315
Cox, Evan E. (GeoSyntec Consultants/CANADA) 6(8):27, 6(9):297
Cox, Jennifer (Clemson University/USA) 6(7):109
Craig, Shannon (Beazer East, Inc./USA) 6(2):239
Crawford, Donald L. (University of Idaho/USA) 6(3):91; 6(9):147

Crecelius, Eric (Battelle/USA) 6(5): 231, 237
Crotwell, Terry (Solutions Industrial & Environmental Services/USA) 6(7):249
Cui, Yanshan (Chinese Academy of Sciences/CHINA) 6(9):113
Cunningham, Al B. (Montana State University/USA) 6(7):117; 6(8):1
Cunningham, Jeffrey A. (Stanford University/USA) 6(7):95
Cutright, Teresa J. (The University of Akron/USA) 6(3):235

da Silva, Marcio Luis Busi (University of Iowa/USA) 6(1):195
Daly, Daniel J. (Energy & Environmental Research Center/USA) 6(5):129
Daniel, Fabien (AEA Technology Environment/UK) 6(10):337
Daniels, Gary (GeoTrans/USA) 6(8):19
Das, K.C. (University of Georgia/USA) 6(9):289
Davel, Jan L. (University of Cincinnati/USA) 6(6):133
Davis, Gregory A. (Microbial Insights Inc./USA) 6(2):97
Davis, Jeffrey L. (U.S. Army/USA) 6(3): 43, 51
Davis, John W. (The Dow Chemical Company/USA) 6(2):89
Davis-Hoover, Wendy J. (U.S. EPA/USA) 6(2):247
De'Ath, Anna M. (Cranfield University/UK) 6(6):329
Dean, Sean (Camp Dresser & McKee. Inc/USA) 6(7):133
DeBacker, Dennis (Battelle/USA) 6(10):145
DeHghi, Benny (Honeywell International Inc./USA) 6(2):39;6(10):283
de Jong, Jentsje (TAUW BV/THE NETHERLANDS) 6(10):253
Del Vecchio, Michael (Envirogen, Inc./USA) 6(9):281
Delille, Daniel (CNRS/FRANCE) 6(2):57
DeLong, George (AIMTech/USA) 6(7):321, 333, 341
Demers, Gregg (ERM/USA) 6(7):45
De Mot, Rene (Catholic University of Leuven/BELGIUM) 6(4):35

Deobald, Lee A. (University of Idaho/USA) 6(9):147
Deschênes, Louise (École Polytechnique de Montréal/CANADA) 6(3):115; 6(9):129
Dey, William S. (Illinois State Geological Survey/USA) 6(9):179
Díaz-Cervantes, Dolores (CINVESTAV-IPN/MEXICO) 6(6):369
Dick, Vincent B. (Haley & Aldrich, Inc./USA) 6(7):27, 281
Diehl, Danielle (The University of New Mexico/USA) 6(9):155
Diehl, Susan V. (Mississippi State University/USA) 6(5):329
Diels, Ludo (VITO/BELGIUM) 6(9):87
DiGregorio, Salvatore (University della Calabria/ITALY) 6(4):131
Di Gregorio, Simona (Universita degli Studi di Verona/ITALY) 6(3):267
Dijkhuis, Edwin (Bioclear/THE NETHERLANDS) 6(5):289
Di Leo, Cristina (EniTecnologie/ITALY) 6(6):173
Dimitriou-Christidis, Petros (Texas A&M University) 6(5):17
Dixon, Robert (Montgomery Watson/ITALY) 6(10):41
Dobbs, Gregory M. (United Technologies Research Center/USA) 6(7):69
Doherty, Amy T. (GZA GeoEnvironmental, Inc./USA) 6(7):165
Dolan, Mark E. (Oregon State University/USA) 6(10):145, 155, 179
Dollhopf, Michael (Michigan State University/USA) 6(8):19
Dondi, Giovanni (Water & Soil Remediation S.r.l./ITALY) 6(6):179
Dong, Yiting (Chinese Academy of Sciences/CHINA) 6(9):113
Dooley, Maureen A. (Regenesis/USA) 6(7):197
Dottridge, Jane (Komex Europe Ltd./UK) 6(4):17
Dowd, John (University of Georgia/USA) 6(9):289
Doughty, Herb (U.S. Navy/USA) 6(10):1

Doze, Jacco (RIZA/THE NETHERLANDS) 6(5):289
Dragich, Brian (California Polytechnic State University/USA) 6(2):1
Drake, John T. (Camp Dresser & McKee Inc./USA) 6(7):273
Dries, Victor (Flemish Public Waste Agency/BELGIUM) 6(7):87
Du, Yan-Hung (National Chung Hsing University/TAIWAN) 6(6):353
Dudal, Yves (École Polytechnique de Montréal/CANADA) 6(3):115
Duffey, J. Tom (Camp Dresser & McKee Inc./USA) 6(5):33
Duffy, Baxter E. (Inland Pollution Services, Inc./USA) 6(7):313
Duijn, Rik (Oostwaardhoeve Co./THE NETHERLANDS) 6(6):59
Durant, Neal D. (GeoTrans, Inc./USA) 6(2):19, 163
Durell, Gregory (Battelle Ocean Sciences/USA) 6(5):231
Dworatzek, S. (University of Toronto/CANADA) 6(8):27
Dwyer, Daryl F. (University of Minnesota/USA) 6(3):219
Dzantor, E. K. (University of Maryland/USA) 6(5):77

Ebner, R. (GMF/GERMANY) 6(2):137
Ederer, Martina (University of Idaho/USA) 6(9):147
Edgar, Michael (Camp Dresser & McKee Inc./USA) 6(7):133
Edwards, Elizabeth A. (University of Toronto/CANADA) 6(8):27
Edwards, Grant C. (University of Guelph/CANADA) 6(6):185
Eggen, Trine (Jordforsk Centre for Soil and Environmental Research/NORWAY) 6(6):157
Eggert, Tim (CDM Federal Programs Corp./USA) 6(6):81
Elberson, Margaret A. (DuPont Co./USA) 6(8):43
Elliott, Mark (Virginia Polytechnic Institute & State University/USA) 6(5):1
Ellis, David E. (Dupont Company/USA) 6(8):43
Ellwood, Derek C. (University of Southampton/UK) 6(9):61
Else, Terri (University of Nevada Las Vegas/USA) 6(9):257
Elväng, Annelie M. (Stockholm University/SWEDEN) 6(3):133
England, Kevin P. (USA) 6(5):105
Ertas, Tuba Turan (San Diego State University/USA) 6(10):301
Escalon, Lynn (U.S. Army Corps of Engineers/USA) 6(3):51
Esparza-Garcia, Fernando (CINVESTAV-IPN/MEXICO) 6(6):219
Evans, Christine S. (University of Westminster/UK) 6(3):75
Evans, Patrick J. (Camp Dresser & McKee, Inc./USA) 6(2):113, 199; 6(8):209

Fabiani, Fabio (EniTecnologie S.p.A./ITALY) 6(6):173
Fadullon, Frances Steinacker (CH2M Hill/USA) 6(3):107
Fang, Min (University of Massachusetts/USA) 6(6):73
Faris, Bart (New Mexico Environmental Department/USA) 6(9):223
Farone, William A. (Applied Power Concepts, Inc./USA) 6(7):103
Fathepure, Babu Z. (Oklahoma State University/USA) 6(8):19
Faust, Charles (GeoTrans, Inc./USA) 6(2):163
Fayolle, Françoise (Institut Français du Pétrole/FRANCE) 6(1):153
Feldhake, David (University of Cincinnati/USA) 6(2):247
Felt, Deborah (Applied Research Associates, Inc./USA) 6(7):125
Feng, Terry H. (Parsons Engineering Science, Inc./USA) 6(2):39; 6(10):283
Fenwick, Caroline (Aberdeen University/UK) 6(2):223
Fernandez, Jose M. (University of Iowa/USA) 6(1):195
Fernández-Sanchez, J. Manuel (CINVESTAV-IPN/MEXICO) 6(6):369

Ferrer, E. (Universidad Complutense de Madrid/SPAIN) 6(4):29
Ferrera-Cerrato, Ronald (Colegio de Postgraduados/MEXICO) 6(6):219
Fiacco, R. Joseph (Environmental Resources Management) 6(7):45
Fields, Jim (University of Georgia/USA) 6(9):289
Fields, Keith A. (Battelle/USA) 6(10):1
Fikac, Paul J. (Jacobs Engineering Group, Inc./USA) 6(6):35
Fischer, Nick M. (Aquifer Technology/USA) 6(8):157, 6(10):15
Fisher, Angela (The Pennsylvania State University/USA) 6(8):201
Fisher, Jonathan (Environment Agency/UK) 6(4):17
Fitch, Mark W. (University of Missouri-Rolla/USA) 6(5):199
Fleckenstein, Janice V. (USA) 6(6):89
Fleischmann, Paul (ZEBRA Environmental Corp./USA) 6(10):139
Fletcher, John S. (University of Oklahoma/USA) 6(5):61
Foget, Michael K. (SHN Consulting Engineers & Geologists, Inc./USA) 6(3):83
Foley, K.L. (U.S. Army Engineer Research & Development Center/USA) 6(5):9
Follner, Christina G. (University of Leipzig/GERMANY) 6(4):81
Fontenot, Martin M. (Syngenta Crop Protection, Inc./USA) 6(6):35
Foote, Eric A. (Battelle/USA) 6(1):137; 6(7):13
Ford, James (Investigative Science Inc./CANADA) 6(2):27
Forman, Sarah R. (URS Corporation/USA) 6(7):321, 333, 341
Fortman, Tim J. (Battelle Marine Sciences Laboratory/USA) 6(3):157
Francendese, Leo (U.S. EPA/USA) 6(3):259
Francis, M. McD. (NOVA Research & Technology Center/CANADA) 6(4):117; 6(5):53,
François, Alan (Institut Français du Pétrole/FRANCE) 6(1):153

Frankenberger, William T. (University of California/USA) 6(9):249
Freedman, David L. (Clemson University/USA) 6(7):109
French, Christopher E. (University of Cambridge/UK) 6(5):69
Friese, Kurt (UFZ Center for Environmental Research/GERMANY) 6(9):43
Frisbie, Andrew J. (Purdue University/USA) 6(3):125
Frisch, Sam (Envirogen Inc./USA) 6(9):281
Frömmichen, René (UFZ Centre for Environmental Research/GERMANY) 6(9):43
Fuierer, Alana M. (New Mexico Institute of Mining & Technology/USA) 6(8):131
Fujii, Kensuke (Obayashi Corporation/JAPAN) 6(10):239
Fujii, Shigeo (Kyoto University/JAPAN) 6(4):149
Furuki, Masakazu (Hyogo Prefectural Institute of Environmental Science/JAPAN) 6(5):321

Gallagher, John R. (University of North Dakota/USA) 6(5):129; 6(6):141
Gambale, Franco (Istituto di Cibernetica e Biofisica/ITALY) 6(5):157
Gambrell, Robert P. (Louisiana State University/USA) 6(5):305
Gandhi, Sumeet (University of Iowa/USA) 6(8):147
Garbi, C. (Universidad Complutense de Madrid/SPAIN) 6(4):29; 6(6):377
García-Arrazola, Roeb (CINVESTAV-IPN/MEXICO) 6(6):369
García-Barajas, Rubén Joel (ESIQIE-IPN/MEXICO) 6(6):369
Garrett, Kevin (Harding ESE/USA) 6(7):205
Garry, Erica (Spelman College/USA) 6(9):53
Gavaskar, Arun R. (Battelle/USA) 6(7):13
Gavinelli, Marco (Ambiente S.p.A./ITALY) 6(6):227
Gebhard, Michael (GeoTrans/USA) 6(8):19

Gec, Bob (Degussa Canada Ltd./CANADA) 6(10):73
Gehre, Matthias (UFZ - Centre for Environmental Research/GERMANY) 6(4):99
Gemoets, Johan (VITO/BELGIUM) 6(4):35; 6(9):87
Gent, David B. (U.S. Army Corps of Engineers/USA) 6(9):241
Gentry, E. E. (Science Applications International Corporation/USA) 6(8):27
Georgiev, Plamen S. (University of Mining & Geology/BULGARIA) 6(9):97
Gerday, Charles (Université de Liège/BELGIUM) 6(2):57
Gerlach, Robin (Montana State University/USA) 6(8):1
Gerritse, Jan (TNO Environmental Sciences/THE NETHERLANDS) 6(2):231; 6(7):61
Gerth, André (BioPlanta GmbH/GERMANY) 6(3):67; 6(5):173
Ghosh, Upal (Stanford University/USA) 6(3):189; 6(6):89
Ghoshal, Subhasis (McGill University/CANADA) 6(9):139
Gibbs, James T. (Battelle/USA) 6(1):137
Gibello, A. (Universidad Complutense/SPAIN) 6(4):29
Giblin, Tara (University of California/USA) 6(9):249
Gilbertson, Amanda W. (University of Missouri-Rolla/USA) 6(5):199
Gillespie, Rick D. (Regenesis/USA) 6(1):107
Gillespie, Terry J. (University of Guelph/CANADA) 6(6):185
Glover, L. Anne (Aberdeen University /UK) 6(2):223
Goedbloed, Peter (Oostwaardhoeve Co./THE NETHERLANDS) 6(6):59
Golovleva, Ludmila A. (Russian Academy of Sciences/RUSSIA) 6(3):75
Goltz, Mark N. (Air Force Institute of Technology/USA) 6(2):173

Gong, Weiliang (The University of New Mexico/USA) 6(9):155
Gossett, James M. (Cornell University/USA) 6(4):125
Govind, Rakesh (University of Cincinnati/USA) 6(5):269; 6(8):35; 6(9):1, 9, 17
Gozan, Misri (Water Technology Center/GERMANY) 6(8):105
Grainger, David (IT Corporation/USA) 6(1):51; 6(2):73
Grandi, Beatrice (Water & Soil Remediation S.r.l./ITALY) 6(6):179
Granley, Brad A. (Leggette, Brashears, & Graham/USA) 6(10):259
Grant, Russell J. (University of York/UK) 6(6):27
Graves, Duane (IT Corporation/USA) 6(2):253; 6(4):109; 6(9):215
Green, Chad E. (University of California/USA) 6(10):311
Green, Donald J. (USAG Aberdeen Proving Ground/USA) 6(7):321, 333, 341
Green, Robert (Alcoa/USA) 6(6):89
Green, Roger B. (Waste Management, Inc./USA) 6(2):247; 6(6):127
Gregory, Kelvin B. (University of Iowa/USA) 6(3):1
Griswold, Jim (Construction Analysis & Management, Inc./USA) 6(1):115
Groen, Jacobus (Vrije Universiteit/THE NETHERLANDS) 6(4):91
Groenendijk, Gijsbert Jan (Hoek Loos bv/THE NETHERLANDS) 6(7):297
Grotenhuis, Tim (Wageningen Agricultural University/THE NETHERLANDS) 6(5):289
Groudev, Stoyan N. (University of Mining & Geology/BULGARIA) 6(9):97
Guarini, William J. (Envirogen, Inc./USA) 6(9):281
Guieysse, Benoît (Lund University/SWEDEN) 6(3):181
Guiot, Serge R. (Biotechnology Research Institute/CANADA) 6(3):165
Gunsch, Claudia (Clemson University/USA) 6(7):109
Gurol, Mirat (San Diego State University/USA) 6(10):301

Ha, Jeonghyub (University of Maryland/USA) 6(10):57
Haak, Daniel (RMT, Inc./USA) 6(10):201
Haas, Patrick E. (Mitretek Systems/USA) 6(7):19, 241, 249; 6(8):73
Haasnoot, C. (Logisticon Water Treatment/THE NETHERLANDS) 6(8):11
Habe, Hiroshi (The University of Tokyo/JAPAN) 6(4):51; 6(6):111
Haeseler, Frank (Institut Français du Pétrole/FRANCE) 6(3):227
Haff, James (Meritor Automotive, Inc./USA) 6(7):173
Haines, John R. (U.S. EPA/USA) 6(9):17
Håkansson, Torbjörn (Lund University/SWEDEN) 6(9):123
Halfpenny-Mitchell, Laurie (University of Guelph/CANADA) 6(6):185
Hall, Billy (Newfields, Inc./USA) 6(5):189
Hampton, Mark M. (Groundwater Services/USA) 6(8):73
Hannick, Nerissa K. (University of Cambridge/UK) 6(5):69
Hannigan, Mary (Mississippi State University) 6(5):329; 6(6):279
Hannon, LaToya (Spelman College/USA) 6(9):53
Hansen, Hans C. L. (Hedeselskabet /DENMARK) 6(2):11
Hansen, Lance D. (U.S. Army Corps of Engineers/USA) 6(3):9, 43, 51; 6(4):59; 6(6):43; 6(7):125; 6(10):115
Haraguchi, Makoo (Sumitomo Marine Research Institute/JAPAN) 6(10):345
Hardisty, Paul E. (Komex Europe, Ltd./ENGLAND) 6(4):17
Harmon, Stephen M. (U.S. EPA/USA) 6(9):17
Harms, Hauke (Swiss Federal Institute of Technology/SWITZERLAND) 6(3):251
Harmsen, Joop (Alterra, Wageningen University and Research Center/THE NETHERLANDS) 6(5):137, 279; 6(6):1, 59

Harper, Greg (TetraTech EM Inc./USA) 6(3):259
Harrington-Baker, Mary Ann (MSE, Inc./USA) 6(9):35
Harris, Benjamin Cord (Texas A&M University/USA) 6(5):17, 25
Harris, James C. (U.S. EPA/USA) 6(6):287, 295
Harris, Todd (Mason and Hanger Corporation/USA) 6(3):35
Harrison, Patton B. (American Airlines/USA) 6(1):121
Harrison, Susan T.L. (University of Cape Town/REP OF SOUTH AFRICA) 6(6):339
Hart, Barry (Monash University/AUSTRALIA) 6(4):1
Hartzell, Kristen E. (Battelle/USA) 6(1):137; 6(10):193
Harwood, Christine L. (Michael Baker Corporation/USA) 6(2):155
Hassett, David J. (Energy & Environmental Research Center/USA) 6(5):129
Hater, Gary R. (Waste Management Inc./USA) 6(2):247
Hausmann, Tom S. (Battelle Marine Sciences Laboratory/USA) 6(3):157
Hawari, Jalal (National Research Council of Canada/CANADA) 6(9):139
Hayes, Adam J. (Triple Point Engineers, Inc./USA) 6(1):183
Hayes, Dawn M. (U.S. Navy/USA) 6(3):107
Hayes, Kim F. (University of Michigan/USA) 6(8):193
Haynes, R.J. (University of Natal/REP OF SOUTH AFRICA) 6(6):101
Heaston, Mark S. (Earth Tech/USA) 6(3):17, 25
Hecox, Gary R. (University of Kansas/USA) 6(4):109
Heebink, Loreal V. (Energy & Environmental Research Center/USA) 6(5):129
Heine, Robert (EFX Systems, Inc./USA) 6(8):19
Heintz, Caryl (Texas Tech University/USA) 6(3):9

Hendrickson, Edwin R. (DuPont Co./USA) 6(8):27, 43
Hendriks, Willem (Witteveen+Bos Consulting Engineers/THE NETHERLANDS) 6(5):289
Henkler, Rolf D. (ICI Paints/UK) 6(2):223
Henny, Cynthia (University of Maine/USA) 6(8):139
Henry, Bruce M. (Parsons Engineering Science, Inc/USA) 6(7):241
Henssen, Maurice J.C. (Bioclear Environmental Biotechnology/THE NETHERLANDS) 6(8):11
Herson, Diane S. (University of Delaware/USA) 6(6):9
Hesnawi, Rafik M. (University of Manitoba/CANADA) 6(6):165
Hetland, Melanie D. (Energy & Environmental Research Center/USA) 6(5):129
Hickey, Robert F. (EFX Systems, Inc./USA) 6(8):19
Hicks, Patrick H. (ARCADIS/USA) 6(1):107
Hiebert, Randy (MSE Technology Applications, Inc./USA) 6(8):79
Higashi, Teruo (University of Tsukuba/JAPAN) 6(9):187
Higgins, Mathew J. (Bucknell University/USA) 6(2):105
Higinbotham, James H. (ExxonMobil Environmental Remediation/USA) 6(8):87
Hines, April (Spelman College/USA) 6(9):53
Hinshalwood, Gordon (Delta Environmental Consultants, Inc./USA) 6(1):43
Hirano, Hiroyuki (The University of Tokyo/JAPAN) 6(6):111
Hirashima, Shouji (Yakult Pharmaceutical Industry/JAPAN) 6(10):345
Hirsch, Steve (Environmental Protection Agency/USA) 6(5):207
Hiwatari, Takehiko (National Institute for Environmental Studies/JAPAN) 6(5):321
Hoag, Rob (Conestoga-Rovers & Associates/USA) 6(1):99

Hoelen, Thomas P. (Stanford University/USA) 6(7):95
Hoeppel, Ronald E. (U.S. Navy/USA) 6(10):245
Hoffmann, Johannes (Hochtief Umwelt GmbH/GERMANY) 6(6):227
Hoffmann, Robert E. (Chevron Canada Resources/CANADA) 6(6):193
Höfte, Monica (Ghent University/BELGIUM) 6(5):223
Holder, Edith L. (University of Cincinnati/USA) 6(2):247
Holm, Thomas R. (Illinois State Water Survey/USA) 6(9):179
Holman, Hoi-Ying (Lawrence Berkeley National Laboratory/USA) 6(4):67
Holoman, Tracey R. Pulliam (University of Maryland/USA) 6(3):205
Hopper, Troy (URS Corporation/USA) 6(2):239
Hornett, Ryan (NOVA Chemicals Corporation/USA) 6(4):117
Hosangadi, Vitthal S. (Foster Wheeler Environmental Corp./USA) 6(9):249
Hough, Benjamin (Tetra Tech EM, Inc./USA) 6(10):293
Hozumi, Toyoharu (Oppenheimer Biotechnology/JAPAN) 6(10):345
Huang, Chin-I (National Chung Hsing University/TAIWAN) 6(10):217
Huang, Chin-Pao (University of Delaware/USA) 6(6):9, 149
Huang, Hui-Bin (DuPont Co./USA) 6(8):43
Huang, Junqi (Air Force Institute of Technology/USA) 6(2):173
Huang, Wei (University of Sheffield/UK) 6(2):207
Hubach, Cor (DHV Noord Nederland/THE NETHERLANDS) 6(8):11
Huesemann, Michael H. (Battelle/USA) 6(3):157
Hughes, Joseph B. (Rice University/USA) 6(5):85; 6(7):19
Hulsen, Kris (University of Ghent/BELGIUM) 6(5):223
Hunt, Jonathan (Clemson University/USA) 6(7):109

Hunter, William J. (U.S. Dept of Agriculture/USA) 6(9):209, 309
Hwang, Sangchul (University of Akron/USA) 6(3):235
Hyman, Michael R. (North Carolina State University/USA) 6(1): 83, 145

Ibeanusi, Victor M. (Spelman College/USA) 6(9):53
Ickes, Jennifer (Battelle/USA) 6(5):231, 237
Ide, Kazuki (Obayashi Corporation Ltd./JAPAN) 6(6):111; 6(10):239
Igarashi, Tsuyoshi (Nippon Institute of Technology/JAPAN) 6(5):321
Infante, Carmen (PDVSA Intevep/VENEZUELA) 6(6):257
Ingram, Sherry (IT Corporation/USA) 6(4):109
Ishikawa, Yoji (Obayashi Corporation/JAPAN) 6(6):249; 6(10):239

Jackson, W. Andrew (Texas Tech University/USA) 6(5):207, 313; 6(9):273
Jacobs, Alan K. (EnSafe, Inc./USA) 6(9):315
Jacques, Margaret E. (Rowan University/USA) 6(5):215
Jahan, Kauser (Rowan University/USA) 6(5):215
James, Garth (MSE Inc./USA) 6(8):79
Jansson, Janet K. (Södertörn University College/SWEDEN) 6(3):133
Japenga, Jan (Alterra/THE NETHERLANDS) 6(5):137
Jauregui, Juan (Universidad Nacional Autonoma de Mexico/MEXICO) 6(6):17
Jensen, James N. (State University of New York at Buffalo/USA) 6(6):89
eon, Mi-Ae (Texas Tech University/USA) 6(9):273
Jerger, Douglas E. (IT Corporation/USA) 6(3):35
Jernberg, Cecilia (Södertörn University College/SWEDEN) 6(3):133
Jindal, Ranjna (Suranaree University of Technology/THAILAND) 6(4):149

Johnson, Dimitra (Southern University at New Orleans/USA) 6(5):151
Johnson, Glenn (University of Utah/USA) 6(5):231
Johnson, Paul C. (Arizona State University/USA) 6(1):11; 6(8):61
Johnson, Richard L. (Oregon Graduate Institute/USA) 6(10):293
Jones, Antony (Komex H_2O Science, Inc./USA) 6(2):223; 6(3):173; 6(10):123
Jones, Clay (University of New Mexico/USA) 6(9):223
Jones, Triana N. (University of Maryland/USA) 6(3):205
Jonker, Hendrikus (Vrije Universiteit/THE NETHERLANDS) 6(4):91
Ju, Lu-Kwang (The University of Akron/USA) 6(6):319

Kaludjerski, Milica (San Diego State University/USA) 6(10):301
Kamashwaran, S. Ramanathen (University of Idaho/USA) 6(3):91
Kambhampati, Murty S. (Southern University at New Orleans/USA) 6(5):145, 151
Kamimura, Daisuke (Gunma University/JAPAN) 6(8):113
Kang, James J. (URS Corporation/USA) 6(1):121; 6(10):223
Kappelmeyer, Uwe (UFZ Centre for Environmental Research/GERMANY) 6(5):337
Karamanev, Dimitre G. (University of Western Ontario/CANADA) 6(10):171
Karlson, Ulrich (National Environmental Research Institute) 6(3):141
Kastner, James R. (University of Georgia/USA) 6(9):289
Kästner, Matthias (UFZ Centre for Environmental Research/GERMANY) 6(4):99; 6(5):337
Katz, Lynn E. (University of Texas/USA) 6(8):139
Kavanaugh, Rathi G. (University of Cincinnati/USA) 6(2):247

Kawahara, Fred (U.S. EPA/USA) 6(9):9
Kawakami, Tsuyoshi (University of Tsukuba/JAPAN) 6(9):187
Keefer, Donald A. (Illinois State Geological Survey/USA) 6(9):179
Keith, Nathaniel (Texas A&M University/USA) 6(5):25
Kelly, Laureen S. (Montana Department of Environmental Quality/USA) 6(6):287
Kempisty, David M. (U.S. Air Force/USA) 6(10):145, 155
Kerfoot, William B. (K-V Associates, Inc./USA) 6(10):33
Keuning, S. (Bioclear Environmental Technology/THE NETHERLANDS) 6(8):11
Khan, Tariq A. (Groundwater Services, Inc./USA) 6(7):19
Khodadoust, Amid P. (University of Cincinnati/USA) 6(5):243, 253, 261
Kieft, Thomas L. (New Mexico Institute of Mining and Technology/USA) 6(8):131
Kiessig, Gunter (WISMUT GmbH/GERMANY) 6(5):173; 6(9):155
Kilbride, Rebecca (CEFAS Laboratory/UK) 6(10):337
Kim, Jae Young (Seoul National University/REPUBLIC OF KOREA) 6(9):195
Kim, Jay (University of Cincinnati/USA) 6(6):133
Kim, Kijung (The Pennsylvania State University/USA) 6(9):303
Kim, Tae Young (Ewha University/REPUBLIC OF KOREA) 6(6):51
Kinsall, Barry L. (Oak Ridge National Laboratory/USA) 6(4):73
Kirschenmann, Kyle (IT Corp/USA) 6(4):109
Klaas, Norbert (University of Stuttgart/GERMANY) 6(2):137
Klecka, Gary M. (The Dow Chemical Company/USA) 6(2):89
Klein, Katrina (GeoTrans, Inc./USA) 6(2):163

Klens, Julia L. (IT Corporation/USA) 6(2):253; 6(9):215
Knotek-Smith, Heather M. (University of Idaho/USA) 6(9):147
Koch, Stacey A. (RMT, Inc./USA) 6(7):181
Koenen, Brent A. (U.S. Army Engineer Research & Development Center/USA) 6(5):9
Koenigsberg, Stephen S. (Regenesis Bioremediation Products/USA) 6(7):197, 257; 6(8):209; 6(10):9, 87
Kohata, Kunio (National Institute for Environmental Studies/JAPAN) 6(5):321
Kohler, Keisha (ThermoRetec Corporation/USA) 6(7):1
Kolhatkar, Ravindra V. (BP Corporation/USA) 6(1):35, 43
Komlos, John (Montana State University/USA) 6(7):117
Komnitsas, Kostas (National Technical University of Athens/GREECE) 6(9):97
Kono, Masakazu (Oppenheimer Biotechnology/JAPAN) 6(10):345
Koons, Brad W. (Leggette, Brashears & Graham, Inc./USA) 6(1):175
Koschal, Gerard (PNG Environmental/USA) 6(1):203
Koschorreck, Matthias (UFZ Centre for Environmental Research/GERMANY) 6(9):43
Koshikawa, Hiroshi (National Institute for Environmental Studies/JAPAN) 6(5):321
Kramers, Jan D. (University of Bern/SWITZERLAND) 6(4):91
Krooneman, Jannneke (Bioclear Environmental Biotechnology/THE NETHERLANDS) 6(7):141
Kruk, Taras B. (URS Corporation/USA) 6(10):223
Kuhwald, Jerry (NOVA Chemicals Corporation/CANADA) 6(5):53
Kuschk, Peter (UFZ Centre for Environmental Research Leipzig/GERMANY) 6(5):337

Laboudigue, Agnes (Centre National de Recherche sur les Sites et Sols Pollués/FRANCE) 6(2):129

LaFlamme, Brian (Engineering Management Support, Inc./USA) 6(10):231

Lafontaine, Chantal (École Polytechnique de Montréal/CANADA) 6(10):171

Laha, Shonali (Florida International University/USA) 6(10):187

Laing, M.D. (University of Natal/REP OF SOUTH AFRICA) 6(9):79

Lamar, Richard (EarthFax Development Corp/USA) 6(6):263

Lamarche, Philippe (Royal Military College of Canada/CANADA) 6(8):95

Lamb, Steven R. (GZA GeoEnvironmental, Inc./USA) 6(7):165

Landis, Richard C. (E.I. du Pont de Nemours & Company/USA) 6(8):185

Lang, Beth (United Technologies Corp./USA) 6(10):41

Langenhoff, Alette (TNO Institute of Environmental Science/THE NETHERLANDS) 6(7):141

LaPat-Polasko, Laurie T. (Parsons Engineering Science, Inc./USA) 6(2):65, 189

Lapus, Kevin (Regenesis/USA) 6(7):257; 6(10):9

LaRiviere, Daniel (Texas A&M University/USA) 6(5):17, 25

Larsen, Lars C. (Hedeselskabet/DENMARK) 6(2):11

Larson, John R. (TranSystems Corporation/USA) 6(7):229

Larson, Richard A. (University of Illinois at Urbana-Champaign/USA) 6(5):181

Lauzon, Francois (Dept of National Defence/CANADA) 6(8):95

Leavitt, Maureen E. (Newfields Inc./USA) 6(1):51; 6(5):189

Lebron, Carmen A. (U.S. Navy/USA) 6(7):95

Lee, B. J. (Science Applications International Corporation) 6(8):27

Lee, Brady D. (Idaho National Engineering & Environmental Laboratory/USA) 6(7):77

Lee, Chi Mei (National Chung Hsing University/TAIWAN) 6(6):353

Lee, Eun-Ju (Louisiana State University/USA) 6(5):313

Lee, Kenneth (Fisheries & Oceans Canada/CANADA) 6(10):337

Lee, Michael D. (Terra Systems, Inc./USA) 6(7):213, 249

Lee, Ming-Kuo (Auburn University/USA) 6(9):105

Lee, Patrick (Queen's University/CANADA) 6(2):121

Lee, Seung-Bong (University of Washington/USA) 6(10):211

Lee, Si-Jin (Kyonggi University/REPUBLIC OF KOREA) 6(1):161

Lee, Sung-Jae (ChoongAng University/REPUBLIC OF KOREA) 6(6):51

Leeson, Andrea (Battelle/USA) 6(10):1, 145, 155, 193

Lehman, Stewart E. (California Polytechnic State University/USA) 6(2):1

Lei, Li (University of Cincinnati/USA) 6(5):243, 261

Leigh, Daniel P. (IT Corporation/USA) 6(3):35

Leigh, Mary Beth (University of Oklahoma/USA) 6(5):61

Lendvay, John (University of San Francisco/USA) 6(8):19

Lenzo, Frank C. (ARCADIS Geraghty & Miller/USA) 6(7):53

Leon, Nidya (PDVSA Intevep/VENEZUELA) 6(6):257

Leong, Sylvia (Crescent Heights High School/CANADA) 6(5):53

Leontievsky, Alexey A. (Russian Academy of Sciences/RUSSIA) 6(3):75

Lerner, David N. (University of Sheffield/UK) 6(1):59; 6(2):207

Lesage, Suzanne (National Water Research Institute/CANADA) 6(7):321, 333, 341

Leslie, Jolyn C. (Camp Dresser & McKee, Inc./USA) 6(2):113
Lewis, Ronald F. (U.S. EPA/USA) 6(5):253, 261
Li, Dong X. (USA) 6(7):205
Li, Guanghe (Tsinghua University/CHINA) 6(7):61
Li, Tong (Tetra Tech EM Inc./USA) 6(10):293
Librando, Vito (Universita di Catania/ITALY) 6(3):149
Lieberman, M. Tony (Solutions Industrial & Environmental Services/USA) 6(7):249
Lin, Cindy (Conestoga-Rovers & Associates/USA) 6(1):99; 6(10):131
Lipson, David S. (Blasland, Bouck & Lee, Inc./USA) 6(10):319
Liu, Jian (University of Nevada Las Vegas/USA) 6(9):265
Liu, Xiumei (Shandong Agricultural University/ CHINA) 6(9):113
Livingstone, Stephen (Franz Environmental Inc./CANADA) 6(6):211
Lizzari, Daniela (Universita degli Studi di Verona/ITALY) 6(3):267
Llewellyn, Tim (URS/USA) 6(7):321, 333, 341
Lobo, C. (El Encin IMIA/SPAIN) 6(4):29
Loeffler, Frank E. (Georgia Institute of Technology/USA) 6(8):19
Logan, Bruce E. (The Pennsylvania State University/USA) 6(9):303
Long, Gilbert M. (Camp Dresser & McKee Inc./USA) 6(6):287
Longoni, Giovanni (Montgomery Watson/ITALY) 6(10):41
Lorbeer, Helmut (Technical University of Dresden/GERMANY) 6(8):105
Lors, Christine (Centre National de Recherche sur les Sites et Sols Pollués /FRANCE) 6(2):129
Lorton, Diane M. (King's College London/UK) 6(2):223; 6(3):173
Losi, Mark E. (Foster Wheeler Environ. Corp./USA) 6(9):249
Loucks, Mark (U.S. Air Force/USA) 6(2):261

Lu, Chih-Jen (National Chung Hsing University/TAIWAN) 6(6):353; 6(10):217
Lu, Xiaoxia (Tsinghua University/CHINA) 6(7):61
Lubenow, Brian (University of Delaware/USA) 6(6):149
Lucas, Mary (Parsons Engineering Science, Inc./USA) 6(10):283
Lundgren, Tommy S. (Sydkraft SAKAB AB/SWEDEN) 6(6):127
Lundstedt, Staffan (Umeå University/SWEDEN) 6(3):181
Luo, Xiaohong (NRC Research Associate/USA) 6(8):167
Luthy, Richard G. (Stanford University/USA) 6(3):189
Lutze, Werner (University of New Mexico/USA) 6(9):155
Luu, Y.-S. (Queen's University/CANADA) 6(2):121
Lynch, Regina M. (Battelle/USA) 6(10):155

Macek, Thomáš (Institute of Chemical Technology/Czech Republic) 6(5):61
MacEwen, Scott J. (CH2M Hill/USA) 6(3):107
Machado, Kátia M. G. (Fund. Centro Tecnológico de Minas Gerais/BRAZIL) 6(3):99
Maciel, Helena Alves (Aberdeen University/UK) 6(1):1
Mack, E. Erin (E.I. du Pont de Nemours & Co./USA) 6(2):81; 6(8):43
Macková, Martina (Institute of Chemical Technology/Czech Republic) 6(5):61
Macnaughton, Sarah J. (AEA Technology/UK) 6(5):305; 6(10):337
Macomber, Jeff R. (University of Cincinnati/USA) 6(6):133
Macrae, Jean (University of Maine/USA) 6(8):139
Madden, Patrick C. (Engineering Consultant/USA) 6(8):87
Madsen, Clint (Terracon/USA) 6(8):157; 6(10):15
Magar, Victor S. (Battelle/USA) 6(1):137; 6(5):231, 237; 6(10):145, 155

Mage, Roland (Battelle
  Europe/SWITZERLAND) 6(6):241;
  6(10):109
Magistrelli, P. (Istituto di Cibernetica e
  Biofisica/ITALY) 6(5):157
Maierle, Michael S. (ARCADIS
  Geraghty & Miller, Inc./USA)
  6(7):149
Major, C. Lee (Jr.) (University of
  Michigan/USA) 6(8):19
Major, David W. (GeoSyntec
  Consultants/CANADA) 6(8):27
Maki, Hideaki (National Institute for
  Environmental Studies/JAPAN)
  6(5):321
Makkar, Randhir S. (University of
  Illinois-Chicago/USA) 6(5):297
Malcolm, Dave (BAE Systems Properties
  Ltd./UK) 6(6):119
Manabe, Takehiko (Hyogo Prefectural
  Fisheries Research Institute/JAPAN)
  6(10):345
Maner, P.M. (Equilon Enterprises,
  LLC/USA) 6(1):11
Maner, Paul (Shell Development
  Company/USA) 6(8):61
Manrique-Ramírez, Emilio Javier
  (SYMCA, S.A. de C.V./MEXICO)
  6(6):369
Marchal, Rémy (Institut Français du
  Pétrole/FRANCE) 6(1):153
Maresco, Vincent (Groundwater &
  Environmental Srvcs/USA)
  6(10):101
Marnette, Emile C. (TAUW BV/THE
  NETHERLANDS) 6(7):297
Marshall, Timothy R. (URS
  Corporation/USA) 6(2):49
Martella, L. (Istituto di Cibernetica e
  Biofisica/ITALY) 6(5):157
Martin, C. (Universidad
  Politecnica/SPAIN) 6(4):29
Martin, Jennifer P. (Idaho National
  Engineering & Environmental
  Laboratory/USA) 6(7):265
Martin, John F. (U.S. EPA/USA)
  6(2):247
Martin, Margarita (Universidad
  Complutense de Madrid/SPAIN)
  6(4):29; 6(6):377

Martinez-Inigo, M.J. (El Encin
  IMIA/SPAIN) 6(4):29
Martino, Lou (Argonne National
  Laboratory/USA) 6(5):207
Mascarenas, Tom (Environmental
  Chemistry/USA) 6(8):157
Mason, Jeremy (King's College
  London/UK) 6(2):223; 6(3):173;
  6(10):123
Massella, Oscar (Universita degli Studi
  di Verona/ITALY) 6(3):267
Matheus, Dacio R. (Instituto de
  Botânica/BRAZIL) 6(3):99
Matos, Tania (University of Puerto Rico
  at Rio Piedras/USA) 6(9):179
Matsubara, Takashi (Obayashi
  Corporation/JAPAN) 6(6):249
Mattiasson, Bo (Lund
  University/SWEDEN) 6(3):181;
  6(6):65; 6(9):123
McCall, Sarah (Battelle/USA)
  6(10):155, 245
McCarthy, Kevin (Battelle Duxbury
  Operations/USA) 6(5):9
McCartney, Daryl M. (University of
  Manitoba/CANADA) 6(6):165
McCormick, Michael L. (The University
  of Michigan/USA) 6(8):193
McDonald, Thomas J. (Texas A&M
  University) 6(5):17
McElligott, Mike (U.S. Air Force/USA)
  6(1):51
McGill, William B. (University of
  Northern British
  Columbia/CANADA) 6(4):7
McIntosh, Heather (U.S. Army/USA)
  6(7):321, 333
McLinn, Eugene L. (RMT, Inc./USA)
  6(5):121
McLoughlin, Patrick W. (Microseeps
  Inc./USA) 6(1):35
McMaster, Michaye (GeoSyntec
  Consultants/CANADA) 6(8):27, 43,
  6(9):297
McMillen, Sara J. (Chevron Research &
  Technology Company/USA)
  6(6):193
Meckenstock, Rainer U. (University of
  Tübingen/GERMANY) 6(4):99
Mehnert, Edward (Illinois State
  Geological Survey/USA) 6(9):179

Meigio, Jodette L. (Idaho National Engineering & Environmental Laboratory/USA) 6(7):77
Meijer, Harro A.J. (University of Groningen/THE NETHERLANDS) 6(4):91
Meijerink, E. (Province of Drenthe/THE NETHERLANDS) 6(8):11
Merino-Castro, Glicina (Inst Technol y de Estudios Superiores/MEXICO) 6(6):377
Messier, J.P. (U.S. Coast Guard/USA) 6(1):107
Meyer, Michael (Environmental Resources Management/BELGIUM) 6(7):87
Meylan, S. (Queen's University/CANADA) 6(2):121
Miles, Victor (Duracell Inc./USA) 6(7):87
Millar, Kelly (National Water Research Institute/CANADA) 6(7):321, 333, 341
Miller, Michael E. (Camp Dresser & McKee, Inc./USA) 6(7):273
Miller, Thomas Ferrell (Lockheed Martin/USA) 6(3):259
Mills, Heath J. (Georgia Institute of Technology/USA) 6(9):165
Millward, Rod N. (Louisiana State University/USA) 6(5):305
Mishra, Pramod Chandra (Sambalpur University/INDIA) 6(9):173
Mitchell, David (AEA Technology Environment/UK) 6(10):337
Mitraka, Maria (Serres/GREECE) 6(6):89
Mocciaro, PierFilippo (Ambiente S.p.A./ITALY) 6(6):227
Moeri, Ernesto N. (CSD-GEOKLOCK/BRAZIL) 6(1):27
Moir, Michael (Chevron Research & Technology Co./USA) 6(1):83
Molinari, Mauro (AgipPetroli S.p.A/ITALY) 6(6):173
Mollea, C. (Politecnico di Torino/ITALY) 6(3):211
Mollhagen, Tony (Texas Tech University/USA) 6(3):9
Monot, Frédéric (Institut Français du Pétrole/FRANCE) 6(1):153

Moon, Hee Sun (Seoul National University/REPUBLIC OF KOREA) 6(9):195
Moosa, Shehnaaz (University of Cape Town/REP OF SOUTH AFRICA) 6(6):339
Morasch, Barbara (University Konstanz/GERMANY) 6(4):99
Moreno, Joanna (URS Corporation/USA) 6(2):239
Morgan, Scott (URS - Dames & Moore/USA) 6(7):321
Morrill, Pamela J. (Camp, Dresser, & McKee, Inc./USA) 6(2):113
Morris, Damon (ThermoRetec Corporation/USA) 6(7):1
Mortimer, Marylove (Mississippi State University/USA) 6(5):329
Mortimer, Wendy (Bell Canada/CANADA) 6(2):27; 6(6):185, 203, 211,
Mossing, Christian (Hedeselskabet/DENMARK) 6(2):11
Mossmann, Jean-Remi (Centre National de Recherche sur les Sites et Sols Pollués/FRANCE) 6(2):129
Moteleb, Moustafa A. (University of Cincinnati/USA) 6(6):133
Mowder, Carol S. (URS/USA) 6(7):321, 333, 341
Moyer, Ellen E. (ENSR International./USA) 6(1):75
Mravik, Susan C. (U.S. EPA/USA) 6(1):167
Mueller, James G. (URS Corporation/USA) 6(2):239
Müller, Axel (Water Technology Center/GERMANY) 6(8):105
Müller, Beate (Umweltschutz Nord GmbH/GERMANY) 6(4):131
Müller, Klaus (Battelle Europe/SWITZERLAND) 6(5):41; 6(6):241
Muniz, Herminio (Hart Crowser Inc./USA) 6(10):9
Murphy, Sean M. (Komex International Ltd./CANADA) 6(4):7
Murray, Cliff (United States Army Corps of Engineers/USA) 6(9):281
Murray, Gordon Bruce (Stella-Jones Inc./CANADA) 6(3):197

Murray, Willard A. (Harding ESE/USA) 6(7):197
Mutch, Robert D. (Brown and Caldwell/USA) 6(2):145
Mutti, Francois (Water & Soil Remediation S.r.l./ITALY) 6(6):179
Myasoedova, Nina M. (Russian Academy of Sciences/RUSSIA) 6(3):75

Nadolishny, Alex (Nedatek, Inc./USA) 6(10):139
Nagle, David P. (University of Oklahoma/USA) 6(5):61
Nam, Kyoungphile (Seoul National University/REPUBLIC OF KOREA) 6(9):195
Narayanaswamy, Karthik (Parsons Engineering Science/USA) 6(2):65
Nelson, Mark D. (Delta Environmental Consultants, Inc./USA) 6(1):175
Nelson, Yarrow (California Polytechnic State University/USA) 6(10):311
Nemati, M. (University of Cape Town/REP OF SOUTH AFRICA) 6(6):339
Nestler, Catherine C. (Applied Research Associates, Inc./USA) 6(4):59, 6(6):43
Nevárez-Moorillón, G.V. (UACH/MEXICO) 6(6):361
Neville, Scott L. (Aerojet General Corp./USA) 6(9):297
Newell, Charles J. (Groundwater Services, Inc./USA) 6(7):19
Nieman, Karl (Utah State University/USA) 6(4):67
Niemeyer, Thomas (Hochtief Umwelt Gmbh/GERMANY) 6(6):227
Nies, Loring (Purdue University/USA) 6(3):125
Nipshagen, Adri A.M. (IWACO/THE NETHERLANDS) 6(7):141
Nishino, Shirley (U.S. Air Force/USA) 6(3):59
Nivens, David E. (University of Tennessee/USA) 6(4):45
Noffsinger, David (Westinghouse Savannah River Company/USA) 6(10):163

Noguchi, Takuya (Nippon Institute of Technology/JAPAN) 6(5):321
Nojiri, Hideaki (The University of Tokyo/JAPAN) 6(4):51; 6(6):111
Noland, Scott (NESCO Inc./USA) 6(10):73
Nolen, C. Hunter (Camp Dresser & McKee/USA) 6(6):287
Norris, Robert D. (Eckenfelder/Brown and Caldwell/USA) 6(2):145; 6(7):35
North, Robert W. (Environ Corporation./USA) 6(7):189
Novak, John T. (Virginia Polytechnic Institute & State University/USA) 6(2):105; 6(5):1
Novick, Norman (Exxon/Mobil Oil Corp/USA) 6(1):35
Nuttall, H. Eric (The University of New Mexico/USA) 6(9): 155, 223
Nuyens, Dirk (Environmental Resources Management/BELGIUM) 6(7):87; 6(9):87
Nzengung, Valentine A. (University of Georgia/USA) 6(9):289

Ochs, L. Donald (Regenesis/USA) 6(10):139
O'Connell, Joseph E. (Environmental Resolutions, Inc./USA) 6(1):91
Odle, Bill (Newfields, Inc./USA) 6(5):189
O'Donnell, Ingrid (BAE Systems Properties, Ltd./UK) 6(6):119
Ogden, Richard (BAE Systems Properties Ltd./UK) 6(6):119
Oh, Byung-Taek (The University of Iowa/USA) 6(8):147, 175
Oh, Seok-Young (University of Delaware/USA) 6(6):149
Omori, Toshio (The University of Tokyo/JAPAN) 6(4):51; 6(6):111
O'Neal, Brenda (ARA/USA) 6(3):43
Oppenheimer, Carl H. (Oppenheimer Biotechnology/USA) 6(10):345
O'Regan, Gerald (Chevron Products Company/USA) 6(1):203
O'Reilly, Kirk T. (Chevron Research & Technology Co/USA) 6(1):83, 145, 203
Oshio, Takahiro (University of Tsukuba/JAPAN) 6(9):187

Author Index

Ozdemiroglu, Ece (EFTEC Ltd./UK) 6(4):17

Padovani, Marco (Centro Ricerche Ambientali/ITALY) 6(4):131
Paganetto, A. (Istituto di Cibernetica e Biofisica/ITALY) 6(5):157
Pahr, Michelle R. (ARCADIS Geraghty & Miller/USA) 6(1):107
Pal, Nirupam (California Polytechnic State University/USA) 6(2):1
Palmer, Tracy (Applied Power Concepts, Inc./USA) 6(7):103
Palumbo, Anthony V. (Oak Ridge National Laboratory/USA) 6(4):73; 6(9):165
Panciera, Matthew A. (University of Connecticut/USA) 6(7):69
Pancras, Tessa (Wageningen University/THE NETHERLANDS) 6(5):289
Pardue, John H. (Louisiana State University/USA) 6(5): 207, 313; 6(9):273
Park, Kyoohong (ChoongAng University/REPUBLIC OF KOREA) 6(6):51
Parkin, Gene F. (University of Iowa/USA) 6(3):1
Paspaliaris, Ioannis (National Technical University of Athens/GREECE) 6(9):97
Paton, Graeme I. (Aberdeen University/UK) 6(1):1
Patrick, John (University of Reading/UK) 6(10):337
Payne, Frederick C. (ARCADIS Geraghty & Miller/USA) 6(7):53
Payne, Jo Ann (DuPont Co./USA) 6(8):43
Peabody, Jack G. (Regenesis/USA) 6(10):95
Peacock, Aaron D. (University of Tennessee/USA) 6(4):73; 6(5):305
Peargin, Tom R. (Chevron Research & Technology Co/USA) 6(1):67
Peeples, James A. (Metcalf & Eddy, Inc./USA) 6(7):173
Pehlivan, Mehmet (Tait Environmental Management, Inc./USA) 6(10):267, 275

Pelletier, Emilien (ISMER/CANADA) 6(2):57
Pennie, Kimberley A. (Stella-Jones, Inc./CANADA) 6(3):197
Peramaki, Matthew P. (Leggette, Brashears, & Graham, Inc./USA) 6(10):259
Perey, Jennie R. (University of Delaware/USA) 6(6):149
Perez-Vargas, Josefina (CINVESTAV-IPN/MEXICO) 6(6):219
Perina, Tomas (IT Corporation/USA) 6(1):51; 6(2):73
Perlis, Shira R. (Rowan University/USA) 6(5):215
Perlmutter, Michael W. (EnSafe, Inc./USA) 6(9):315
Perrier, Michel (École Polytechnique de Montréal/CANADA) 6(4):139
Perry, L.B. (U.S. Army Engineer Research & Development Center/USA) 6(5):9
Persico, John L. (Blasland, Bouck & Lee, Inc./USA) 6(10):319
Peschong, Bradley J. (Leggette, Brashears & Graham, Inc./USA) 6(1):175
Peters, Dave (URS/USA) 6(7):333
Peterson, Lance N. (North Wind Environmental, Inc./USA) 6(7):265
Petrovskis, Erik A. (Geotrans Inc./USA) 6(8):19
Peven-McCarthy, Carole (Battelle Ocean Sciences/USA) 6(5):231
Pfiffner, Susan M. (University of Tennessee/USA) 6(4):73
Phelps, Tommy J. (Oak Ridge National Laboratory/USA) 6(4):73
Pickett, Tim M. (Applied Biosciences Corporation/USA) 6(9):331
Pickle, D.W. (Equilon Enterprises LLC/USA) 6(8):61
Pierre, Stephane (École Polytechnique de Montréal/CANADA) 6(10):171
Pijls, Charles G.J.M. (TAUW BV/THE NETHERLANDS) 6(10):253
Pirkle, Robert J. (Microseeps, Inc./USA) 6(1):35
Pisarik, Michael F. (New Fields/USA) 6(1):121

Piveteau, Pascal (Institut Français du Pétrole/FRANCE) 6(1):153
Place, Matthew (Battelle/USA) 6(10):245
Plata, Nadia (Battelle Europe/SWITZERLAND) 6(5):41
Poggi-Varaldo, Hector M. (CINVESTAV-IPN/MEXICO) 6(3):243; 6(6):219
Pohlmann, Dirk C. (IT Corporation/USA) 6(2):253
Pokethitiyook, Prayad (Mahidol University/THAILAND) 6(10):329
Polk, Jonna (U.S. Army Corps of Engineers/USA) 6(9):281
Pope, Daniel F. (Dynamac Corp/USA) 6(1):129
Porta, Augusto (Battelle Europe/SWITZERLAND) 6(5):41; 6(6):241; 6(10):109
Portier, Ralph J. (Louisiana State University/USA) 6(5):305
Powers, Leigh (Georgia Institute of Technology/USA) 6(9):165
Prandi, Alberto (Water & Soil Remediation S.r.l/ITALY) 6(6):179
Prasad, M.N.V. (University of Hyderabad/INDIA) 6(5):165
Price, Steven (Camp Dresser & McKee, Inc./USA) 6(9):303
Priester, Lamar E. (Priester & Associates/USA) 6(10):65
Pritchard, P. H. (Hap) (U.S. Navy/USA) 6(7):125
Profit, Michael D. (CDM Federal Programs Corporation/USA) 6(6):81
Prosnansky, Michal (Gunma University/JAPAN) 6(9):201
Pruden, Amy (University of Cincinnati/USA) 6(1):19
Ptacek, Carol J. (University of Waterloo/CANADA) 6(9):71

Radosevich, Mark (University of Delaware/USA) 6(6):9
Radtke, Corey (INEEL/USA) 6(3):9
Raetz, Richard M. (Global Remediation Technologies, Inc./USA) 6(6):311
Rainwater, Ken (Texas Tech University/USA) 6(3):9

Ramani, Mukundan (University of Cincinnati/USA) 6(5):269
Raming, Julie B. (Georgia-Pacific Corp./USA) 6(1):183
Ramírez, N. E. (ECOPETROL-ICP/COLOMBIA) 6(6):319
Ramsay, Bruce A. (Polyferm Canada Inc./CANADA) 6(2):121; 6(10):171
Ramsay, Juliana A. (Queen's University/CANADA) 6(2):121; 6(10):171
Rao, Prasanna (University of Cincinnati/USA) 6(9):1
Ratzke, Hans-Peter (Umweltschutz Nord GMBH/GERMANY) 6(4):131
Reardon, Kenneth F. (Colorado State University/USA) 6(8):53
Rectanus, Heather V. (Virginia Polytechnic Institute & State University/USA) 6(2):105
Reed, Thomas A. (URS Corporation/USA) 6(8):157; 6(10):15, 95
Rees, Hubert (CEFAS Laboratory/UK) 6(10):337
Rehm, Bernd W. (RMT, Inc./USA) 6(2):97; 6(10):201
Reinecke, Stefan (Franz Environmental Inc./CANADA) 6(6):211
Reinhard, Martin (Stanford University/USA) 6(7):95
Reisinger, H. James (Integrated Science & Technology Inc/USA) 6(1):183
Rek, Dorota (IT Corporation/USA) 6(2):73
Reynolds, Charles M. (U.S. Army Engineer Research & Development Center/USA) 6(5):9
Reynolds, Daniel E. (Air Force Institute of Technology/USA) 6(2):173
Rice, John M. (RMT, Inc./USA) 6(7):181
Richard, Don E. (Barr Engineering Company/USA) 6(3):219; 6(5):105
Richardson, Ian (Conestoga-Rovers & Associates/USA) 6(10):131
Richnow, Hans H. (UFZ-Centre for Environmental Research/GERMANY) 6(4):99

Rijnaarts, Huub H.M. (TNO Institute of Environmental Science/THE NETHERLANDS) 6(2):231
Ringelberg, David B. (U.S. Army Corps of Engineers/USA) 6(5):9; 6(6):43; 6(10):115
Ríos-Leal, E. (CINVESTAV-IPN/MEXICO) 6(3):243
Ripp, Steven (University of Tennessee/USA) 6(4):45
Ritter, Michael (URS Corporation/USA) 6(2):239
Ritter, William F. (University of Delaware/USA) 6(6):9
Riva, Vanessa (Parsons Engineering Science, Inc./USA) 6(2):39
Rivas-Lucero, B.A. (Centro de Investigacion en Materiales Avanzados/MEXICO) 6(6):361
Rivetta, A. (Universita degli Studi di Milano/ITALY) 6(5):157
Robb, Joseph (ENSR International/USA) 6(1):75
Robertiello, Andrea (EniTecnologie S.p.A./ITALY) 6(6):173
Robertson, K. (Queen's University/CANADA) 6(2):121
Robinson, David (ERM, Inc./USA) 6(7):45
Robinson, Sandra L. (Virginia Polytechnic Institute & State University/USA) 6(5):1
Rockne, Karl J. (University of Illinois-Chicago/USA) 6(5):297
Rodríguez-Vázquez, Refugio (CINVESTAV-IPN/MEXICO) 6(3):243; 6(6):219, 369
Römkens, Paul (Alterra/THE NETHERLANDS) 6(5):137
Rongo, Rocco (University della Calabria/ITALY) 6(4):131
Roorda, Marcus L. (Rowan University/USA) 6(5):215
Rosser, Susan J. (University of Cambridge/UK) 6(5):69
Rowland, Martin A. (Lockheed-Martin Michoud Space Systems/USA) 6(7):1
Royer, Richard (The Pennsylvania State University/USA) 6(8):201

Ruggeri, Bernardo (Politecnico di Torino/ITALY) 6(3):211
Ruiz, Graciela M. (University of Iowa/USA) 6(1):195
Rupassara, S. Indumathie (University of Illinois at Urbana-Champaign/USA) 6(5):181

Sacchi, G.A. (Universita degli Studi di Milano/ITALY) 6(5):157
Sahagun, Tracy (U.S. Marine Corps./USA) 6(10):1
Sakakibara, Yutaka (Waseda University/JAPAN) 6(8):113; 6(9):201
Sakamoto, T. (Queen's University/CANADA) 6(10):171
Salam, Munazza (Crescent Heights High School/CANADA) 6(5):53
Salanitro, Joseph P. (Equilon Enterprises, LLC/USA) 6(1):11; 6(8):61
Salvador, Maria Cristina (CSD-GEOKLOCK/BRAZIL) 6(1):27
Samson, Réjean (École Polytechnique de Montréal/CANADA) 6(3):115; 6(4):139; 6(9):27
San Felipe, Zenaida (Monash University/AUSTRALIA) 6(4):1
Sánchez, F.N. (ECOPETROL-ICP/COLOMBIA) 6(6):319
Sánchez, Gisela (PDVSA Intevep/VENEZUELA) 6(6):257
Sánchez, Luis (PDVSA Intevep/VENEZUELA) 6(6):257
Sanchez, M. (Universidad Complutense de Madrid/SPAIN) 6(4):29; 6(6):377
Sandefur, Craig A. (Regenesis/USA) 6(7):257; 6(10):87
Sanford, Robert A. (University of Illinois at Urbana-Champaign/USA) 6(9):179
Santangelo-Dreiling, Theresa (Colorado Dept. of Transportation/USA) 6(10):231
Saran, Jennifer (Kennecott Utah Copper Corp./USA) 6(9):323
Sarpietro, M.G. (Universita di Catania/ITALY) 6(3):149

Sartoros, Catherine (Université du Québec à Montréal/CANADA) 6(3):165
Saucedo-Terán, R.A. (Centro de Investigacion en Materiales Avanzados/MEXICO) 6(6):361
Saunders, James A. (Auburn University/USA) 6(9):105
Sayler, Gary S. (University of Tennessee/USA) 6(4):45
Scalzi, Michael M. (Innovative Environmental Technologies, Inc./USA) 6(10):23
Scarborough, Shirley (IT Corporation/USA) 6(2):253
Schaffner, I. Richard (GZA GeoEnvironmental, Inc./USA) 6(7):165
Scharp, Richard A. (U.S. EPA/USA) 6(9):9
Schell, Heico (Water Technology Center/GERMANY) 6(8):105
Scherer, Michelle M. (The University of Iowa/USA) 6(3):1
Schipper, Mark (Groundwater Services) 6(8):73
Schmelling, Stephen (U.S. EPA/USA) 6(1):129
Schnoor, Jerald L. (University of Iowa/USA) 6(8):147
Schoefs, Olivier (École Polytechnique de Montréal/CANADA) 6(4):139
Schratzberger, Michaela (CEFAS Laboratory/UK) 6(10):337
Schulze, Susanne (Water Technology Center/GERMANY) 6(2):137
Schuur, Jessica H. (Lund University/SWEDEN) 6(6):65
Scrocchi, Susan (Conestoga-Rovers & Associates/USA) 6(1):99; 6(10):131
Sczechowski, Jeff (California Polytechnic State University/USA) 6(10):311
Seagren, Eric A. (University of Maryland/USA) 6(10):57
Sedran, Marie A. (University of Cincinnati/USA) 6(1):19
Seifert, Dorte (Technical University of Denmark/DENMARK) 6(2):11
Semer, Robin (Harza Engineering Company, Inc./USA) 6(7):157

Semprini, Lewis (Oregon State University/USA) 6(10):145, 155, 179
Seracuse, Joe (Harding ESE/USA) 6(7):205
Serra, Roberto (Centro Ricerche Ambientali/ITALY) 6(4):131
Sewell, Guy W. (U.S. EPA/USA) 6(1):167; 6(7):125; 6(8):167
Sharma, Pawan (Camp Dresser & McKee Inc./USA) 6(7):305
Sharp, Robert R. (Manhattan College/USA) 6(7):117
Shay, Devin T. (Groundwater & Environmental Services, Inc./USA) 6(10):101
Shelley, Michael L. (Air Force Institute of Technology/USA) 6(5):95
Shen, Hai (Dynamac Corporation/USA) 6(1): 129, 167
Sherman, Neil (Louisiana-Pacific Corporation/USA) 6(3):83
Sherwood Lollar, Barbara (University of Toronto/CANADA) 6(4):91, 109
Shi, Jing (EFX Systems, Inc./USA) 6(8):19
Shields, Adrian R.G. (Komex Europe/UK) 6(10):123
Shiffer, Shawn (University of Illinois/USA) 6(9):179
Shin, Won Sik (Lousiana State University/USA) 6(5):313
Shiohara, Kei (Mississippi State University/USA) 6(6):279
Shirazi, Fatemeh R. (Stratum Engineering Inc./USA) 6(8):121
Shoemaker, Christine (Cornell University/USA) 6(4):125
Sibbett, Bruce (IT Corporation/USA) 6(2):73
Silver, Cannon F. (Parsons Engineering Science, Inc./USA) 6(10):283
Silverman, Thomas S. (RMT, Inc./USA) 6(10):201
Simon, Michelle A. (U.S. EPA/USA) 6(10):293
Sims, Gerald K. (USDA-ARS/USA) 6(5):181
Sims, Ronald C. (Utah State University/USA) 6(4):67; 6(6):1
Sincock, M. Jennifer (ENVIRON International Corp./USA) 6(7):189

Sittler, Steven P. (Advanced Pollution Technologists, Ltd./USA) 6(2):215
Skladany, George J. (ERM, Inc./USA) 6(7):45, 213
Skubal, Karen L. (Case Western Reserve University/USA) 6(8):193
Slenders, Hans (TNO-MEP/THE NETHERLANDS) 6(7):289
Slomczynski, David J. (University of Cincinnati/USA) 6(2):247
Slusser, Thomas J. (Wright State University/USA) 6(5):95
Smallbeck, Donald R. (Harding Lawson/USA) 6(10):231
Smets, Barth F. (University of Connecticut/USA) 6(7):69
Smith, Christy (North Carolina State University/USA) 6(1):145
Smith, Colin C. (University of Sheffield/UK) 6(2):207
Smith, John R. (Alcoa Inc./USA) 6(6):89
Smith, Jonathan (The Environment Agency/UK) 6(4):17
Smith, Steve (King's College London/UK) 6(2):223; 6(3):173; 6(10):123
Smyth, David J.A. (University of Waterloo/CANADA) 6(9):71
Sobecky, Patricia (Georgia Institute of Technology/USA) 6(9):165
Sola, Adrianna (Spelman College/USA) 6(9):53
Sordini, E. (EniTechnologie/ITALY) 6(6):173
Sorensen, James A. (University of North Dakota/USA) 6(6):141
Sorenson, Kent S. (Idaho National Engineering and Environmental Laboratory./USA) 6(7):265
South, Daniel (Harding ESE/USA) 6(7):205
Spain, Jim (U.S. Air Force/USA) 6(3):59; 6(7):125
Spasova, Irena Ilieva (University of Mining & Geology/BULGARIA) 6(9):97
Spataro, William (University della Calabria/ITALY) 6(4):131

Spinnler, Gerard E. (Equilon Enterprises, LLC/USA) 6(1):11; 6(8):61
Springael, Dirk (VITO/BELGIUM) 6(4):35
Srinivasan, P. (GeoTrans, Inc./USA) 6(2):163
Stansbery, Anita (California Polytechnic State University/USA) 6(10):311
Starr, Mark G. (DuPont Co./USA) 6(8):43
Stehmeier, Lester G. (NOVA Research Technology Centre/CANADA) 6(4):117; 6(5):53
Stensel, H. David (University of Washington/USA) 6(10):211
Stordahl, Darrel M. (Camp Dresser & McKee Inc./USA) 6(6):287
Stout, Scott (Battelle/USA) 6(5):237
Strand, Stuart E. (University of Washington/USA) 6(10):211
Stratton, Glenn (Nova Scotia Agricultural College/CANADA) 6(3):197
Strybel, Dan (IT Corporation/USA) 6(9):215
Stuetz, R.M. (Cranfield University/UK) 6(6):329
Suarez, B. (ECOPETROL-ICP/COLOMBIA) 6(6):319
Suidan, Makram T. (University of Cincinnati/USA) 6(1):19; 6(5):243, 253, 261; 6(6):133,
Suthersan, Suthan S. (ARCADIS Geraghty & Miller/USA) 6(7):53
Suzuki, Masahiro (Nippon Institute of Technology/JAPAN) 6(5):321
Sveum, Per (Deconterra AS/NORWAY) 6(6):157
Swallow, Ian (BAE Systems Properties Ltd./UK) 6(6):119
Swann, Benjamin M. (Camp Dresser & McKee Inc./USA) 6(7):305
Swannell, Richard P.J. (AEA Technology Environment/UK) 6(10):337

Tabak, Henry H. (U.S. EPA/USA) 6(5):243, 253, 261, 269; 6(9):1, 17
Takai, Koji (Fuji Packing/JAPAN) 6(10):345

Talley, Jeffrey W. (University of Notre Dame/USA) 6(3):189; 6(4):59; 6(6):43; 6(7):125; 6(10):115
Tao, Shu (Peking University/CHINA) 6(7):61
Taylor, Christine D. (North Carolina State University/USA) 6(1):83
Ter Meer, Jeroen (TNO Institute of Environmental Science/THE NETHERLANDS) 6(2):231; 6(7):289
Tétreault, Michel (Royal Military College of Canada/CANADA) 6(8):95
Tharpe, D.L. (Equilon Enterprises LLC/USA) 6(8):61
Theeuwen, J. (Grontmij BV/THE NETHERLANDS) 6(7):289
Thomas, Hartmut (WASAG DECON GMbH/GERMANY) 6(3):67
Thomas, Mark (EG&G Technical Services, Inc./USA) 6(10):49
Thomas, Paul R. (Thomas Consultants, Inc./USA) 6(5):189
Thomas, Robert C. (University of Georgia/USA) 6(9):105
Thomson, Michelle M. (URS Corporation/USA) 6(2):81
Thornton, Steven F. (University of Sheffield/UK) 6(1):59, 6(2):207
Tian, C. (University of Cincinnati/USA) 6(8):35
Tiedje, James M. (Michigan State University/USA) 6(7):125; 6(8):19
Tiehm, Andreas (Water Technology Center/GERMANY) 6(2):137; 6(8):105
Tietje, David (Foster Wheeler Environmental Corportation/USA) 6(9):249
Timmins, Brian (Oregon State University/USA) 6(10):179
Togna, A. Paul (Envirogen Inc/USA) 6(9):281
Tolbert, David E.(U.S. Army/USA) 6(9):281
Tonnaer, Haimo (TAUW BV/THE NETHERLANDS) 6(7):297; 6(10):253
Toth, Brad (Harding ESE/USA) 6(10):231

Tovanabootr, Adisorn (Oregon State University/USA) 6(10):145
Travis, Bryan (Los Alamos National Laboratory/USA) 6(10):163
Trudnowski, John M. (MSE Technology Applications, Inc./USA) 6(9):35
Truax, Dennis D. (Mississippi State University/USA) 6(9):241
Trute, Mary M. (Camp Dresser & McKee, Inc./USA) 6(2):113
Tsuji, Hirokazu (Obayashi Corporation Ltd./JAPAN) 6(6):111, 249; 6(10):239
Tsutsumi, Hiroaki (Prefectural University of Kumamoto/JAPAN) 6(10):345
Turner, Tim (CDM Federal Programs Corp./USA) 6(6):81
Turner, Xandra (International Biochemicals Group/USA) 6(10):23
Tyner, Larry (IT Corporation/USA) 6(1):51; 6(2):73

Ugolini, Nick (U.S. Navy/USA) 6(10):65
Uhler, Richard (Battelle/USA) 6(5):237
Unz, Richard F. (The Pennsylvania State University/USA) 6(8):201
Utgikar, Vivek P. (U.S. EPA/USA) 6(9):17

Valderrama, Brenda (Universidad Nacional Autónoma de México/MEXICO) 6(6):17
Vallini, Giovanni (Universita degli Studi di Verona/ITALY) 6(3):267
van Bavel, Bert (Umeå University/SWEDEN) 6(3):181
van Breukelen, Boris M. (Vrije University/THE NETHERLANDS) 6(4):91
VanBroekhoven, K. (Catholic University of Leuven/BELGIUM) 6(4):35
Vandecasteele, Jean-Paul (Institut Français du Pétrole/FRANCE) 6(3):227
VanDelft, Frank (NOVA Chemicals/CANADA) 6(5):53
van der Gun, Johan (BodemBeheer bv/THE NETHERLANDS) 6(5):289

van der Werf, A. W. (Bioclear Environmental Technology/THE NETHERLANDS) 6(8):11
van Eekert, Miriam (TNO Environmental Sciences /THE NETHERLANDS) 6(2):231; 6(7):289
Van Hout, Amy H. (IT Corporation/USA) 6(3):35
Van Keulen, E. (DHV Environment and Infrastructure/THE NETHERLANDS) 6(8):11
Vargas, M.C. (ECOPETROL-ICP/COLOMBIA) 6(6):319
Vazquez-Duhalt, Rafael (Universidad Nacional Autónoma de México/MEXICO) 6(6):17
Venosa, Albert (U.S. EPA/USA) 6(1):19
Verhaagen, P. (Grontmij BV/THE NETHERLANDS) 6(7):289
Verheij, T. (DAF/THE NETHERLANDS) 6(7):289
Vidumsky, John E. (E.I. du Pont de Nemours & Company/USA) 6(2):81; 6(8):185
Villani, Marco (Centro Ricerche Ambientali/ITALY) 6(4):131
Vinnai, Louise (Investigative Science Inc./CANADA) 6(2):27
Visscher, Gerolf (Province of Groningen/THE NETHERLANDS) 6(7):141
Voegeli, Vincent (TranSystems Corporation/USA) 6(7):229
Vogt, Bob (Louisiana-Pacific Corporation/USA) 6(3):83
Volkering, Frank (TAUW bv/THE NETHERLANDS) 6(4):91
von Arb, Michelle (University of Iowa) 6(3):1
Vondracek, James E. (Ashland Inc./USA) 6(5):121
Vos, Johan (VITO/BELGIUM) 6(9):87
Voscott, Hoa T. (Camp Dresser & McKee, Inc./USA) 6(7):305
Vough, Lester R. (University of Maryland/USA) 6(5):77

Waisner, Scott A. (TA Environmental, Inc./USA) 6(4):59; 6(10):115

Walecka-Hutchison, Claudia M. (University of Arizona/USA) 6(9):231
Wall, Caroline (CEFAS Laboratory/UK) 6(10):337
Wallace, Steve (Lattice Property Holdings Plc./UK) 6(4):17
Wallis, F.M. (University of Natal/REP OF SOUTH AFRICA) 6(6):101; 6(9):79
Walton, Michelle R. (Idaho National Engineering & Environmental Laboratory/USA) 6(7):77
Walworth, James L. (University of Arizona/USA) 6(9):231
Wan, C.K. (Hong Kong Baptist University/CHINA) 6(6):73
Wang, Chuanyue (Rice University/USA) 6(5):85
Wang, Qingren (Chinese Academy of Sciences/CHINA [PRC]) 6(9):113
Wani, Altaf (Applied Research Associates, Inc./USA) 6(10):115
Wanty, Duane A. (The Gillette Company/USA) 6(7):87
Warburton, Joseph M. (Parsons Engineering Science/USA) 6(7):173
Watanabe, Masataka (National Institute for Environmental Studies/JAPAN) 6(5):321
Watson, James H.P. (University of Southampton/UK) 6(9):61
Wealthall, Gary P. (University of Sheffield/UK) 6(1):59
Weathers, Lenly J. (Tennessee Technological University/USA) 6(8):139
Weaver, Dallas E. (Scientific Hatcheries/USA) 6(1):91
Weaverling, Paul (Harding ESE/USA) 6(10):231
Weber, A. Scott (State University of New York at Buffalo/USA) 6(6):89
Weeber, Philip A. (Geotrans/USA) 6(10):163
Wendt-Potthoff, Katrin (UFZ Centre for Environmental Research/GERMANY) 6(9):43
Werner, Peter (Technical University of Dresden/GERMANY) 6(3):227; 6(8):105

West, Robert J. (The Dow Chemical Company/USA) *6*(2):89
Westerberg, Karolina (Stockholm University/SWEDEN) *6*(3):133
Weston, Alan F. (Conestoga-Rovers & Associates/USA) *6*(1):99; *6*(10):131
Westray, Mark (ThermoRetec Corp/USA) *6*(7):1
Wheater, H.S. (Imperial College of Science and Technology/UK) *6*(10):123
White, David C. (University of Tennessee/USA) *6*(4):73; *6*(5):305
White, Richard (EarthFax Engineering Inc/USA) *6*(6):263
Whitmer, Jill M. (GeoSyntec Consultants/USA) *6*(9):105
Wick, Lukas Y. (Swiss Federal Institute of Technology/SWITZERLAND) *6*(3):251
Wickramanayake, Godage B. (Battelle/USA) *6*(10):1
Widada, Jaka (The University of Tokyo/JAPAN) *6*(4):51
Widdowson, Mark A. (Virginia Polytechnic Institute & State University/USA) *6*(2):105; *6*(5):1
Wieck, James M. (GZA GeoEnvironmental, Inc./USA) *6*(7):165
Wiedemeier, Todd H. (Parsons Engineering Science, Inc./USA) *6*(7):241
Wiessner, Arndt (UFZ - Centre for Environmental Research/GERMANY) *6*(5):337
Wilken, Jon (Harding ESE/USA) *6*(10):231
Williams, Lakesha (Southern University at New Orleans/USA) *6*(5):145
Williamson, Travis (Battelle/USA) *6*(10):245
Willis, Matthew B. (Cornell University/USA) *6*(4):125
Willumsen, Pia Arentsen (National Environmental Research Institute/DENMARK) *6*(3):141
Wilson, Barbara H. (Dynamac Corporation/USA) *6*(1):129
Wilson, Gregory J. (University of Cincinnati/USA) *6*(1):19

Wilson, John T. (U.S. EPA/USA) *6*(1):43, 167
Wiseman, Lee (Camp Dresser & McKee Inc./USA) *6*(7):133
Wisniewski, H.L. (Equilon Enterprises LLC/USA) *6*(8):61
Witt, Michael E. (The Dow Chemical Company/USA) *6*(2):89
Wong, Edwina K. (University of Guelph/CANADA) *6*(6):185
Wong, J.W.C. (Hong Kong Baptist University/CHINA) *6*(6):73
Wood, Thomas K. (University of Connecticut/USA) *6*(5):199
Wrobel, John (U.S. Army/USA) *6*(5):207

Xella, Claudio (Water & Soil Remediation S.r.l./ITALY) *6*(6):179
Xing, Jian (Global Remediation Technologies, Inc./USA) *6*(6):311

Yamamoto, Isao (Sumitomo Marine Research Institute/JAPAN) *6*(10):345
Yamazaki, Fumio (Hyogo Prefectural Institute of Environmental Science/JAPAN) *6*(5):321
Yang, Jeff (URS Corporation/USA) *6*(2):239
Yerushalmi, Laleh (Biotechnology Research Institute/CANADA) *6*(3):165
Yoon, Woong-Sang (Sam) (Battelle/USA) *6*(7):13
Yoshida, Takako (The University of Tokyo/JAPAN) *6*(4):51; *6*(6):111
Yotsumoto, Mizuyo (Obayashi Corporation Ltd./JAPAN) *6*(6):111
Young, Harold C. (Air Force Institute of Technology/USA) *6*(2):173

Zagury, Gérald J. (École Polytechnique de Montréal/CANADA) *6*(9): 27, 129
Zahiraleslamzadeh, Zahra (FMC Corporation/USA) *6*(7):221
Zaluski, Marek H. (MSE Technology Applications/USA) *6*(9):35
Zappi, Mark E. (Mississippi State University/USA) *6*(9):241

Author Index

Zelennikova, Olga (University of Connecticut/USA) 6(7):69
Zhang, Chuanlun L. (University of Missouri/USA) 6(9):165
Zhang, Wei (Cornell University/USA) 6(4):125
Zhang, Zhong (University of Nevada Las Vegas/USA) 6(9):257
Zheng, Zuoping (University of Oslo/NORWAY) 6(2):181
Zocca, Chiara (Universita degli Studi di Verona/ITALY) 6(3):267
Zwick, Thomas C. (Battelle/USA) 6(10):1

KEYWORD INDEX

This index contains keyword terms assigned to the articles in the ten-volume proceedings of the Sixth International In Situ and On-Site Bioremediation Symposium (San Diego, California, June 4-7, 2001). Ordering information is provided on the back cover of this book.

In assigning the terms that appear in this index, no attempt was made to reference all subjects addressed. Instead, terms were assigned to each article to reflect the primary topics covered by that article. Authors' suggestions were taken into consideration and expanded or revised as necessary. The citations reference the ten volumes as follows:

6(1): Magar, V.S., J.T. Gibbs, K.T. O'Reilly, M.R. Hyman, and A. Leeson (Eds.), *Bioremediation of MTBE, Alcohols, and Ethers*. Battelle Press, Columbus, OH, 2001. 249 pp.

6(2): Leeson, A., M.E. Kelley, H.S. Rifai, and V.S. Magar (Eds.), *Natural Attenuation of Environmental Contaminants*. Battelle Press, Columbus, OH, 2001. 307 pp.

6(3): Magar, V.S., G. Johnson, S.K. Ong, and A. Leeson (Eds.), *Bioremediation of Energetics, Phenolics, and Polycyclic Aromatic Hydrocarbons*. Battelle Press, Columbus, OH, 2001. 313 pp.

6(4): Magar, V.S., T.M. Vogel, C.M. Aelion, and A. Leeson (Eds.), *Innovative Methods in Support of Bioremediation*. Battelle Press, Columbus, OH, 2001. 197 pp.

6(5): Leeson, A., E.A. Foote, M.K. Banks, and V.S. Magar (Eds.), *Phytoremediation, Wetlands, and Sediments*. Battelle Press, Columbus, OH, 2001. 383 pp.

6(6): Magar, V.S., F.M. von Fahnestock, and A. Leeson (Eds.), *Ex Situ Biological Treatment Technologies*. Battelle Press, Columbus, OH, 2001. 423 pp.

6(7): Magar, V.S., D.E. Fennell, J.J. Morse, B.C. Alleman, and A. Leeson (Eds.), *Anaerobic Degradation of Chlorinated Solvents*. Battelle Press, Columbus, OH, 2001. 387 pp.

6(8): Leeson, A., B.C. Alleman, P.J. Alvarez, and V.S. Magar (Eds.), *Bioaugmentation, Biobarriers, and Biogeochemistry*. Battelle Press, Columbus, OH, 2001. 255 pp.

6(9): Leeson, A., B.M. Peyton, J.L. Means, and V.S. Magar (Eds.), *Bioremediation of Inorganic Compounds*. Battelle Press, Columbus, OH, 2001. 377 pp.

6(10): Leeson, A., P.C. Johnson, R.E. Hinchee, L. Semprini, and V.S. Magar (Eds.), *In Situ Aeration and Aerobic Remediation*. Battelle Press, Columbus, OH, 2001. 391 pp.

A

abiotic/biotic dechlorination 6(8):193
acenaphthene 6(5):253
acetate as electron donor 6(3):51; 6(9):297
acetone 6(2):49
acid mine drainage, (*see also* mine tailings) 6(9):1, 9, 27, 35, 43, 53
acrylic vessel 6(5):321
actinomycetes 6(10):211
activated carbon biomass carrier 6(6):311; 6(8):113

activated carbon **6(8)**:105
adsorption **6(3)**:243; **6(5)**:253; **6(6)**:377;
　6(7):77; **6(8)**:131; **6(9)**:86
advanced oxidation **6(1)**:121; **6(10)**:33
aerated submerged **6(10)**:329
aeration **6(6)**:203
anaerobic/aerobic treatment **6(6)**:361;
　6(7):229
age dating **6(5)**:231, 237
air sparging **6(1)**:115, 175; **6(2)**:239;
　6(9):215; **6(10)**:1, 9, 41, 49, 65, 101,
　115, 123, 163, 223
alachlor **6(6)**:9
algae **6(5)**:181
alkaline phosphatase **6(9)**:165
alkane degradation **6(5)**:313
alkylaromatic compounds **6(6)**:173
alkylbenzene **6(2)**:19
alkylphenolethoxylate **6(5)**:215
Amaranthaceae **6(5)**:165
Ames test **6(6)**:249
ammonia **6(1)**:175; **6(5)**:337
amphipod toxicity test **6(5)**:321
anaerobic **6(1)**:35, 43; **6(3)**:91; 205;
　6(5):17, 25, 261, 297, 313; **6(6)**:133;
　6(7):249, 297; **6(9)**:147, 303
anaerobic biodegradation **6(1)**:137;
　6(5):1; **6(8)**:167
anaerobic bioventing **6(3)**:9
anaerobic petroleum degradation **6(5)**:25
anaerobic sparging **6(7)**:297
aniline **6(6)**:149
Antarctica **6(2)**:57
anthracene **6(3)**:165, 251; **6(6)**:73
aquatic plants **6(5)**:181
arid-region soils **6(9)**:231
aromatic dyes **6(6)**:369
arsenic **6(2)**:239, 261; **6(5)**:173; **6(9)**: 97,
　129
atrazine **6(5)**:181; **6(6)**:9
azoaromatic compounds **6(6)**:149
Azomonas **6(6)**:219

B

bacterial transport **6(8)**:1
barrier technologies **6(1)**:11; **6(3)**:165;
　6(7):289; **6(8)**:61, 79, 87, 105, 121;
　6(9):27, 71, 195, 209, 309
basidiomycete **6(6)**:101
benthic **6(10)**:337

benzene **6(1)**:1, 67, 75, 145, 167, 203;
　6(4):91,117; **6(8)**:87; **6(10)**:123
benzene, toluene, ethylbenzene, and
　xylenes (BTEX) **6(1)**:43, 51, 59, 107,
　129, 167, 195; **6(2)**:11, 19, 137, 215,
　223, 270; **6(4)**:99; **6(5)**:33; **6(7)**:133;
　6(8):105; **6(10)**: 1, 23, 49, 65, 95,
　123, 131
benzo(a)pyrene **6(3)**:149; **6(6)**:101
benzo(e)pyrene **6(3)**:149
BER, *see* biofilm-electrode reactor
bioassays **6(3)**:219
bioaugmentation **6(1)**:11; **6(3)**:133;
　6(4):59; **6(6)**:9, 43, 111; **6(7)**:125;
　6(8):1, 11, 19, 27, 43, 53, 61, 147,
　175
bioavailability **6(3)**:115, 157, 173, 189,
　51; **6(4)**:7; **6(5)**:253, 279, 289; **6(6)**:1
bioavailable FeIII assay **6(8)**:209
biobarrier **6(1)**:11; **6(3)**:165; **6(7)**:289;
　6(8):61, 79, 105, 121; **6(9)**:27, 71,
　209, 309
BIOCHLOR model **6(2)**:155
biocide **6(7)**:321, 333
biodegradability **6(6)**:193
biodegradation **6(1)**:19,153; **6(3)**:165,
　181, 205, 235; **6(10)**:187
biofilm **6(3)**:251; **6(4)**:149; **6(8)**:79;
　6(9):201, 303
biofilm-electrode reactor (BER) **6(9)**:201
biofiltration **6(4)**:149
biofouling **6(7)**:321, 333
bioindicators **6(1)**:1; **6(3)**:173; **6(5)**:223
biological carbon regeneration **6(8)**:105
bioluminescence **6(1)**:1; **6(3)**:173; **6(4)**:45
biopile **6(6)**:81, 127, 141, 227, 249, 287
bioreactors **6(1)**:91; **6(6)**:361; **6(8)**:11, 35;
　6(9):1, 265, 281, 303, 315; **6(10)**:171,
　211
biorecovery of metals **6(9)**:9
bioreporters **6(4)**:45
biosensors **6(1)**:1
bioslurping **6(10)**:245, 253, 267, 275
bioslurry and bioslurry reactors **6(3)**:189;
　6(6):51, 65
biosparging **6(10)**:115, 163
biostabilization **6(6)**:89
biostimulation **6(6)**:43
biosurfactant **6(3)**:243; **6(7)**:53
bioventing **6(10)**:109, 115, 131
biphasic reactor **6(3)**:181

Keyword Index

biological oxygen demand (BOD) **6(10)**:311
BTEX, *see* benzene, toluene, ethylbenzene, and xylenes
Burkholderia cepacia **6(1)**:153; **6(7)**:117; **6(8)**:53
butane **6(1)**:137, 161
butyrate **6(7)**:289

C

cadmium **6(3)**:91; **6(9)**:79, 147
carAa, see carbazole 1,9a-dioxygenase gene
carbazole-degrading bacterium **6(6)**:111
carbazole 1,9a-dioxygenase gene (*carAa*) **6(4)**:51
Carbokalk **6(9)**:43
carbon isotope **6(4)**:91, 99, 109, 117; **6(10)**:115
carbon tetrachloride (CT) **6(2)**:81, 89; **6(5)**:113; **6(7)**:241; **6(8)**:185, 193
cesium-137 **6(5)**:231
CF, *see* chloroform
charged coupled device camera **6(2)**:207
chelators addition (EDGA, EDTA) **6(5)**:129, 137, 145, 151; **6(9)**:123, 147
chemical oxidation **6(7)**:45
chicken manure **6(9)**:289
chlorinated ethenes **6(7)**:27, 61, 69, 109; **6(10)**:163, 201, 231
chlorinated solvents **6(2)**:145; **6(7)**:all; **6(8)**:19; **6(10)**:231
chlorobenzene **6(8)**:105
chloroethane **6(2)**:113; **6(7)**:133, 249
chloroform (CF) **6(2)**:81; **6(8)**:193
chloromethanes **6(8)**:185
chlorophenol **6(3)**:75, 133
chlorophyll fluorescence **6(5)**:223
chromated copper arsenate **6(9)**:129
chromium (Cr[VI]) **6(8)**:139, 147; **6(9)**:129, 139, 315
chrysene **6(6)**:101
citrate and citric acid **6(5)**:137; **6(7)**:289
cleanup levels **6(6)**:1
coextraction method **6(4)**:51
Coke Facility waste **6(2)**:129
combined chemical toxicity (*see also* toxicity) **6(5)**:305
cometabolic air sparging **6(10)**:145, 155, 223

cometabolism **6(1)**:137, 145, 153, 161; **6(2)**:19; **6(6)**:81, 141; **6(7)**:117; **6(10)**:145, 155, 163, 171, 179, 193, 201, 211, 217, 223, 231; 239
competitive inhibition **6(2)**:19
composting **6(3)**:83; **6(5)**:129, **6(6)**:73, 119, 165, 257; **6(7)**:141
constructed wetlands **6(5)**:173, 329
contaminant aging **6(3)**:157, 197
contaminant transport **6(3)**:115
copper **6(9)**:79, 129
cosolvent effects **6(1)**:175, 195, 203, 243
cosolvent extraction **6(7)**:125
cost analyses and economics of environmental restoration **6(1)**:129; **6(4)**:17; **6(8)**:121; **6(9)**:331; **6(10)**:65, 211
Cr(VI), *see* chromium
creosote **6(3)**:259; **6(4)**:59; **6(5)**:1, 237, 329; **6(6)**:81, 101, 141, 295
cresols **6(10)**:123
crude oil **6(5)**:313; **6(6)**:193, 249; **6(10)**:329
CT, *see* carbon tetrachloride
cyanide **6(9)**:331
cytochrome P-450 **6(6)**:17

D

2,4-DAT, *see* diaminotoluene
DCA, *see* dichloroethane
1,1-DCA, *see* 1,1-dichloroethane
1,2-DCA, *see* 1,2-dichloroethane
DCE, *see* dichloroethene
1,1-DCE, *see* 1,1-dichloroethene
1,2-DCE, *see* 1,2-dichloroethene
c-DCE, *see* cis-dichloroethene
DCM, *see* dichloromethane
DDT, *see also* dioxins *and* pesticides **6(6)**:157
2,4-DNT, *see* dinitrotoluene
dechlorination kinetics **6(2)**:105; **6(7)**:61
dechlorination **6(2)**:231; **6(3)**:125; **6(5)**:95; **6(7)**:13, 61, 165, 173, 333; **6(8)**:19, 27, 43
DEE, *see* diethyl ether
Dehalococcoides ethenogenes **6(8)**:19, 43
dehalogenation **6(8)**:167
denaturing gradient gel electrophoresis (DGGE) **6(1)**:19; **6(4)**:35

denitrification **6(2)**:19; **6(4)**:149; **6(5)**:17, 261; **6(8)**:95; **6(9)**:179, 187, 195, 201, 209, 223, 309
dense, nonaqueous-phase liquid (DNAPL) **6(7)**:13, 19, 35, 181; **6(10)**:319
depletion rate **6(1)**:67
desorption **6(3)**:235, 243; **6(5)**:253; **6(6)**:377; **6(7)**:53, 77; **6(8)**:131
DGGE, *see* denaturing gradient gel electrophoresis
DHPA, *see* dihydroxyphenylacetate
dialysis sampler **6(5)**:207
diaminotoluene (2,4-DAT) **6(6)**:149
dibenzofuran-degrading bacterium **6(6)**:111
dibenzo-p-dioxin **6(6)**:111
dibenzothiophene **6(3)**:267
dichlorodiethyl ether **6(10)**:301
dichloroethane (DCA) **6(2)**:39; **6(7)**:289
1,1-dichloroethane (1,1-DCA; 1,2-DCA) **6(2)**:113; **6(5)**:207; **6(7)**:133, 165
1,2-dichloroethane (1,2-DCA) **6(5)**:207
dichloroethene, dichloroethylene **6(2)**:97, 155; **6(4)**:125; **6(5)**:105,113; **6(7)**:157, 197
cis-dichloroethene, *cis*-dichloroethylene (*c*-DCE) **6(2)**:39, 65, 73; 105, 173; **6(5)**:33, 95, 207; **6(7)**:1, 13, 61, 133, 141, 149, 165, 173, 181, 189, 205, 213, 221, 249, 273, 281, 289, 297, 305; **6(8)**:11, 19, 27, 43, 73, 105, 157, 209; **6(10)**:41, 145, 155, 179, 201
1,1-dichloroethene, 1,1-dichloroethylene (1,1-DCE) **6(2)**:39; **6(7)**:165, 229; **6(8)**:157; **6(10)**:231
1,2-dichloroethene and 1,2-dichloroethylene (1,2-DCE) **6(2)**:113
dichloromethane (DCM) **6(2)**:81; **6(8)**:185
diesel fuel **6(1)**:175; **6(2)**:57; **6(5)**:305; **6(6)**:81, 141, 165; **6(10)**:9
diesel-range organics (DRO) **6(10)**:9
diethyl ether (DEE) **6(1)**:19
dihydroxyphenylacetate (DHPA) **6(4)**:29
diisopropyl ether (DIPE) **6(1)**:19, 161
1,3-dinitro-5-nitroso-1,3,5-triazacyclohexane (MNX) (*see also* explosives *and* energetics) **6(3)**:51; **6(8)**:175
dinitrotoluene (2,4-DNT) **6(3)**:25, 59; **6(6)**:127, 149
dioxins **6(6)**:111

DIPE, *see* diisopropyl ether
dissolved oxygen **6(2)**:189, 207
16S rDNA sequencing **6(8)**:19
DNAPL, *see* dense, nonaqueous-phase liquid
DNX, *see* explosives and energetics
DRO, *see* diesel-range organics
dual porosity aquifer **6(1)**:59
dyes **6(6)**:369

E

ecological risk assessment **6(4)**:1
ecotoxicity, (*see also* toxicity) **6(1)**:1; **6(4)**:7
ethylenedibromide (EDB) **6(10)**:65
EDGA, *see* chelate addition
EDTA, *see* chelate addition
effluent **6(4)**:1
electrokinetics **6(9)**:241, 273
electron acceptors and electron acceptor processes **6(2)**:1, 137, 163, 231; **6(5)**:17, 25, 297; **6(7)**:19
electron donor amendment **6(3)**:25, 35, 51, 125; **6(7)**:69, 103,109, 141, 181, 249, 289, 297; **6(8)**:73; **6(9)**:297, 315
electron donor delivery **6(7)**:19, 27, 133, 173, 213, 221, 265, 273, 281, 305
electron donor mass balance **6(2)**:163
electron donor transport **6(4)**:125; **6(7)**:133; **6(9)**:241
embedded carrier **6(9)**:187
encapsulated bacteria **6(5)**:269
enhanced aeration **6(10)**:57
enhanced desorption **6(7)**:197
environmental stressors **6(4)**:1
enzyme induction **6(6)**:9; **6(10)**:211
ERIC sequences **6(4)**:29
ethane **6(2)**:113; **6(7)**:149
ethanol 6(1):19,167,175, 195, 203; **6(5)**:243; **6(6)**:133; **6(9)**:289
ethene and ethylene **6(2)**:105,113; **6(5)**:95; **6(7)**:1, 95, 133, 141, 205, 281, 297, 305; **6(8)**:11, 43, 167, 175, 209
ethylene dibromide **6(10)**:193
explosives and energetics **6(3)**:9, 17, 25, 35, 43, 51, 67; **6(5)**:69; **6(6)**:119, 127, 133; **6(7)**:125

F

fatty acids *6*(5):41
Fe(II), *see* iron
Fenton's reagent *6*(6):157
fertilizer *6*(5):321; *6*(6):35; *6*(10):337
fixed-bed and fixed-film reactors *6*(5):221, 337; *6*(6):361; *6*(9):303
flocculants *6*(6):279
flow sensor *6*(10):293
fluidized-bed reactor *6*(1):91; *6*(6):133, 311; *6*(9):281
fluoranthene *6*(3):141; *6*(6):101
fluorogenic probes *6*(4):51
food safety *6*(9):113
formaldehyde *6*(6):329
fractured shale *6*(10):49
free-product recovery *6*(6):211
Freon *6*(2):49
fuel oil *6*(5):321
fungal remediation *6*(3):75, 99; *6*(5):61, 279; *6*(6):17, 101, 157, 263, 319, 329, 369
Funnel-and-Gate™ *6*(8):95

G

gas flux *6*(6):185
gasoline *6*(1):35, 75, 161, 167, 195; *6*(10):115
gasoline-range organics (GRO) *6*(10):9
manufactured gas plants and gasworks *6*(2):137; *6*(10):123
GCW, *see* groundwater circulating well
gel-encapsulated biomass *6*(8):35
GEM, see genetically engineered microorganisms
genetically engineered microorganisms (GEM) *6*(4):45; *6*(5):199; *6*(7):125
genotoxicity, (*see also* toxicity) *6*(3):227
Geobacter *6*(3):1
geochemical characterization *6*(4):91
geographic information system (GIS) *6*(2):163
geologic heterogeneity *6*(2):11
germination index 6(3):219; *6*(6):73
GFP, *see* green fluorescent protein
GIS, *see* geographic information system
glutaric dialdehyde dehydrogenase *6*(4):81
Gordonia terrae *6*(1):153
green fluorescent protein (GFP) *6*(5):199
GRO, see gasoline-range organics

groundwater *6*(3):35; *6*(8): 35, 87, 121; *6*(10):231
groundwater circulating well (GCW) *6*(7):229, 321; *6*(10):283, 293

H

H_2 gas, *see* hydrogen
H_2S, *see* hydrogen sulfide
halogenated hydrocarbons *6*(9):61
halorespiration *6*(8):19
heavy metal *6*(2):239; *6*(5):137, 145, 157, 165, 173; *6*(6):51; *6*(9):53, 61, 71, 79, 86, 97, 113, 129, 147
herbicides *6*(5):223; *6*(6):35
hexachlorobenzene *6*(3):99
hexane *6*(3):181, *6*(6):329
HMX, *see* explosives and energetics
hollow fiber membranes *6*(5):269
hopane *6*(6):193; *6*(10):337
hornwort *6*(5):181
HRC® (a proprietary hydrogen-release compound) *6*(3):17, 25, 107; *6*(7):27, 103, 157, 189, 197, 205, 221, 257, 305, *6*(8):157, 209
^{2}H-tetradecane (*see also* tetradecane) *6*(2):27
humates *6*(1):99
hybrid treatment *6*(10):311
hydraulic containment *6*(8):79
hydraulically facilitated remediation *6*(2):239
hydrocarbon *6*(6):235; *6*(10):329
hydrogen (H_2 gas) *6*(2):199; *6*(9):201
hydrogen injection, in situ *6*(7):19
hydrogen isotope *6*(4):91
hydrogen peroxide *6*(1):121; *6*(6):353; *6*(10):33
hydrogen release compound, see HRC®
hydrogen sulfide (H_2S) *6*(9):123
hydrogen *6*(2):231, *6*(7):61, 305
hydrolysis *6*(1):83
hydrophobicity *6*(3):141
hydroxyl radical *6*(1):121
hydroxylamino TNT intermediates *6*(5):85

I

immobilization *6*(8):53
immobilized cells *6*(8):121
immobilized soil bioreactor *6*(10):171

in situ oxidation **6(7)**:1
industrial effluents **6(6)**:303, 361
inhibition **6(9)**:17
injection strategies, in situ **6(7)**:19, 133, 173, 213, 221, 265, 273, 305, 313; **6(9)**:223; **6(10)**:23, 163
insecticides **6(6)**:27
intrinsic biodegradation **6(2)**:89, 121
intrinsic remediation, *see* natural attenuation
ion migration **6(9)**:241
iron (Fe[II]) **6(5)**:1
iron barrier **6(8)**:139, 147, 157, 167
iron oxide **6(3)**:1
iron precipitation **6(3)**:211
iron-reducing processes **6(2)**:121; **6(3)**:1; **6(5)**:1, 17, 25; **6(6)**:149; **6(8)**:193, 201, 209; **6(9)**:43, 323
IR-spectroscopy **6(4)**:67
isotope analyses **6(2)**:27; **6(4)**:91; **6(8)**:27
isotope fractionation **6(4)**:99, 109, 117

J

jet fuel **6(10)**:95, 139

K

KB-1 strain **6(8)**:27
kerosene **6(6)**:219
kinetics **6(8)**:131, **6(1)**:1, 19, 27, 167; **6(2)**:11, 19, 105; **6(3)**:173; **6(4)**:131; **6(7)**:61
Klebsiella oxytoca **6(7)**:117
Kuwait **6(6)**:249

L

laccase **6(3)**:75; **6(6)**:319
lactate and lactic acid **6(7)**:103, 109, 165, 181, 213, 265, 281, 289; **6(8)**:139; **6(9)**:155, 273
lagoons **6(6)**:303
land treatment units (LTU) **6(6)**:1; **6(6)**:81, 141, 287, 295
landfarming **6(3)**:259; **6(4)**:59; **6(5)**:53, 279; **6(6)**:1, 43, 59, 179, 203, 211, 235
landfills **6(2)**:145, 247; **6(4)**:91; **6(8)**:113
leaching **6(9)**:187
lead **6(5)**:129, 145, 151, 157

lead-210 **6(5)**:231
light, nonaqueous-phase liquids (LNAPL) **6(1)**:59; **6(4)**:35; **6(10)**:57, 109, 245, 253, 275
lindane, (*see also* pesticides) **6(5)**:189
linuron (*see also* herbicides) **6(5)**:223
LNAPL, *see* light, nonaqueous-phase liquids
Lolium multiflorum **6(5)**:9
LTU, *see* land treatment units
lubricating oil **6(6)**:173
luciferase **6(3)**:133
lux **6(4)**:45

M

mackinawite **6(9)**:155
macrofauna **6(10)**:337
magnetic separation **6(9)**:61
magnetite **6(3)**:1; **6(8)**:193
manganese **6(2)**:261
manufactured gas plant (MGP) **6(2)**:19; **6(3)**:211, 227; **6(10)**:123
mass balance **6(2)**:163
mass transfer limitation **6(3)**:157
mass transfer **6(1)**:67
MC-100, see mixed culture
media development **6(9)**:147
Meiofauna **6(5)**:305; **6(10)**:337
membrane **6(5)**:269; **6(9)**:1, 265
metabolites **6(3)**:227
metal reduction **6(8)**:1
metal precipitation **6(9)**:9, 165
metals, biorecovery of **6(9)**:9
metals speciation **6(9)**:129
metal toxicity (*see also* toxicity) **6(9)**:17, 129
metals **6(5)**:129, 305; **6(8)**:1; **6(9)**:9, 17, 27, 105, 123, 129, 155, 165
methane oxidation **6(10)**:171, 187, 193, 201, 223, 231
methane **6(1)**:183; **6(8)**:113
methanogenesis **6(1)**:35, 43, 183; **6(3)**:205; **6(9)**:147
methanogens **6(3)**:91
methanol **6(1)**:183; **6(7)**:141, 289, 297
methanotrophs **6(10)**:171, 187, 201
methylene chloride **6(2)**:39; **6(10)**:231
Methylosinus trichosporium **6(10)**:187
methyl *tert*-butyl ether *or* methyl *tertiary*-butyl ether (MTBE) **6(1)**:1, 11, 19, 27, 35, 43, 51, 59, 67, 75, 83, 91, 107

115, 121, 129, 137, 145, 153,161, 195, **6(2)**:215; **6(8)**:61; **6(10)**:1, 65
MGP, *see* manufactured gas plant
microbial heterogeneity **6(4)**:73
microbial isolation **6(3)**:267
microbial population dynamics **6(4)**:35
microbial regrowth **6(2)**:253; **6(7)**:1, 13; **6(10)**:319
microcosm studies **6(7)**:109; **6(10)**:179
microencapsulation **6(8)**:53
microfiltration **6(9)**:201
microporous membrane **6(9)**:265
microtox assay **6(3)**:227
mine tailings (*see also* acid mine drainage) **6(5)**:173; **6(9)**:27, 71
mineral oil **6(5)**:279, 289; **6(6)**:59
mineralization **6(2)**:121; **6(3)**:165; **6(6)**:165; **6(8)**:175; **6(9)**:139, 155
MIP, *see* membrane interface probe
mixed culture **6(8)**:61
mixed wastes **6(3)**:91; **6(7)**:133; **6(9)**:139
MNX, *see* 1,3-dinitro-5-nitroso-1,3,5-triazacyclohexane
modeling **6(1)**:51; **6(2)**:105, 155, 181, **6(4)**:125, 131, 139, 149; **6(6)**:339, 377; **6(8)**:185; **6(9)**:27, 105; **6(10)**:163
moisture content **6(2)**:247
molasses as electron donor **6(3)**:35; **6(7)**:53, 103, 149, 173; **6(9)**:315
monitored natural attenuation (*see also* natural attenuation) **6(1)**:183, **6(2)**:11, 163, 199, 223, 253, 261
monitoring techniques **6(2)**:27,189, 199, 207; **6(4)**:59
motor oil **6(5)**:53
MPE, *see* multiphase extraction
multiphase extraction (MPE) well design **6(10)**:245, 259
MTBE, *see* methyl *tert*-butyl ether
multiphase extraction **6(10)**:245, 253, 259, 267, 275
municipal solid waste **6(2)**:247
Mycobacterium sp. IFP 2012 **6(1)**:153
Mycobacterium adhesion **6(3)**:251
mycoremediation **6(6)**:263

NAPL, *see* nonaqueous-phase liquid
natural attenuation **6(1)**:27, 35, 43, 51, 59, 75, 83, 183, 195; **6(2)**:1,39, 73, 81, 89, 97, 105, 137, 145, 173, 181, 215; **6(4)**:91, 99, 117; **6(5)**:33, 189, 321; **6(8)**:185, 209; **6(9)**:179; **6(10)**:115, 163
natural gas **6(10)**:193
natural organic carbon **6(2)**:261
natural organic matter **6(2)**:81, 97; **6(8)**:201
natural recovery **6(5)**:132, 231
nitrate contamination **6(9)**:173
nitrate reduction **6(3)**:51; **6(5)**:25; **6(9)**:331
nitrate utilization efficiency **6(6)**:353
nitrate **6(2)**:1; **6(3)**:17, 43; **6(6)**:353; **6(8)**:95, 147; **6(9)**:179, 187, 195, 209, 223, 257
nitrification **6(4)**:149; **6(5)**:337; **6(9)**:215
nitroaromatic compounds (*see also* explosives and energetics) **6(3)**:59, 67; **6(6)**:149
nitrobenzene, *see also* explosives and energetics **6(6)**:149
nitrocellulose, *see also* explosives and energetics **6(6)**:119
nitrogen fixation **6(6)**:219
nitrogen utilization **6(9)**:231
nitrogenase **6(6)**:219
nitroglycerin, *see also* explosives and energetics **6(5)**:69
nitrotoluenes, *see also* explosives and energetics **6(6)**:127
nitrous oxide **6(8)**:113
^{13}C-NMR, *see* nuclear magnetic resonance spectroscopy
nonaqueous-phase liquids (NAPLs) **6(1)**:67, 203; **6(3)**:141; **6(7)**:249
nonylphenolethoxylates **6(5)**:215
nuclear magnetic resonance spectroscopy (^{13}C-NMR) **6(4)**:67
nutrient augmentation **6(3)**:59; **6(5)**:329; **6(6)**:257; **6(7)**:313; **6(9)**:331; **6(10)**:23
nutrient injection **6(10)**:101
nutrient transport **6(9)**:241

N

naphthalene **6(1)**:1; **6(2)**:121; **6(3)**:173, 227; **6(5)**:1, 253; **6(6)**:51; **6(8)**:95, **6(9)**:139; **6(10)**:123

O

oily waste **6(4)**:35; 6(6):257; **6(10)**:337, 345
oil-coated stones **6(10)**:329

optimization **6(5)**:279
ORC® (a proprietary oxygen-release compound) **6(1)**:99,107; **6(2)**:215; **6(3)**:107; **6(7)**:229; **6(10)**:9, 15, 87, 95, 139
organic acids **6(2)**:39
organophosphorus **6(6)**:17, 27
advanced oxidation **6(6)**:157, **6(10)**:311
oxygen-release compound, *see* ORC®
oxygen-release material **6(10)**:73
oxygen respiration **6(9)**:231; **6(10)**:57
oxygenation **6(1)**:107, 145
ozonation **6(1)**:121; **6(10)**:33, 149, 301

P

packed-bed reactors **6(9)**:249; **6(10)**:329
PAHs, *see* polycyclic aromatic hydrocarbons
paper mill waste **6(4)**:1
paraffins **6(3)**:141
partitioning **6(9)**:129
PCBs, *see* polychlorinated biphenyls
PCP toxicity (*see also* toxicity) **6(3)**:125
PCP, *see* pentachlorophenol
PCR analysis, *see* polymerase chain reaction
pentachlorophenol (PCP) **6(3)**:83, 91, 99, 107, 115, 125; **6(5)**:329; **6(6)**:279, 287, 295, 329
percarbonate **6(10)**:73
perchlorate **6(9)**:249, 257, 265, 273, 281, 289, 297, 303, 309, 315
perchloroethene, perchloroethylene **6(7)**:53
permeable reactive barriers **6(3)**:1; **6(8)**: 73, 87, 95, 121, 139, 147, 157, 167, 175, 185; **6(9)**:71, 309, 323; **6(10)**:95
pesticides **6(5)**:189; **6(6)**:9, 17, 35
PETN reductase **6(5)**:69
petroleum hydrocarbon degradation **6(4)**:7; **6(5)**:9, 17, 25; **6(8)**:131; **6(10)**: 65, 101, 245, 345
phenanthrene **6(2)**:121; **6(3)**:227, 235, 243; **6(6)**:51, 65, 73
phenol **6(6)**:303, 319, 329
phenolic waste **6(6)**:311
phenol-oxidizing cultures **6(10)**:211, 217, 239
phenyldodecane **6(2)**:27
phosphate precipitation **6(9)**:165
PHOSter **6(10)**:65

photocatalysis **6(10)**:311
physical/chemical pretreatment **6(1)**:1, 51; **6(2)**:253; **6(3)**:149; **6(5)**:9, 33, 41, 53, 61, 69, 77, 85,105, 113, 121, 129,137, 145, 151, 157, 165, 189, 199, 207, 279, 337; **6(6)**:59, 157, 241; **6(7)**:1, 13; **6(9)**:113, 173; **6(10)**:239, 311, 319
phytotoxicity (*see also* toxicity) **6(5)**:41, 223
phytotransformation **6(5)**:85
pile-turner **6(6)**:249
PLFA, *see* phospholipid fatty acid analysis
polychlorinated biphenyls (PCBs) **6(2)**:39,105,173; **6(5)**:33, 61, 95, 113, 231, 289; **6(6)**:89, **6(7)**:13, 61, 69, 95, 109, 125, 133, 141, 149, 165, 181, 189, 197, 205, 213, 241, 249, 273, 297, 305; **6(8)**:11,19, 27, 43, 157, 167, 193, 209; **6(10)**:33, 41, 231, 283
polycyclic aromatic hydrocarbons (PAHs) **6(2)**:19, 121, 129, 137; **6(3)**:141, 149, 157, 165, 173, 181, 189, 197, 205, 211, 219, 227, 235, 243; **6(4)**:35, 45, 59, 67; **6(5)**:1, 9, 17, 41, 237, 243, 251, 253, 261, 269, 279, 289, 305, 329; **6(6)**:43, 51, 59, 65, 73, 81, 89, 101, 279, 295, 297; **6(7)**:125; **6(8)**:95; **6(9)**:139; **6(10)**:33, 123
polymerase chain reaction (PCR) analysis **6(4)**:29, 35, 51; **6(8)**:43
polynuclear aromatic hydrocarbons, *see* polycyclic aromatic hydrocarbons
poplar lipid fatty acid analysis (PLFA) **6(3)**:189
poplar trees **6(5)**:113, 121, 189
potassium permanganate **6(2)**:253; **6(7)**:1
precipitation **6(9)**:105; **6(10)**:301
pressurized-bed reactor **6(6)**:311
propane utilization **6(1)**:137; **6(10)**:145, 155, 179, 193
propionate **6(7)**:265, 289
Pseudomonas fluorescens **6(3)**:173
pyrene **6(3)**:165, 235; **6(4)**:67; **6(6)**: 65, 73, 101
pyridine **6(4)**:81

R

RABITT, *see* reductive anaerobic biological in situ treatment technology
radium **6(5)**:173
rapeseed oil **6(6)**:65
RDX, *see* research development explosive
rebound **6(10)**:1
recirculation well **6(7)**:333, 341; **6(10)**:283
redox measurement and control **6(1)**:35; **6(2)**:11, 231; **6(5)**:1; **6(9)**:53
reductive anaerobic biological in situ treatment technology (RABITT) **6(7)**:109
reductive dechlorination **6(2)**:39, 65, 97, 105, 145, 173; **6(4)**:125; **6(7)**:45, 53, 87, 103, 109, 133, 141, 149, 157, 181, 197, 205, 213, 221, 249, 257, 265, 273, 289, 297; **6(8)**:11, 73, 105, 157, 209
reductive dehalogenation **6(7)**:69
reed canary grass **6(5)**:181
research development explosive (RDX) **6(3)**:1, 9, 17, 25, 35, 43, 51; **6(6)**:133; **6(8)**:175
respiration and respiration rates **6(2)**:129; **6(4)**:59; **6(6)**:185, 227
respirometry **6(6)**:127; **6(10)**:217
rhizoremediation **6(5)**:9, 61, 199
Rhodococcus opacus **6(4)**:81
risk assessment **6(2)**:215; **6(4)**:1
16S rRNA sequencing **6(8)**:43; **6(9)**:147
rock-bed biofiltration **6(4)**:149
rotating biological contactor **6(9)**:79
rototiller **6(6)**:203
RT3D **6(10)**:163

S

salinity **6(9)**:257
salt marsh **6(5)**:313
SC-100, *see* single culture
Sea of Japan **6(5)**:321
sediments **6(3)**:91; **6(5)**:231, 237, 253, 261, 269, 279, 289, 297, 305; **6(6)**:51, 59; **6(9)**:61
selenium **6(9)**:323, 331
semivolatile organic carbon (SVOC) **6(2)**:113
sheep dip **6(6)**:27
Shewanella putrefaciens **6(8)**:201
silicon oil **6(3)**:141, 181
single culture **6(8)**:61
site characterization **6(10)**:139
site closure **6(2)**:215
slow-release fertilizer **6(2)**:57
sodium glycine **6(9)**:273
soil treatment **6(3)**:181; **6(6)**:1
soil washing **6(5)**:243; **6(6)**:241
soil-vapor extraction (SVE) **6(1)**:183; **6(10)**:1, 41, 131, 223
solids residence time **6(10)**:211
sorption **6(5)**:215, 253; **6(6)**:377; **6(8)**:131; **6(9)**:79, 105
source zone **6(7)**:13, 19, 27, 181; **6(10)**:267
soybean oil **6(7)**:213
sparging **6(10)**:33, 145, 155
stabilization **6(6)**:89
substrate delivery **6(7)**:281
sulfate reduction **6(1)**:35; **6(3)**:43, 91; **6(5)**:261, 313; **6(6)**:339; **6(7)**:69, 95; **6(8)**:139, 147, 193; **6(9)**:1, 9, 17, 27, 35, 43, 61, 71, 86, 105, 123, 147
sulfide precipitation **6(9)**:123
surfactants **6(5)**:215; **6(6)**:73; **6(7)**:213, 321, 333; **6(8)**:131
sustainability **6(6)**:1
SVE, *see* soil vapor extraction
SVOC, *see* semivolatile organic carbon
synthetic pyrethroid **6(6)**:27

T

TCA, see trichlorethane
1,1,1-TCA, *see* 1,1,1-trichloroethane
1,1,2-TCA, *see* 1,1,2-trichloroethane
2,4,6-TCP, *see* 2,4,6-trichlorophenol
1,1,1,2-TeCA,*see* tetrachloroethane
1,1,2,2-TeCA, *see* tetrachloroethane
1,3,5-TNB, *see* 1,3,5-trinitrobenzene
TAME, *see* tertiary methyl-amyl ether
TBA, *see* tertiary butyl alcohol
TBF, *see* tertiary butyl formate
TCE oxidation, *see* trichloroethene, trichloroethylene
TCE, *see* trichloroethene
TCP, *see* trichlorophenol
t-DCE, *see* trans-dichloroethene, trans-dichloroethylene
technology comparisons **6(7)**:45; **6(9)**:323
terrazyme **6(10)**:345

tertiary butyl alcohol (TBA) **6(1)**:19, 27, 35, 51, 59, 91, 145, 153, 161
tertiary butyl formate (TBF) **6(1)**:145, 161
tertiary methyl-amyl ether (TAME) **6(1)**:59, 161
tetrachloroethane (1,1,1,2-TeCA, 1,1,2,2-TeCA) **6(5)**:207; **6(7)**:321, 341; **6(8)**:193
tetradecane (see also ^{2}H-tetradecane) **6(3)**:181
thermal desorption **6(3)**:189, **6(6)**:35
TNB, *see* trinitrobenzene
TNT, see trinitrotoluene
TNX, *see* 1,3,5-trinitroso-1,3,5-triazacyclohexane
tobacco plant **6(5)**:69
toluene **6(1)**:145; **6(2)**:181; **6(7)**:95; **6(8)**:35, 131
total petroleum hydrocarbons (TPH) **6(2)**:1; **6(5)**:9; **6(6)**:127, 173, 179, 193, 227, 241, 249; **6(10)**:15, 73, 115, 337
toxicity **6(1)**:1; **6(3)**:67, 189, 227; **6(4)**:7; **6(5)**:41, 61, 223, 305; **6(9)**:17, 129
TPH, *see* total petroleum hydrocarbons
trace gas emissions **6(6)**:185
trans-dichloroethene, trans-dichloroethylene **6(5)**:95, 207; **6(7)**:165
transgenic plants **6(5)**:69
transpiration **6(5)**:189
Trecate oil spill **6(6)**:241; **6(10)**:109
trichloroethane (TCA) **6(7)**:241, 281
1,1,1-trichloroethane (1,1,1-TCA; 1,1,2-TCA) **6(2)**:39, 113, 464; **6(5)**:207; **6(7)**:87,165, 281
1,1,2-trichloroethane (1,1,2-TCA) **6(5)**:207
trichloroethene, trichloroethylene (TCE) **6(2)**:39, 65, 73, 97, 105, 113, 155, 173, 253; **6(4)**:125; **6(5)**:33, 95, 105, 113, 207; **6(7)**:1, 13, 53, 61, 69, 77, 87, 109, 117, 133, 141, 149, 157, 181, 189, 197, 205, 213, 221, 241, 249, 265, 273, 281, 297, 305, **6(8)**:11, 19, 27, 35, 43, 53, 73, 105,147, 157, 193, 209; **6(10)**:41, 131, 145, 155, 163, 171, 179, 187, 201, 211, 217, 223, 231, 239, 283, 319
2,4,6-trichlorophenol (2,4,6-TCP) **6(3)**:75; **6(8)**:121
trichlorotrifluoroethane **6(2)**:49

trinitrobenzene (TNB) **6(3)**:9, 25
1,3,5-trinitroso-1,3,5-triazacyclohexane (TNX) **6(8)**:175
trinitrotoluene (TNT) **6(3)**:35, 67; **6(5)**:69, 77, 85; **6(6)**:133

U

underground storage tank (UST) **6(1)**:67, 129
uranium **6(5)**:173; **6(7)**:77; **6(9)**:155, 165
UST, *see* underground storage tank

V

vacuum extraction **6(1)**:115
vadose zone **6(1)**:183; **6(2)**:39, 65, 97, 105, 113, 155, 173; **6(3)**:9; **6(5)**:33, 105; **6(7)**:1,13, 61, 133, 141, 197, 205, 213, 249, 273, 281, 305; **6(8)**:11,19, 43, 73, 157, 209; **6(10)**:41, 163
vegetable oil **6(6)**:65; **6(7)**103, 213, 241, 249
vinyl chloride **6(2)**:73; **6(4)**:109; **6(5)**:95; **6(7)**:95,149, 157, 165, 173, 289, 297, **6(10)**:231
vitamin B_{12} **6(7)**:321, 333, 341
VOCs, *see* volatile organic carbons
volatile fatty acid **6(7)**:61
volatile organic carbons (VOCs) **6(2)**:113, 189; **6(5)**:113, 121

W

wastewater treatment **6(5)**:215; **6(6)**:149; **6(9)**:173
water potential **6(9)**:231
weathering **6(4)**:7
wetlands **6(5)**:33, 95, 105, 313, 329; **6(9)**:97
white rot fungi, (*see also* fungal remediation) **6(3)**:75, 99; **6(6)**:17, 157, 263
windrow **6(6)**:81, 119, 141
wood preservatives **6(3)**:83, 259; **6(4)**:59; **6(6)**:279

X

xylene **6(1)**:67

Y
yeast extract *6*(7):181

Z
zero-valent iron *6*(8):157, 167; *6*(9):71
zinc *6*(4):91; *6*(9):79